Geographic Information Systems in Business

James B. Pick
University of Redlands, USA

IDEA GROUP PUBLISHING
Hershey • London • Melbourne • Singapore

Acquisitions Editor:	Mehdi Khosrow-Pour
Senior Managing Editor:	Jan Travers
Managing Editor:	Amanda Appicello
Development Editor:	Michele Rossi
Copy Editor:	Alana Bubnis
Typesetter:	Jennifer Wetzel
Cover Design:	Lisa Tosheff
Printed at:	Yurchak Printing Inc.

Published in the United States of America by
Idea Group Publishing (an imprint of Idea Group Inc.)
701 E. Chocolate Avenue, Suite 200
Hershey PA 17033 USA
Tel: 717-533-8845
Fax: 717-533-8661
E-mail: cust@idea-group.com
Web site: http://www.idea-group.com

and in the United Kingdom by
Idea Group Publishing (an imprint of Idea Group Inc.)
3 Henrietta Street
Covent Garden
London WC2E 8LU
Tel: 44 20 7240 0856
Fax: 44 20 7379 3313
Web site: http://www.eurospan.co.uk

Library of Congress Cataloging-in-Publication Data

Geographic information systems in business / James B. Pick, editor.
p. cm.
Includes bibliographical references and index.
ISBN 1-59140-399-5 (hardcover) -- ISBN 1-59140-400-2 (pbk.) -- ISBN 1-59140-401-0 (ebook)
1. Management--Geographic information systems. 2. Business--Geographic information systems. I. Pick, James B.
HD30.213.G46 2005
910'.285--dc22

2004003754

British Cataloguing in Publication Data
A Cataloguing in Publication record for this book is available from the British Library.

All work contributed to this book is new, previously-unpublished material. The views expressed in this book are those of the authors, but not necessarily of the publisher.

Dedication

This book is dedicated with appreciation to my wife, Dr. Rosalyn M. Laudati, who was always patient and supportive with the long hours and deadlines of editing that sometimes intruded on family time.

Geographic Information Systems in Business

Table of Contents

Section II: Conceptual Frameworks

Section III: Applications and the Future

Foreword

By: Jack Dangermond
President, ESRI Inc.

Throughout my career I have been convinced that the use of geographic information systems (GIS) technology by businesses would result in better decision-making, increased efficiency, significant cost benefits, and improved customer satisfaction. Although GIS is very widely used by local, state, and federal governments and utilities, most of the business community has been slow to embrace this technology. One reason for the slow adoption of spatial technologies has been the lack of educational opportunities to learn about GIS in our business schools. In recent years, the business community has discovered GIS and the advantages of spatial analysis. But still, GIS is rarely taught in business schools. Part of the reason for the dearth of GIS in business schools is the lack of research books on GIS with a focus on the business side, good textbooks, and usable case studies on GIS applications to business processes. I expect that this book will help change that by making available a valuable resource for educators and researchers.

This book brings together North American and European leaders of thought in the use of GIS for business applications. The contributors to this book are a veritable "Who's Who" from the academic world of GIS and business. The book covers a broad range of topics and business applications, from agriculture to real estate to health care. The chapters address and expand on important business-related methods and concepts including spatial decision support systems, the design of enterprise wide GIS systems, a software design approach to GIS-based knowledge discovery using qualitative reasoning, the role of GIS in systems that include a wide variety of geospatial data sources, conceptual models of e-geobusiness applications, the relationship of GIS to mobile technology and location based services, and emerging technologies.

As we fully enter the Information Age, we are experiencing an overwhelming flood of data. We need tools to help us sift through and organize the data to find useful information that can better inform business processes. Geographic information systems provide us with a powerful tool for organizing and searching data within geographies.

This book is useful to business school professors who want to offer their students the best of the new techniques, business school students looking for marketable skills, business leaders looking for an edge in a highly competitive business environment, and individuals looking to improve their skill set to better compete for jobs in a high-tech world.

I believe that this book will help us move toward a more spatially literate society, a world in which the business schools are providing comprehensive education that includes an understanding of the spatial sciences and how to use the powerful tools for analysis of geographic data.

Preface

The Growth and Development of GIS in Business

Geographical information systems (GISs) access spatial and attribute information, analyze it, and produce outputs with mapping and visual displays. An early definition stated: GIS is "an information system that is designed to work with data referenced by spatial or geographic coordinates. In other words, a GIS is both a database system with specific capabilities for spatially-referenced data, as well as a set of operations for working with the data" (Star & Estes, 1990).

GIS in business has grown as a significant part of this subject. It has been stimulated by the rapid expansion of GIS use in the private sector during the 1990s and early 21st century. Companies are utilizing this technology for a variety of applications, including marketing, retail, real estate, health care, energy, natural resources, site location, logistics, transportation, and supply chain management. GIS can be combined with global positioning systems, remote sensing, and portable wireless devices to provide location-based services in real-time. GIS is more and more being delivered over the Internet. Increasingly, it constitutes a strategic resource for firms.

This book fills a gap in the scholarly literature on GIS. Although books and journals are devoted to GIS in general (Longley et al., 2000; Clarke, 2003) and to its practical applications in business (Grimshaw, 2000; Boyles, 2002), there has not been a book solely focused on research for GIS in business. As Chapter II points out, there is a deficit of peer-reviewed research on GIS in business, which means this book can be helpful in bringing forward a compendium of current research. Also, by its two literature review chapters and references throughout, this volume can serve to direct interested persons to diverse and sometimes scattered sources of existing scholarship.

The early developments leading to GIS stem from the mid-20th century (Clarke, 2003). Swedish weather mapping was computer-based in the mid-1950s (Longley et al., 2000). In the late 1950s in the UK, Terry Coppock performed geographical analysis of a half million agricultural census records (Longley et al., 2000). At this time, GIS was conceptualized by Waldo Tobler (Tobler, 1959), who foresaw the role of map input, map analysis, and map output (Clarke, 2003). Batch computer programs for GIS were produced in the 1960s by several groups (Clarke, 2003). The early uses of GIS were in government, at the federal, state, and local levels. Canadian governments were especially significant early adopters of GIS. This is not surprising, since Canada is an advanced nation having extensive land area and natural resources, which could benefit by improved public management. In the mid-1960s, Ralph Tomlinson and others utilized computers to perform intensive mapping of the Canada Land Inventory. He led in producing the Canada Geographic Information System (CGIS), which many regard as the first GIS (Longley et al., 2000). In the same period, the Harvard University's Laboratory for Computer Graphics and Spatial Analysis designed and developed software leading to an improved GIS program, Odyssey (Clarke, 2003). Commercial programs became available in the late 1960s by companies such as ESRI Inc. and others. Like other information technologies, early GIS uses were constrained by computers' low disk storage capacity, slow processor speeds, and bulky sizes. GIS was more constrained than the average range of IS applications, because of the additional need to store spatially referenced boundary files. In the late 1960s and early 1970s, remote sensing, i.e., photographs of the earth's surface, was developed and later linked with GIS (Longley et al., 2000).

One of the underlying enablers of GIS over the past 35 years has been the rapid increase in both computer storage capacities and processing speed. As seen in *Table 1*, the ratio of transistors per silicon chip increased at a rate that doubled approximately every one and a half years, a phenomenon known as Moore's Law (for Gordon Moore, who formulated it in 1965). The rate has increased at that amount during the past 40 years. The GISs that ran on bulky mini-computers in the mid-1980s with processing speeds of around 16 megahertz today run on small laptops with speeds of 4 gigahertz (4 billion Hz) or more. Although some have questioned whether Moore's Law and other growth rates will continue in the long range, all prognosticators are indicating storage densities will grow in the mid-term.

For GIS, the faster speeds have allowed much more refined databases, analysis, modeling, visualization, mapping features, and user interfaces. GIS applications and its user base grew rapidly in the 1990s and early 21st century. It has become connected with global positioning systems, the Internet, and mobile technologies. With multiplying applications, it continues to find new uses every year. Datatech projected that the sum of revenues for GIS core-business will be \$1.75 billion in 2003, an 8 percent increase from 2002 (*Directions Magazine*, 2003). The GIS software vendor sales totaled \$1.1 billion, two thirds of the total, while services accounted for 24 percent (*Directions Magazine*, 2003).

Concomitant with the increase in chip capacity has been a dramatic fall in price per transistor (Intel, 2003). From one dollar per transistor in 1968, the price has fallen to a cost of \$0.0000005 per transistor in 2002 (Intel, 2003).

At the level of large-sized systems and applications, expanded computing power, combined with the Internet and modern telecommunications infrastructure, allows GIS to be deployed across an organizations as a worldwide enterprise system. In enterprise applications, the GIS processing is centered in specialized groups of servers that are interconnected through middleware to the client-based end users. The development of enterprise GIS resembles the trend towards enterprise resource planning systems (ERP).

Sometimes they are merged; in fact, many ERP systems allow for interconnections to GIS software.

A number of other technology trends have led to the expanding use of GIS. They include more sophisticated and robust GIS software, evolving database design, improved visualization display — both hardware and software — and, since 1992, the growth of the commercial Internet (Longley et al., 2000). Like other information systems applications, GIS has benefited notably from the Internet. As a consequence, GIS applications are available as web services, and, in some cases, a single map server responds to millions of requests per week. This area of GIS is rapidly expanding. GIS is utilized in location-based applications refers to applications where small portable devices are connected by the Internet to send and receive data to and from centralized computing resources. Hand-held GIS devices such as ArcPad (ESRI, 2003), coupled with other mobile devices, support these applications.

Another group of related technologies has been more specifically advantageous to GIS in business. Some of the more important ones are given in the attached table.

These associated technologies are discussed in many of the chapters. They have added to the momentum of GIS use in business.

From the standpoint of academia, GIS originated in the 1960s and 1970s in landscape architecture, geography, cartography, and remote sensing (Longley et al., 2000). During the last 20 years, it has branched into other academic disciplines, notably computer science (Longley et al., 2000), statistics, and more particularly geostatistics (Getis, 2000), land administration (Dale & McLaren, 2000), urban planning, public policy (Greene, 2000), social sciences, medicine (Khan, 2003), and the humanities (Gregory, Kemp, & Mostern, 2002).

In the 1990s, it began to spill over into the business disciplines including management (Huxhold & Levinsohn, 1995), information systems (Grimshaw, 2000), organizational

Table 1. Moore's Law — Transistor Capacity of Intel Processor Chips, 1971-2000

Year of Introduction	Chip	No. of Transistors per chip	MIPS*
1971	4004	2,250	0.06
1972	8008	2,500	
1974	8080	5,000	0.64
1978	8086	29,000	0.75
1982	286	120,000	2.66
1985	386	275,000	5.00
1989	486	1,180,000	20.00
1993	Pentium	3,100,000	66.00
1997	Pentium II	7,500,000	1,000.00
1999	Pentium III	24,000,000	
2000	Pentrium IV	55,000,000	14,000.00

* *millions of instructions per second*
Source: Intel (2003)

studies (Reeve & Petch, 1999), real estate (Thrall, 2002), retail management (Longley et al., 2003), and telecommunications (Godin, 2001).

In the early 21st century, some business schools have recognized the importance of GIS by including it as a required course or degree emphasis: for instance, the elective GIS course at University of California Berkeley's Haas School of Business, and University of Redlands' MBA emphasis in GIS (UCGIS, 2003). Several business schools have established centers for GIS research, such as Wharton Geographic Information Systems Laboratory. University College London established the interdisciplinary Centre for Advanced Spatial Analysis (CASA), which is an initiative to combine spatial technologies in several disciplines that deal with geography, location, business, and the built environment. The interest of business schools in GIS is just getting started, but is likely to be stimulated by the rapid growth in industry of GIS and location-based services.

Another set of developments contributing to the study of GIS in business consists of its concepts, methodologies, and theories. Geographic information systems utilize methods and techniques drawn from many disciplines, including geography, cartography, spatial information science, information systems, statistics, economics, and business. It is typical of new fields to draw on referent disciplines, eventually combining concepts to form a core for the field. Some of the concepts and theories for GIS in business and their referent disciplines are shown in *Table 3*. Some of them are referred to and elaborated on in chapters of this book. They include decision support systems (from information systems), remote sensing (from geography and spatial information science), geostatistics (from spatial information science and statistics), marketing theories (from marketing), and cost-benefit analysis (from economics and business), and spatial analysis (from geography). The latter two are discussed here as examples of the conceptual origins for business GIS.

Table 2. Examples of Technologies Closely Associated with GIS for Business

Technology	*Importance for GIS in Business*
Global positioning systems	GPS combined with GIS allows real-time locational information to be applied for business purposes.
RFID	Allows portable products of any type to be spatially registered and to carry data that can be accessed and updated remotely. Useful in business because its supply chains and inventories consist of goods that are moved around and can benefit by being tracked (Richardson, 2003).
Spatial features built into leading relational databases, such as Oracle	Makes large-scale GIS applications easier and more efficient to realize. GIS software packages have specific add-ons to link to the database spatial features. Applies to business because enterprise applications are mostly adopted by businesses
Mobile wireless communications	Allows field deployment of GIS technologies in mobile commerce. Useful in supporting the real-time field operations of businesses (Mennecke & Strader, 2003). Combines GIS, GPS, and wireless technologies.
Hand-held GIS, such as ArcPad	A new type of product that is equivalent to PDAs, cell phones, and other mobile devices. It contains GPS and scaled-down versions of standard GIS software. Gives businesses field flexibility in inputting, modifying, and utilizing data. Important in business sectors, such as retail, that have substantial field force (ESRI, 2003).
Map server software	Specialized software to support servers that deliver GIS over the internet. The software converts maps from conventional GIS storage form into versions that are coded and optimized for web delivery

Cost-benefit (C-B) analysis was developed by economists originally, and applied to justify a wide variety of public sector and private sector projects. It takes concepts from economics including the time value of money, the influence of markets on C-B analysis, and determination of break-even point. Business disciplines adopted it and farther refined it for business problems. The information systems discipline in particular expanded the theory to analyze the costs and benefits of information systems (King & Schrems, 1978). The information systems field added the related concepts of the productivity paradox, which analyzes investment in IS and the returns on investment (Brynjolfsson, 1993; Lucas, 1999; Strassmann, 1999; Devaraj & Kohli, 2002). These theories and concepts apply to GIS in business because they form the principal methods and theories for decision-makers to decide whether to adopt and deploy GISs.

Spatial analysis stemmed originally from developments in geography and regional science in the early 1960s (Fischer, 2000). It includes "methods and techniques to analyze the pattern and form of geographical objects, ... the inherent properties of geographical space, ... spatial choice processes, and the spatial-temporal evolution of complex spatial systems" (Fischer, 2000). A simple example of spatial analysis is the overlay, which juxtaposes two or more map layers on top of each another: the positions of spatial objects can be compared between layers, for instance highways on one layer crossing the boundaries of marketing territories on a second layer.

Chapter III on techniques and methods by Greene & Stager discusses some spatial analysis methods, as well as two more elaborate case studies. Spatial analysis techniques differ from ordinary database functions by involving computations on spatial attributes (such as points, lines, and polygons), rather than just data attributes (such as numbers and characters). Advanced applications of spatial analysis involve elaborate spatial simulation, modeling, and visualization (Longley & Batty, 2003). This side of GIS is less familiar to scholars in the business disciplines. For this reason, some of

Table 3. Referent Disciplines for Concepts and Theories of GIS

Concept or Theory in GIS in Business	*Referent Discipline*
Spatial Analysis	Geography, Regional Science
Location Theory	Geography
Gravity Model	Geography
Remote Sensing	Geography, Earth Sciences
Decision Support Systems	Information Systems
Knowledge-Based Discovery	Information Systems
Data Mining	Information Systems
Location Based Services	Information Systems
Value of IT Investment	Information Systems, Economics
Electronic Business	Information Systems, Economics
Networking Configuration	Telecommunications
Visualization	Computer Science
Geostatistics	Statistics
Customer Relationship Management	Marketing, Information Systems
Adoption/Diffusion Theory	Marketing
Market Segmentation	Marketing
CAMA and AVM Models	Real Estate
Cost-Benefit Analysis	Economics, Business
Organizational Theory	Management, Sociology

its elements are included in the Greene & Stager chapter. Other sections in this volume refer to spatial analysis, including in Chapters VI, VIII, and XII.

Organization of the Book

This book is divided into three parts: Section I: Foundation and Research Literature, Section II: Conceptual Frameworks, and Section III: Applications and the Future. Section I examines the development of the field of GIS in business, summarizes its research literature, and provides a foundation for analytical methods and techniques of GIS in business. Section II examines conceptual frameworks for GIS as seen in the context of information systems and other business discipline. Section III analyzes GIS business applications in the real world, including health care services, marketing, retail, real estate, the power industry, and agriculture. The section and book ends with discussion of future applications of GIS.

Section I: Foundation & Research Literature

The four chapters in Section I examine the body of scholarly research literature on GIS in business, survey techniques and methods of GIS for business, and analyze its costs and benefits. This part critically reviews the body of knowledge available for this field, as well as presenting some of its fundamental business blocks.

Chapter I. GIS in business as a scholarly field developed over the past four decades, drawing from and relating to information systems and other business disciplines, as well as to the real world. In the first chapter, "Concepts and Theories of GIS in Business," Peter Keenan delineates the growth of this field's body of knowledge, referencing and linking together key studies in the literature. The role of GIS has progressed from information reporting to spatially enabled databases and to spatial decision support systems. This paralleled the movement generally of the IS field towards decision support and strategic systems. The literature and key concepts for important areas of business application of GIS are reviewed, notably logistical support, operational support, marketing, service, trends in spatial decision support systems (SDSS), electronic commerce, and mobile commerce. In service, for instance, the movement towards customer relationship management (CRM) systems is further reinforced by GIS. Customers' spatial relationships can be utilized to provide better service. For consumer electronic commerce, GIS supports the delivery logistics. In mobile services, GIS, combined with wireless and GPS, customizes service at the customer location. The chapter later refers to the classical Nolan stage theories of IS growth (Nolan, 1973). It suggests that GIS in the business world today is entering the expansion/contagion stage. GIS will be helpful in the subsequent stage of data integration. However, the data administration stage may pose for GIS problems due to its complexity. The author asserts GIS to have yet unrealized potential in business. This chapter is informative of the growth and maturation of the field's body of knowledge and the diverse literature that supports it.

Chapter II. This chapter, "GIS and Decision Making in Business: A Literature Review," by Esperanza Huerta, Celene Navarrete, and Terry Ryan, focuses on the extent of research during the past 12 years in one area within business GIS, namely GIS and decision support systems. The authors perform a comprehensive and in-depth literature review of leading information systems journals and conference proceedings, predominantly in information systems along with some from the GIS field. Over the dozen years, the 20 publications contained merely nine articles on GIS and decision support! A well-known model of decision support by Todd & Benbasat (2000) is utilized to classify the articles by area, which showed a deficit of studies on "desired effect" and "decision strategy." The paucity of peer-reviewed research in the GIS-DSS area suggests an overall lack of research on GIS in business, underscoring the importance of bringing forward the contributions in this book.

Chapter III. "Techniques and Methods of GIS for Business" focuses on spatial methods that are commonplace for GISs and can be applied in the business world. The chapter starts with rudimentary elements, such as spatial databases, spatial queries, mapping classifications, table operations, buffers and overlays. It provides simple instances of how those operations can be applied to business. The chapter ends with two case studies of more sophisticated spatial analyses, one on industrial specialization and location quotient analysis in an urban labor market, and the second on trade area analysis, based on the gravity model, which examines the specific instance of opera houses in the Midwest. The chapter is somewhat introductory, and will benefit the reader having limited knowledge of spatial analysis.

Chapter IV. In anticipating applying GIS in an organization, a crucial aspect is to assess the costs and benefits. The chapter on "Costs and Benefits of GIS in Business" examines the key factors and methods for assessing costs and benefits. Cost-benefit (C-B) analysis for GIS differs from C-B analysis in non-spatial IS in two ways. First, GIS software tends to be linked with other technologies and software, such as GPS, wireless technologies, RFID, statistical software, and modeling packages. This need to link up may result in added costs as well as benefits. Second, GIS data and data management must deal with both attribute and spatial data, which influence C-B differently. Third, the visualization aspect of GIS is hard to quantify and therefore adds to intangible costs and benefits. The costs and benefits are related to the organizational hierarchy of an organization. There is a long-term trend for GIS business applications to move up this hierarchy, i.e., from the operational to managerial to strategic levels. At the higher levels, benefits become more difficult to assess. A related topic considered with respect to GIS is the productivity paradox. The productivity paradox refers to studies that have had ambiguous results on whether IT investments lead to added value. The productivity paradox and value of IT investment literature is discussed as it relates to assessing the payoff of GIS.

Section II: Conceptual Frameworks

This part of the book includes studies that expand on and contribute to conceptual frameworks drawn mostly from the information systems field.

Chapter V. Scholars and industry specialists tend to be familiar with desktop or laptop GIS, but less so with enterprise deployments of GIS. Those have a variety of architectures, comprising spatial processors, databases, networking, and interconnecting components such as middleware. In "Spatial Data Repositories: Design, Implementation, and Management Issues," Julian Ray presents a new taxonomy for the architectures of large-scale GIS, and analyzes the design, implementation, and management issues related to this taxonomy. Special attention is given to how spatial data repositories (SDR) function in these enterprise arrangements. The design issues include how databases perform, physical storage, provision of real-time data, how to update data, and the integration of multi-vendor products. Implementation considers the formats of spatial data, steps to load spatial data, and the compatibility of spatial data within SDRs. Enterprise GIS systems raise management issues that are discussed, notably the costs, staffing, licensing, and security of SDRs. The future movement is towards real-time systems and subscription-based web services. The chapter will be useful to companies planning enterprise-wide geographic information systems, and to scholars studying them.

Chapter VI. Knowledge discovery, or the process of extracting data from large datasets, has undergone thorough study for non-spatial relational databases. On the other hand, knowledge discovery spatial databases have been little investigated. "Mining Geo-Referenced Databases: A Way to Improve Decision-Making," by Maribel Yasmina Santos and Luis Alfredo Amaral, presents a model and application of spatial knowledge discovery. It is based on a new model of qualitative relations between spatial attributes, which retains standard data-mining features as well. The model includes qualitative spatial relations of three types — direction, distance, and topology. The model is expressed in tables that apply these relations singly or in sequence. The authors have designed and built a working prototype system, PADRÃO, for knowledge discovery in spatial databases (KDSD). PADRÃO is built on top of the components of Microsoft Access, the Clementine data-mining package, and the GIS software Geomedia Professional. PADRÃO prototypes an application to regional banking credit decisions in Portugal. The KDSD approach draws on and leverages from existing literature about knowledge discovery to provide a conceptual base, logic, algorithms, and software to give convincing results for its spatial rendition. Besides academics, industry designers and other practitioners will benefit from the chapter.

Chapter VII. The movement of GIS upward in organizational level has occurred over the past 30 years and has paralleled similar steps in development in conventional ISs from transaction processing to MIS to decision support systems. "GIS as Spatial Decision Support Systems," by Suprasith Jarupathirun and Fatemeh Zahedi, centers on the decision-support role of GIS; it analyzes what is unique about spatial decision support systems (SDSS) vs. DSS. Besides SDSS's wide range of applications, SDSS has spatial analytical tools that go beyond ordinary DSSs and include standard zoom, buffer, overlay, and other spatial functions, many reviewed in Chapter III. It also has advanced, specialized functions for special purposes that are both spatial and analytical including, for example, 3-D visualization, statistical modeling, and network analysis. The authors dig deeper on visualization by identifying through the literature the unique visualization features of SDSS that include the dynamic nature of map visualization, visual thinking, and the behavioral impact on decision makers. Given all this, how can the efficacy of an SDSS be evaluated and tested? The authors present a conceptual model of SDSS that can constitute a basis for testing and evaluation. The model includes technology, problem tasks, and behavioral abilities, and the resultant task-technology fit, as well as incentives, goals, performance, and utilization. Future enhancements of SDSS may include use of 3-D, animation, and intelligent agents. A chapter rich in its literature references, it advances understanding of the properties of SDSS and enlarges its conceptual theory. SDSS is at the core of why GIS is essential to real-world decision makers, so practitioners should be interested as well.

Chapter VIII. Although 80 percent of business data is potentially spatially-referenced, opportunities to utilize its spatial aspects are often missed in industry. However, managers possessing spatial mindsets can tap into considerably more of the spatial potential and bring new types of spatial data, such as remotely-sensed data, to bear on improved decision-making. Spatiotemporal data, i.e., spatial data that is not from a single time slice but extending over time, can enhance business decisions. In "The Value of Using GIS and Geospatial Data to Support Organizational Decision Making," W. Lee Meeks and Subhasish Dasgupta emphasize the data side of spatial decision-making models. Where do the data come from? What is the data's accuracy and utility for the problems at hand? Have all available sources of data been looked into? Can automated tools such as search engines ease the challenge of identifying the right spatial data? Once the spatially-referenced data are available, do managers have the mindset to take advantage of it? The chapter starts with the conventional SDSS model, but enlarges it to include data sources and the ability to comprehend/use the data. It expands the range of sources of spatial data from maps, scanning, and GPS to include remotely-sensed data. The potential of remotely-sensed data is growing, since satellites' spectral resolution, spatial resolution, and accuracy have increased. Managers in industry need to be open to including remotely-sensed data for decision-making. The chapter forms a complement to Chapter VII, since it elaborates greatly on the data side of the SDSS model, whereas Chapter VII emphasizes decision-making and visualization.

Chapter IX. There is potential for spatially-enabled business, or geo-business as this chapter's authors refer to it, to advance from physical to digital to virtual applications. However, reaching the state of virtual application depends on appropriate business conditions in which the spatially-enabled virtual business is justified to be beneficial. In the chapter "Strategic Positioning of Location Applications for Geo-Business," Gary Hackbarth and Brian Mennecke present conceptual models that help to understand whether the spatially-enabled virtual business is appropriate or not. The first model, the net-enablement business innovation cycle (NEBIC), modified from Wheeler (2002), consists of the steps of identifying appropriate net technologies, matching them with economic opportunities, executing business innovations internally, and taking the innovation to the external market. The process consumes time and resources, and depends on organizational learning feedback. The second model, modified from Choi et al. (1997), classifies geo-business applications into 27 cells in three dimensions, consisting of virtual products, processes and agents. Each dimension has three categories: physical, digital, and virtual. The authors discuss examples of spatially-enabled applications that fall into certain cells of this model. The model is helpful in seeing both the potential and limitations for net-enabled applications. A final model classifies spatially-enabled applications by operational, managerial, and individual levels. Examples are given that demonstrate spatial applications at each level. The chapter helps to establish frameworks for virtual geo-business applications, which include evolving stages over time of e-enablement; a classification of physical-digital-virtual processes, products, and agents; and the differences in spatial applications at the operational, managerial, and individual levels of decision-making. These models are useful in not perceiving geo-business applications as all or nothing in virtual enablement, but rather as located somewhere across a complex multidimensional range.

Section III: Applications & the Future

This part of the book examines GIS applications in a number of sectors. It is not intended to be comprehensive, but to give in-depth analyses of several varied areas. It finishes with a teaching case of GIS in agriculture and a study that considers the future of GIS in the business world.

Chapter X. Chapter X begins Section III of the book on Applications and the Future by addressing GIS in health care services. The authors Brian Hilton, Thomas Horan, & Bengisu Tulu emphasize the variety of health care uses, presenting the results of three case studies at the operational, managerial, and strategic levels. "Geographic Information Systems in Health Care Services" refers to Anthony's classical theory of organizational levels and illustrates its relevance with three cases, the first at the operational level of a health care company operating a spatially-enabled system for making physician appointments for claimants with disabilities. In a managerial level case, government providers of emergency medical services need to provide spatial technologies to connect with mobile devices accessing the emergency 911 system. At the strategic level, spatial technologies are utilized to support the display of epidemiological data on SARS as part of the large-scale National Electronic Disease Surveillance System (NEDSS). The authors analyze the solutions and outcomes of these case studies, as well as future issues that need to be addressed by the management of the case organizations — for instance, the health care company needs to better integrate its spatial and non-spatial databases. This chapter is helpful in its analysis and comparison of the successes of three varied cases of GIS in healthcare services.

Chapter XI. Marketing that includes spatial analysis has enhanced utility. For instance, a marketing study of a person's residential location can indicate his/her likely consumption pattern. Nanda Viswanathan, in "GIS in Marketing," considers key constructs of the marketing field and how GIS and spatial science have the potential to enlarge the dimensions of marketing and increase its efficiency. The chapter begins by considering marketing in terms of space, time, and demographics. These three components are nearly always present for real-world marketing problems.

GIS supports marketing models of both space and time that include demographics as attributes. The chapter examines spatially-enabled strategies for products, pricing, promotions, and distribution. For instance, the product life cycle traditionally is applied to the whole economy. For instance, a car product is marketed differently at initial roll-out, versus its peak sales time, versus as a mature product. GIS allows product-life-cycles models to be disaggregated into small geographic areas, with the tapestry of differences revealed through mapping and spatial analysis. For distribution, the supply chain can be modeled spatially. A further enhancement is to add real-time, location-based information to achieve a dynamic view of the supply chain. What are the locations and destinations of certain products at this moment and how can their movement and deliveries be spatially-optimized?

Another chapter topic is GIS to support marketing analysis and strategy. Spatial models can support market segmentation, customer relationship management, competitive analysis, and simulating dynamic markets. For example, competitive analysis of products can be done for small areas, for instance census tracts. The interaction effects of competition in one small zone influencing other small zones can be included in spatial competition models. Mapping and visualization can inform marketers of fine differences in competition by location. A final chapter segment cautions that the combined spatial marketing techniques of GIS, GPS, mobile devices, and the Internet may pose serious privacy and ethical issues. The author recommends that the American Marketing Association's ethical codes for Internet marketing be extended to GIS and location-based services. As costs decrease and data-availability expands, marketers can realize the diverse uses suggested in this chapter.

Chapter XII. Retailing is inherently spatial. Stores, customers, and advertising have intertwined physical locations that underpin business outcomes. In "The Geographical Edge: Spatial Analysis of Retail Loyalty Program Adoption," spatial analysis is utilized to spatially-enhance a traditional production diffusion model, which is illustrated for a single store of a major retailer. Authors Arthur Allaway, Lisa Murphy, and David Berkowitz discuss in detail a prototype of a cutting-edge marketing technique. Data recorded in the store's POS system from the loyalty card data that customers entered is supplemented with census and other community data. The customer addresses are geocoded, in order to obtain X-Y coordinate locations. Other data on the loyalty adoption cards include the products purchased, time and date of purchase, previous adoptions, and spending behavior. This is supplemented by adding in U.S. Census sociodemographic data at the block group level.

The ensuing database contains records on 18,000 loyalty-program adopters in the store's territory. Spatial diffusion results show the particular influence of early innovators on their neighborhoods and the entire course of adoption and diffusion. Three distinct spatial diffusion stages are evident. Furthermore, the location of the store and the billboards advertising the loyalty program are influential. The authors demonstrate that the billboards can be manipulated experimentally to test assumptions. The chapter reinforces a common point in the book that there is potentially much more spatially-enabled data than people recognize, and that new, innovative uses are waiting to be discovered.

Chapter XIII. Real estate valuation can be done for large samples of properties encompassing whole municipalities and regions. With the increasing affordability of GIS software, spatial analysis can be added to traditional non-spatial estimation methods, increasing their predictive accuracy. Susan Wachter, Michelle Thompson, and Kevin Gillen, in "Geospatial Analysis for Real Estate Valuation Models," give theoretical background on models that include spatial variables, and then illustrate the Automated Valuation Model (AVM) with a case study of a community in southern California. The traditional Computer Assisted Mass Appraisal (CAMA) model estimates real estate values based on prior prices, while the classic, non-spatial hedonic model estimates values from housing characteristics of the immediate area. The authors combine the hedonic and spatial models in the form of a linear regression. The spatial part of this model consists of real-estate prices at particular radial distances from the property being estimated. Their results for Yucca Valley, California, demonstrate substantial improvement in regression significance and predictive power for the mixed hedonic-spatial model, compared to hedonic alone or spatial alone. The real estate industry and local and regional governments are beginning to adopt such mixed models. This chapter substantiates the benefit of including spatial components in real estate valuation models. It also suggests that there is future potential to build valuation models with more spatial dimensions, enhancing their significance and accuracy.

Chapter XIV. Large-sized power systems are essential elements for advanced societies. Their software support systems need to be reliable, well-maintained and able to respond to emergency situations. Although these large systems are mostly taken for granted by consumers, system failures such as the widespread U.S.-Canadian electrical grid failure in the summer of 2003, raise questions and concerns. "Monitoring and Analysis of Power Line Failures: An Example of the Role of GIS," by Oliver Fritz and Petter Skerfving, explains the role of GIS in these multilayered and geographically-distributed software systems. The chapter starts by explaining software support systems for power lines. The systems function at the operational level to support line monitoring and maintenance, while the management level, they support optimization of the system, as well as capacity and economic planning of the network, such as pricing and estimates of customer base.

GIS is a modular component that offers advantages to these software systems. At a low level, it can provide basic mapping of fault locations, to assist in emergency repair. Other benefits appear post-incident since fault maps can be overlaid with weather and topographic maps, assisting experts to analyze of the causes of outages. At a higher level, GIS displays and analysis can assist in investment planning of new lines and other assets. An aspect of GIS of profound significance is its integrative role in encouraging cross-department applications and managing the power line systems. The authors present a case study that combines Power System Monitoring (PSM) software for fault detection with GIS for map display. The chapter emphasizes the role of GIS in the power industry, as one modular component within large-scale monitoring, maintenance, and analysis of software systems.

Chapter XV. In "GIS in Agriculture," Anne Mims Adrian, Chris Dillard, and Paul Mask delineate modern precision agriculture and explain the role of GIS. Precision agriculture utilizes measurements of soil type, crop yield, and remote sensing data to pinpoint micro-areas for special treatments. Farm equipment can be automated to deliver exact amounts of fertilizers and chemicals to particular micro-areas. Since the movement of farm vehicles can be detected precisely, GIS and GPS together sense exactly where the micro-areas are and inform automated systems when to effect precision treatment. The systems yield large amounts of information. Unfortunately, farmers and agricultural managers may not be able to process more than a small fraction of it. The authors suggest that farmers need to become better trained in these technologies, and to gain greater confidence and motivation to utilize them. Until now, adoption rates for GIS have been slow. One reason is that farmers struggle with economically justifying the new technologies. There is potential that a higher percentage of farms will adopt GIS and GPS technologies. GIS in agriculture has so far been primarily at the levels of supporting operations on the ground, but the time is ripe for expanding the use of spatial decision support systems by farmers.

Chapter XVI. "Isobord's Geographic Information System Solution," by Derrick Neufeld and Scott Griffith, is an educational case study of a GIS adoption decision confronting a small Canadian firm, Isobord. The firm was later acquired by Dow Bioproducts. The case pertains to many issues raised in this book. Isobord is a small particleboard firm operating on the Canadian prairies in Manitoba that has discovered an environmentally sound approach to acquiring its materials, namely to substitute straw instead of wood. However, since it doesn't make economic sense for farmers to deliver the straw, Isobord had to develop its own pick-up service over a large area with a radius of 50 miles. However, pick-up is very difficult in the flat prairie landscape, which lacks markers and has rough roads.

The answer was to utilize a combination of GIS and GPS to pinpoint pick-up locations. The case details how Isobord begin with its own local software solutions and then graduated to the use of commercial packages. At the end of the case, the firm is at the point of deciding on one of three alternative software solutions, each offering a different platform, software, and servicing. The case raises the issues of GIS costs and benefits, planning, human resources, outsourcing, and project scope. The firm differs from most other cases in this book in its small size and budget, and its limited training and experience with GIS. The chapter can be useful to teachers, researchers, and practitioners.

Chapter XVII. How are spatial technologies and GIS moving towards the future? What changes in hardware, software, platforms, delivery, and applications are anticipated? The book's final chapter, "GIS and the Future in Business IT," by Joseph Francica, identifies areas of rapid enhancements and changes, and extrapolates trends into the future. The chapter is practitioner-grounded, since the author is familiar with the cutting-edge in industry.

Several factors underlying anticipated changes are the declining prices of GIS products, database products that are spatially enhanced, location-based services, and web delivery of spatial data and services. Price reductions have contributed to making GIS products ever more widely available, while the inclusion of spatial components in standard databases expands spatial analysis capabilities to a much broader customer group of general-purpose database users. The chapter examines the future trends of web services, wireless location-based services, open-source GIS, further database spatial enhancements, scalable vector graphics, and spatially-empowered XML. Open source refers to software products for which the source code is freely and readily available. It is a software industry-wide trend that offers pluses and minuses that apply as much to GIS as to other technologies. For GIS, open-source offers affordability and ability to change code, but brings along problems of software quality and robustness, standards, and maintenance.

Some examples of future applications are examined, including truck fleet management and field service, and customer relationship management (CRM) to identify and understand the relative locations of customers, suppliers, and the sales/marketing force. CRM can be implemented alongside an enterprise resource planning systems (ERP).

Another future scenario is GIS accessing satellite-based remote imagery combined with the widespread and rich government databases available in the U.S. and some other nations. The e-environment will profoundly affect GIS use, since non-technical users will be able to easily access sophisticated spatial web services that will provide everything a traditional desktop GIS offers, and much more.

Conclusions

In conclusion, the chapters in this volume add to the foundation of research on geographic information systems in business. The authors provide substantial review of the literature, offer revised and updated conceptual frameworks to unify and weave together geographic information science with conceptual theories in academic business disciplines, and give examples of empirical investigations and case studies that test or challenge the concepts. The book should complement other publications that have focused on applied aspects of GIS in business.

It is hoped that the readers will regard this volume as a starting base, from which to expand the theories and empirical testing. As GIS and its related technologies continue to become more prevalent and strategic for enterprises, a growing academic base of knowledge can provide useful ideas to the wider group of real-world practitioners, and vice versa. It is hoped this volume will stimulate further opportunities for researchers on GIS in Business to develop what is today a limited research area into a full-fledged scholarly field, linked to business practice.

References

Boyles, D. (2002). *GIS means business* (Vol. 2). Redlands, CA: ESRI Press.

Brynjolfsson, E. (1993). The productivity paradox of information technology. *Communications of the ACM, 36*(12), 67.

Castle III, G.H. (1998). *GIS in real estate: Integrating, analyzing, and presenting locational information.* Chicago, IL: Appraisal Institute.

Choi, S.Y., Stahl, D.O., & Whinston, A.B. (1997). The economics of electronic commerce. Indianapolis, IN: Macmillan Technical Publishing.

Clarke, K. (2003). *Getting started with geographic information systems.* Upper Saddle River, NJ: Prentice Hall.

Dale, P.F., & McLaren, R.A. (2000). GIS in land administration. In P.A. Longley, M.F. Goodchild, D.J. Maguire, & D. W. Rhind (Eds.), *Geographical information systems* (Vol. 1, pp. 859-875). New York: John Wiley & Sons.

Devaraj, S., & Kohli, R. (2002). *The IT payoff: Measuring the business value of information technology investments.* New York: Pearson Education.

Directions Magazine. (2003, August 9). Datatech reports GIS revenues forecast to grow 8% of $1.75 billion in 2003: Utilities and Government Increase Spending. Retrieved November 2003: www.directionsmag.com.

ESRI. (2003). *ArcPad: Mapping and GIS for mobile systems.* Redlands, CA: ESRI Inc. Retrieved November 2003: http://www.esri.com/software/arcpad.

Fischer, M.M. (2000). Spatial analysis: Retrospect and prospect. In P.A. Longley, M. F. Goodchild, D. J. Maguire, & D. W. Rhind (Eds.), *Geographical Information Systems* (Vol. 1, pp. 283-292). New York: John Wiley & Sons.

Getis, A. (2000). Spatial statistics. In P.A. Longley, M. F. Goodchild, D. J. Maguire, & D. W. Rhind (Eds.), *Geographical Information Systems* (Vol. 1, pp. 239-251). New York: John Wiley & Sons.

Godin, L. (2001). *GIS in telecommunications*. Redlands, CA: ESRI Press.

Greene, R.W. (2000). *GIS in public policy*. Redlands, CA: ESRI Press.

Gregory, I., Kemp, K.K., & Mostern, R. (2002). Geographical information and historical research: Integrating quantitative and qualitative methodologies. *Humanities and Computing*.

Grimshaw, D. (2000). *Bringing geographical information systems into business* (2nd ed.). New York: John Wiley & Sons.

Harder, C. (1997). *ArcView GIS means business*. Redlands, CA: ESRI Press.

Harder, C. (1999). *Enterprise GIS for energy companies*. Redlands, CA: ESRI Press.

Huxhold, W. E. & Levinsohn, A.G. (1995). *Managing geographic information system projects*. New York: Oxford University Press.

Intel. (2003). Moore's Law. Retrieved December 2003: http://www.intel.com.

King, J.L., & Schrems, E.L. (1978). Cost-benefit analysis in information systems development and operation. *ACM Computing Surveys*, *10*(1), 19-34.

Longley, P.A. & Batty, M. (2003). *Advanced spatial analysis: The CASA book of GIS*. Redlands, CA: ESRI Press.

Longley, P.A., Boulton, C., Greatbatch, I., & Batty, M. (2003). Strategies for integrated retail management using GIS. In P.A. Longley & M. Batty (Eds.), *Advanced spatial analysis: The CASA book of GIS* (pp. 211-231).

Longley, P.A., Goodchild, M.F., Maguire, D.J. & Rhind, D.W. (2000). Introduction. In P.A. Longley, M.F. Goodchild, D.J. Maguire, & D.W. Rhind (Eds.), *Geographical Information Systems* (Vol. 1, pp. 1-20). New York: John Wiley & Sons.

Longley, P.A., Goodchild, M.F., Maguire, D.J., & Rhind, D.W. (eds.). (2000). *Geographical Information Systems* (two vols.). New York: John Wiley & Sons.

Mennecke, B.E., & Strader, T.J. (2003). *Mobile commerce: Technology, theory, and applications*. Hershey, PA: Idea Group Publishing.

Nolan, R.L. (1973). Managing the crises in Data Processing. *Harvard Business Review*, *57*(2), 115-126.

Richardson, H.L. (2003). Tuning in RFID. *World Trade*, *16*(11), 46-47.

Star, J.L., & Estes, J.E. (1990). *Geographic information systems: Socioeconomic applications* (2nd ed). London: Routledge.

Strassmann, P.A. (1999). *Information productivity: Assessing the information management costs of U.S. industrial corporations*. New Canaan, CT: The Information Economics Press.

Tobler, W.R. (1959). Automation and Cartography. *Geographical Review*, *49*, 526-534.

Todd, P., & Benbasat, I. (2000). The impact of information technology on decision making: A cognitive perspective. In R.W. Zmud (Ed.), *Framing the domains of IT management* (pp. 1-14). Cincinnati, OH: Pinnaflex Education Resources.

UCGIS. (2003). University consortium for geographic information science home page. Retrieved November 2003: http://www.ucgis.org. Leesburg, VA: University Consortium for Geographic Information Science. Note: this home page indicates where GIS is taught in the member group of universities.

Wheeler, B.C. (2002). NEBIC: A dynamic capabilities theory for assessing net-enablement. *Information Systems Research, 13*(2), 125-146.

James B. Pick
University of Redlands, USA

Acknowledgments

This book could not have been accomplished without the support, cooperation, and collaboration of many persons and institutions. The first acknowledgment goes to the chapter authors, with whom it has been a remarkably easy process to work. Each chapter was reviewed anonymously by three reviewers. They worked hard and added a lot to the book and acknowledgment is expressed to each of them. With several exceptions, the chapter authors contributed reviews of other chapters, and deserve recognition. In addition, the following external reviewers examined one or more of the chapters: Rob Burke, Rafat Fazeli, Jon Gant, Murray Jennex, Mahmoud Kaboudan, Dick Lawrence, Wilson Liu, Doug Mende, Monica Perry, Mike Phoenix, and Vijay Sugurmaran.

At University of Redlands, appreciation is expressed to campus leaders, including President James Appleton for fostering spatial information science on the campus over many years. Although I arrived on the campus as an applied GIS researcher, it was the university's atmosphere and proximity to ESRI Inc. that helped me grow as a teacher and fuller researcher in this field. I thank the university's Information Technology Services for technology support and to the School of Business Faculty Support Services for a variety of assistance at many stages.

Early discussions of the project with Rob Burke and Tony Burns from ESRI Inc. were helpful in formulating idea and scope of the book, and late discussion with Mike Phoenix of ESRI Inc. was a stimulus to wrapping it up. Acknowledgment is expressed to them, as well as to ESRI President Jack Dangermond for his interest and forward to the book. I would also like to acknowledge the Association for Information Systems, which has sponsored a GIS track for quite a few years at its annual conference, and which stimulated contacts and ideas for this book.

At Idea Group, special thanks to the book's support team, especially Michele Rossi, Development Editor, Jan Travers, Senior Managing Editor, Mehdi Khosrow-Pour, Senior Academic Editor, and Jennifer Sundstrom, Assistant Marketing Manager. They were cooperative, helpful, and offered insights and expertise that improved the book.

James B. Pick
University of Redlands, USA

Section I

Foundation & Research Literature

Chapter I

Concepts and Theories of GIS in Business

Peter Keenan, University College Dublin, Ireland

Abstract

This chapter looks at the concepts and theories underlying the application of GIS in business. It discusses the role of information technology in business generally and how GIS is related to other business systems. Different views of GIS use are introduced and the chapter suggests that decision support applications of GIS are more relevant to most businesses than purely operational applications. Porter's value chain approach is used to assess the potential of GIS to contribute to management. GIS is seen as an emerging technology that will increase importance in business in the future.

Introduction

Information technology (IT) has had a powerful impact on the business world in the last 50 years. IT has facilitated the transformation of business and has allowed new business forms to come into existence. This transformation has reflected the potential of IT both as a cost saving mechanism and as a tool for supporting business decision-making. New developments such as the Internet and mobile applications have an important ongoing impact on business, continuing the process of transformation started by the punched card 50 years earlier. Geographic information systems (GISs) are an area of IT application with a significantly different history from other types of information system. GIS-based applications are now becoming widespread in business, playing a role that reflects both the similarity of GIS to other forms of IT and the distinct characteristics of spatial applications.

Origins of Geographic and Business Use of IT

Business use of information technology started in the 1950s in payroll, billing and invoice processing applications. These applications exploited data processing techniques that had been previously used by government agencies such as the U.S. Census Bureau. GIS has its origins in the use of IT for geographic related activities in North America in the same period. These early applications were typically government orientated, such as transport planning in Detroit and Chicago and the Canada Geographic Information System (CGIS) (Coppock & Rhind, 1991).

Early business applications of IT employed relatively simple processing that could be automated using the comparatively crude computer technology of the period. One example was payroll processing, where only four or five simple calculations were required for each individual. This computerization of simple numeric processing was an automation of clerical work, analogous to the automation of manufacturing in the earlier part of the 20th century. The high cost of computing in this period meant that this type of application was mainly confined to large organizations with a high volume of transactions. While these early data processing applications were relatively unsophisticated, they had a significant impact as they concerned activities critical to business. Data processing techniques allowed these critical operations to be performed faster, more accurately and, above all, more cheaply than manual methods. Despite the relatively high cost of computing at this time, significant cost reductions could be achieved by this automation of the clerical processes required for the day-to-day operation of all businesses. Consequently, early business applications of IT had a widespread impact on routine accounting operations, but were initially much less important in other departments of the organization. In a similar way, the early applications of geographic computer processing were only of interest to the small number of companies involved in map-making, surveying or similar geography-based activities. For example, in the oil industry GIS had a role in exploration at an early stage, but would not have been used in marketing in this sector until much more recently. Many early private sector organizations provided consultancy services or GIS software to the public sector. One example would be Tomlinson Associates, set up in 1977 in Ottawa, Canada by Roger Tomlinson, one of the pioneers of GIS. Another example of an early GIS commercial organization would be the Environmental Systems Research Institute established in 1969. This later became ESRI, which is now the main player in the GIS software market.

Development of IT Towards Decision Applications

As IT became more capable and less expensive, business use of computing moved from the automation of clerical processes to decision support applications. This change exploited the superior interaction made possible by time-sharing computers, and the developments in data organization made possible by developments in database management software. The data available in organizations was initially used to produce regular reports in the form of a Management Information System (MIS). The introduction of improved user interfaces in the 1970s facilitated the introduction of Decision Support

Systems (DSSs). These systems constitute a flexible user-friendly interface linked to problem databases and specific models. As the name suggests, DSSs aim to support, rather than replace, the decision-maker (Sprague, 1980). This form of IT became of interest to managers throughout the organization, as these systems could support decision-making in diverse business functions such as marketing or human resource planning. IT use therefore began to spread throughout all of the business functions, a trend facilitated by the introduction of user-friendly personal computers in the 1980s. Improved networking allowed these machines to be connected together, and this has allowed access from a variety of applications to centralized resources such as databases. Modern business applications continue to exploit the rapidly increasing computational power of the computer; but also derive increasing benefits from the ability of IT to store and organize data (databases), distribute the information derived (networking), and present that information in an interactive format (interfaces). This trend also found expression in the development of systems such as Executive Information Systems (EIS) that provide executive management with an overview of business activity within the organization and of competitive forces on the outside.

A similar sequence of developments occurred within GIS, although largely independently from other forms of IT. The distinct development of GIS was partially a consequence of the much larger amounts of data required for spatial applications when compared to business data processing. This meant that the evolution from automation applications to decision support applications was delayed by 10 to 15 years for GIS when compared to traditional business systems (Densham, 1991). Nevertheless, as computer technology became more powerful, the functionality of GIS software greatly increased. This trend, combined with the lower cost of GIS hardware, has facilitated more ambitious spatial applications. Modern GIS provides distinctive database techniques, specialized data processing and a sophisticated interface for dealing with spatial data.

Consequently, interest in decision support in the GIS field grew in the 1980s when the concept of a Spatial Decision Support System (SDSS) was introduced (Armstrong, Densham, & Rushton, 1986). SDSS was built around GIS with the inclusion of appropriate decision models. By the end of the 1980s, SDSS was a recognized area within the GIS community (Densham, 1991). Over time, decision support applications have found increasing acceptance as an application of GIS and spatial applications have come to constitute an increasing proportion of DSS applications (Keenan, 2003). These applications typically require the synthesis of spatial techniques with other business orientated decision-making approaches based on accounting, financial or operations research techniques.

Initially GIS software was run on mainframe computers, then on relatively expensive graphics workstations. However, as computer performance improved in the 1990s, it has become possible to run GIS software on standard personal computers. This meant that the machines commonly used in businesses were sufficiently powerful to do some useful work with spatial data. Powerful GIS software is now readily available on the Microsoft Windows platform, which is widely used in business and is familiar to business users. GIS vendors have also recognized the market potential of business applications and GIS software has evolved to meet the needs of this broader set of users, facilitating the design of business applications.

Table 1. Computerized Support for Decision Making (Adapted from Turban and Aronson, 2001, pg. 22)

Phase	Description	Traditional Tools	Spatial Tools
Early	Compute, "crunch numbers," summarize, organize	Early computer programs, management science models 1950s - 1960s	Computerized cartography 1960s - 1970s
Intermediate	Find, organize and display decision relevant information	Database management systems, MIS 1970s	Workstation GIS 1980s
Current	Perform decision relevant computations on decision relevant information; organize and display the results. Query based and user-friendly approach. "What if" analysis	Financial models, spreadsheets, trend exploration, operations research models, decision support systems. 1980s - 1990s	Spatial decision support systems 1990s
Just beginning	Complex and fuzzy decision situations, expanding to collaborative decision making and machine learning	Group support systems, neural computing, knowledge management, fuzzy logic, intelligent agents	Group SDSS, Intelligent spatial interfaces, evolutionary techniques for spatial problems, Geolibraries

These developments show a clear trend. Early applications of IT had a cost reduction role, similar to other forms of mass production. However, it was quickly realized that computer technology has a dual nature: it can be used to automate, but as a by-product of this automation it can also produce large amounts of information about the process being automated. In a widely cited book, Zuboff (1988) coins the term *informate* to describe the ability of technology to provide information about processes as well as automating them. GIS has also been seen as an *informating* technology (Madon & Sahay, 1997; Snellen, 2001), as it moves from data processing applications to decision oriented applications. The *informating* role of GIS is particularly evident in a business context, where decision-makers value the problem visualization provided by a map, rather than the map itself.

Within the GIS research community, there has been ongoing debate whether GIS is just another information system or whether it has unique characteristics that separate it from other systems. Maguire (1991) conducts a review of the definitions of GIS and suggests that GIS can be seen as a form of IS, with a distinctive orientation towards spatial data and processing. Maguire identifies three views of GIS, with each view focusing on one functional aspect of GIS technology. The *map* view sees GIS as a map processing or display system. The *database* view is concerned with simple analysis, such as overlaying, buffering. The *spatial analysis* view focuses on more complex analytical functions such as modeling and decision-making. While these views have something in common with the use of IS for data processing, database management and more elaborate DSS applications, there are also some differences. The map view of GIS includes techniques not widely used in business applications, such as map production using raster operations. The distinction between a map view and database view of a GIS is less clear in business mapping, as these applications generally involve at least a simple database structure to allow the storage of attribute data in addition to geographic data. The spatial analysis view of a GIS implies that the GIS provides models providing analysis of interest to a decision maker. In the business context, appropriate analysis usually requires the addition of specific business models. In this case the GIS is a platform which can be developed into an analysis system with the addition of appropriate models (Hess, Rubin, & West, 2004; Keenan, 1996). Nevertheless, the development of GIS can be seen as approximating to the phases of development of other forms of IS (*Table 2*). Presentation mapping, although much more sophisticated, can be related to the fixed format reporting of MIS. The database view of GIS, which allows onscreen query, can be compared to modern EIS systems.

Spatial Visualization

The vast majority of modern GIS applications are characterized by sophisticated graphics, and this capacity for visualization allows GIS to provide effective support for problem representation in spatial problems. Long before computer technology was introduced, users gained an improved understanding of spatial problems by the use of maps. While maps were usually initially prepared by governments for political or military reasons, these could also be used for business applications. An important early map, the 1815 geological map of England by Smith (Winchester, 2002), also facilitated business projects such as coal mining and canal construction. In the same period, British Admiralty charts were also seen as an important advantage for British merchant ships trading in distant parts of the world. Early government maps could also be used to assess business potential; one example of this was the 19th century "Atlas to accompany the second report of the Irish Railway Commissioners," which showed population, traffic flow, geology, and topography all displayed on the same map (Gardner, Griffith, Harness, & Larcom, 1838). This allowed easy understanding of the feasibility of proposed railway routes planned by the private railway companies of that period.

Table 2. Views of GIS

GIS View (Maguire, 1991)	GIS operations	Comparable IS System
Map	Map creation	Data processing
	Map presentation	MIS Business Graphics
Database	Simple analysis Visualization	EIS
Spatial analysis	Specialized analysis	DSS

The growth of IS has seen the introduction of new information representation paradigms. As technology has advanced, users' ability to work with information has been enhanced by innovations such as graphical user interfaces. Even relatively simple concepts, such as the representation of multiple spreadsheet tables as tabbed worksheets, or the use of hypertext have greatly enhanced the usability of computer systems. The rapid pace of change in technology has provided scope for the use of new problem representations. However, it takes some time for interface design to take advantage of these developments, as suitable references must be found to assist in the design of new information representation paradigms.

One of the most important strategies in interface design is the use of a visual problem representation to improve user interaction. The area of visual modeling (Bell, 1994) is a recognized part of management support systems. Visual modeling is based on the concept that it is easier to interact with a visual representation of a model than its mathematical equivalent. Geographical techniques have been identified as being relevant to the general field of computer graphics, which has had an important influence on business use of IT for decision-making by facilitating visualization applications. Researchers from the IS tradition have noted that computer technology is especially appropriate for the display of mapping data. Ives (1982) suggested that maps were too difficult to produce manually for most business applications, and that computerized techniques would make this form of representation much more widely available. Cartography has been seen as being an important source of principles for the design of business graphics (DeSanctis, 1984); this reflects the fact that many decision makers are accustomed to using maps, although this may not be true in all cultures (Sahay & Walsham, 1996; Walsham & Sahay, 1999). Speier (2003) noted that information visualization techniques have been widely applied in science and geography, but have only been recently integrated into business applications. Tegarden (1999) uses the example of the 1854 map of the incidence of Cholera by John Snow to illustrate the power of visualization. This map is frequently cited as the ancestor of computerized GIS.

As decision-makers in many business sectors are used to the concept of a map, the display of onscreen maps has long been incorporated in computer-based DSS and EIS systems. Many areas of DSS application are concerned with geographic data, an influential early example being the Geodata Analysis and Display system (GADS) (Grace, 1977). GADS was used to build a DSS for the planning of patrol areas for the police department in San Jose, California. This system allowed a police officer to display a map outline and to call up data by geographical zone, showing police calls for service, activity levels, service time, etc. The increasingly widespread use in business of GIS-based systems for map creation and display since GADS reflects the importance of visualization in human information processing.

In the business context, visualization in GIS poses a challenge to interface designers to provide facilities that meet the problem representation needs of users, while also providing convenient ways of interacting with that representation. Computer interface design generally has yet to take full advantage of the increased power of computing and the richer set of possibilities that this offers for user interaction. The complex nature of spatial data requires GIS to use sophisticated visualization techniques to represent information. It is therefore quite challenging for GIS to also to provide an interactive interface on the same screen. Consequently, GIS applications can especially benefit from better designed human-computer interfaces which meet their specific needs (Hearnshaw & Medyckyj-Scott, 1993). Visualization has been recognized in the GIS community as an important aspect of GIS (Buttenfield & Mackaness, 1991). This may reflect support for the map view of GIS. One limitation of GIS interface designs is that they are seen to provide a means for visualizing results only, rather than providing a comprehensive problem representation for all stages of the problem (Blaser, Sester, & Egenhofer, 2000). A more comprehensive system would allow problem specification using interactive techniques. One example is the Tolomeo system (Angehrn & Lüthi, 1990; Angehrn, 1991). In this case, the user can sketch their problem in a geographical context and the Tolomeo system will try to infer the appropriate management science model to use to solve the problem outlined by the user. Another example of sketching might be a real estate agent who could use a GIS interface capable of interpreting a sketch of a customer's preferences for location (Blaser et al., 2000). In this case the system might interpret the districts where the customer wanted to live and whether they wanted to be close to the sea or other features.

Views of GIS Use

Spatial Data

The spread of GIS technology has been accompanied by simultaneous growth in the amount of digital data available. Extensive collections of spatial data now exist for most developed countries. The same geographical data sets may be used by many different organizations, as many businesses will operate in the same geographic region. Most of the data used in traditional IT applications is sourced within an organization and concerns customers, suppliers, employees, etc. Data of interest in a GIS may include information on existing customers, but will also include data on shared transportation networks and demographic data on people who are not yet customers. Consequently, GIS is somewhat unusual when compared to other business IT applications, in that many users typically outsource both their software and a large part of their data. As the business use of IT moves from internal data processing applications to EIS applications, external data is of increasing importance, and this needs to be effectively linked to external GIS data. The availability and pricing of spatial data is an important factor in the widespread use of GIS, as a significant amount of geographic data is sourced outside the organizations using it.

Geographic data may be collected by the government and made available at little or no cost to organizations that want to use it; this is the case in the U.S. On the other hand, European governments generally seek to recover the cost of spatial data collection from users. Any assessment of the potential of the GIS field to business must take account of the cost and availability of the common data, as well as software and hardware.

Use of GIS as Automation Tool

GIS is of interest to a wide range of businesses. These organizations use IT in very different ways and this influences their adoption of GIS. Reflecting Zuboff's concepts of *automating* and *informating,* GIS may be seen as a means to automate spatial operations or as a tool for obtaining better information about business operations. Map automation is most relevant where traditional paper maps were used; this arises only in specialist roles in most business organizations. One example is the field of Facilities Management (FM), which makes use of computer assisted design (CAD) approaches to record factory layouts, locations of pipe networks, etc. Typically these layouts were superimposed on maps, therefore GIS can be used to better integrate this data and to produce appropriate integrated maps in a less expensive and timelier way.

Utility companies, such as electricity, gas, or water companies, can also exploit GIS to support routine maintenance of pipe, cable, and power networks. For these organizations, the ability to locate quickly a pipe or cable is critical to their ability to continue to provide service to their customers. Traditional approaches suffered from missing data, for example, where a map was lost, and inadequate indexing of the data available. GIS-based technology can be used to automate the search procedure for pipe location, thereby making operations more efficient. Just as data processing allowed simple checks on the integrity of data, GIS-based applications can improve the quality of spatial data used. The productivity gains alone from this type of application may be sufficient to justify the use of GIS, just as productivity gains can justify the use of data processing in business generally.

GIS as an Information Reporting Tool

While automation applications of GIS are not of direct interest to most businesses, applications with the capacity to *informate* are potentially of much wider interest. The simplest forms of information-based applications are those where a map is produced with some graphical information on attribute values superimposed. Presentation mapping has been identified as the dominant requirement of the business use of GIS-based technology (Landis, 1993). Presentation mapping creates a one-way report; the user cannot query the map presented, instead the user assimilates the information provided and indirectly manipulates the data. For example, a map may be displayed on screen with superimposed bar charts on each region showing sales for an organization's products. This is similar to other graphics and charts produced in business software; the graphic provides a report, not an interface. The use of maps as an extension of business graphics is facilitated by the inclusion of a simple mapping add-on in Microsoft Excel. This allows the creation of a form of chart where simple graphics can be associated with spatial entities. A choropleth map (thematic map) displays attribute data, in this case population, associated with relevant spatial units. One example can be seen in *Figure 1*, which allows the user to identify the states in Australia with faster population growth. This type of simple graphic can make the visualization of areas of potential demand easier than a traditional table format. Other simple graphic maps allow the display of bar or pie charts for each spatial entity on the map.

Figure 1. Text View and Map View of Population Change in Australia (Generated Using Microsoft Map in Excel)

Australian State	Population Change 2000-2001
New South Wales	1.1%
Victoria	1.3%
Queensland	1.9%
South Australia	0.5%
Western Australia	1.3%
Tasmania	0.2%
Northern Territory	0.7%
Australia Capital Territory	1.0%

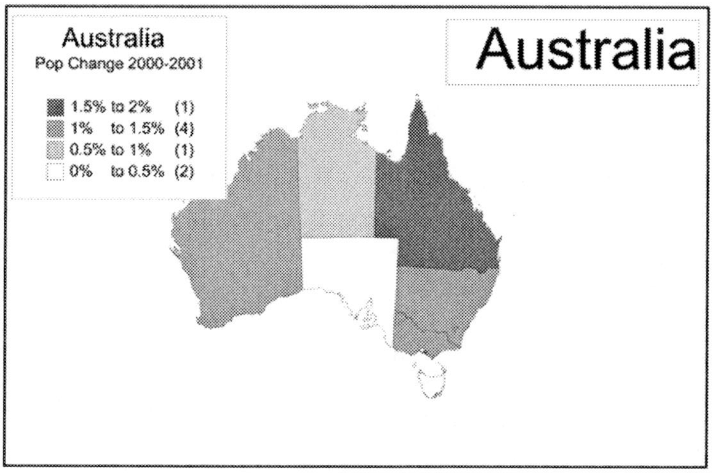

Modern desktop GIS software, such as ArcGIS or Mapinfo, can be regarded as much more than presentation mapping software. This software can better be regarded as illustrating the database or spatial analysis view of GIS. However, in addition this type of software also provides comprehensive presentation facilities. These facilities include the ability to generate thematic maps using a variety of shading techniques, bar and pie charts, graduated symbols, and dot density maps. Modern presentation mapping software allows three-dimensional representations to be used, with the capability to extrude areas on the map to represent particular attribute values.

GIS as a Database

A GIS interface can be used to query a database, although this requires a more sophisticated interface with the ability to formulate a query using the interactive commands. As IT has developed, a limited level of database functionality has become common in almost all software applications. This trend has also been seen in GIS where modern desktop packages, such as Mapinfo, ArcGIS, or Maptitude, have sophisticated database functionality. Database capability allows queries be generated in the GIS to show only areas selected by attribute value, e.g., sales value. This type of software also

Figure 2. Selection of Part of a Geographic Database

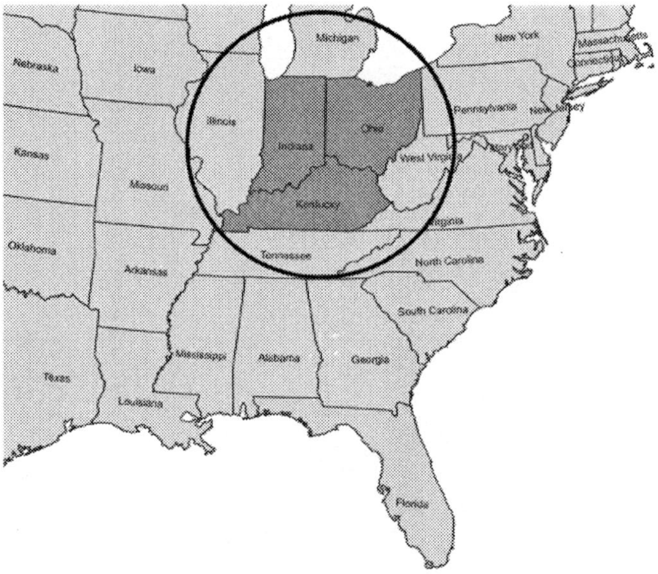

allows simple spatial database queries, such as selection of a particular region (Figure 2), and operations such as buffering or overlay.

In the business world, information systems have continued to evolve towards the introduction of large databases, which extensive networking then make available throughout the organization. This evolution has led to the introduction of EIS; these systems need to facilitate information retrieval from traditional forms of non-spatial data and a variety of types of data outside the organization. A limited map presentation capability is a recognized feature of EIS-type applications and the use of map representation can reduce the information overload that might arise in the use of an EIS. These maps form the basis of an interface for querying data; this facility can include the ability to conduct spatial operations. Spatial data is increasingly becoming a standard part of corporate databases, as evidenced by the alliances between the GIS market leader ESRI and organizations like IBM, Oracle and SAP (Good, 1999).

Spatial Decision Support

IT applications generally have moved from automation applications to decision support applications and GIS is following the same path (see *Table 1*). In most cases, spatial data is only one form of data relevant to business users, since many business sectors have existing non-GIS based DSS systems. Traditional users of DSS include fields such as marketing and routing (Eom, Lee, & Kim, 1993) with obvious scope for the use of GIS. While the growth of traditional IS has already made an important contribution to the management of these fields, it has not yet fully catered for the spatial component of decisions. The ability to handle both spatial and non-spatial data appropriately is required for better support for management decision-making in a range of applications. Effective decision support is characterized by the use of specialized models directed at the specific business decision being made. These need to be closely integrated with GIS techniques to enhance business decision-making. In this book, the literature on spatial decision support is elaborated on in the chapter by Huerta, Ryan, & Navarrete, while its theoretical aspects are examined in the chapter by Jarupathirun & Zahedi.

Business Applications of GIS

Contribution of IT to Business

Business organizations operate in an ever changing and challenging environment, in which competitive forces require that information technology be exploited to the full. One widely cited model of business, the Value Chain model (Porter, 1985), identifies five primary business activities. These are (1) inbound logistics (inputs), (2) operations, (3) outbound logistics (outputs), (4) marketing and sales, and (5) service. Porter argues that the ability to perform effectively particular activities, and to manage the linkages between these activities, is a source of competitive advantage. An organization exists to deliver a product or service, for which the customer is willing to pay more than the sum of the costs of all activities in the value chain. Consequently, management should be concerned with ways to improve these activities.

Information technology can contribute to the efficient organization of all of these primary business activities. As the business environment becomes increasingly competitive, the use of IT becomes an important component of business strategy. Importantly, spatial techniques can have a major role in this contribution. In addition to the basic issues raised by the value chain model, other developments in business provide further opportunities for the use of spatial techniques. There is increasing concern about the natural environment and companies are anxious to be seen to respond to these concerns. Issues such as pollution control often have a spatial dimension and planning for the location of new facilities requires the use of spatial techniques to address public concern over issues such as traffic impact.

Logistics Support

Business logistics has an inherent spatial dimension, as goods must be moved from one point to another. Modern businesses have sophisticated supply chains, with goods being moved around the world on a just-in-time basis. However, these supply chains are vulnerable to disruption due to political events, bad weather and natural disasters, and unforeseen events such as quarantine due to disease. In these circumstances, it is important to be aware of the spatial location of parties involved and to be able to plan rapidly alternative routes to resolve any difficulties. It is therefore not surprising that routing and location analysis are some of the most important areas of application of spatial techniques, a good example being the comprehensive restructuring of a Proctor and Gamble's logistics (Camm et al., 1997).

Logistics applications are therefore of considerable importance to business and a field where the contribution of quantitative approaches has long been recognized. In fields such as routing (Bodin, Golden, Assad, & Ball, 1983) and location analysis (Church, 2002), technical analysis has long had a role in management planning. Early DSS design for this class of applications has been driven predominantly by the quantitative techniques used (Keenan, 1998). However, such model driven systems often had very limited database or interface components and the DSS provided little contextual information to the user. The limitations of the technology meant that early systems were unable to fully incorporate geographic information. Consequently, users often continued to use paper maps to complement their use of computerized techniques.

With the availability of less expensive GIS software and associated hardware, these systems have tended to evolve by initially providing presentation mapping to show the solutions generated, with later systems allowing query operations through the map interface (Reid, 1993). However, the full potential for logistics support can only be reached when new interactions between non-spatial models and GIS techniques are fully exploited.

Operational Support

Organizations with substantial use of spatial data for logistics form one group of potential users of GIS techniques. Other organizations will focus on the use of spatial techniques for different operational applications. Information technology continues to be of critical importance to the routine operation of many businesses, which rely on systems such as airline booking systems, point of sale systems and bank networks to facilitate their routine operations. The initial role of IT in these organizations is one of increasing efficiency and cost reduction. However, as technology has moved towards *informating* applications, the scope of these sectors has been changed by the use of technology. For example, the complex pricing models found in the airline industry would be difficult to sustain without IT. While many operational applications of GIS lie in the government sector, these often involve private contractors. For example, road networks may be publicly owned, but may be constructed and maintained by the private sector. The use of GIS should lead to greater efficiencies in this type of application and ultimately to new procedures and processes for the allocation of this type of work.

However, as with other business applications, the collection of large amounts of data for operational purposes can provide data for use in decision-oriented applications. Business data processing produces low-level transaction data that can be aggregated and processed for EIS applications. In a similar way, those organizations using spatial data for operational reasons have the opportunity to exploit their spatial data resources for strategic management purposes. This will mean a move towards spatial decision support applications and the incorporation of spatial data in EIS systems. The synthesis of EIS and spatial techniques is most promising where there is already a large volume of operational spatial data in the organization, as well as a requirement for access to spatial data outside the organization. However, if managers are to take advantage of the inclusion of spatial data in EIS, and other GIS applications in business, they must be aware of contribution of spatial techniques.

Marketing

In disciplines such as marketing, additional possibilities for analysis are provided by the availability of increasing amounts of reasonably priced spatial data. Demographic data is of particular importance to business (Mennecke, 1997) and basic census information is now available for use in GIS throughout the western world. The relevance of GIS to this type of work is becoming widely recognized (Fung & Remsen, 1997). The availability of user-friendly SDSS to manipulate this type of data will lead to additional decision possibilities being examined which are difficult to evaluate without the use of such techniques (Grimshaw, 2000). This is reflected in increasing interest in spatial applications for sectors such as retailing (Nasirin & Birks, 2003) which may not have used this form of technical analysis in the past.

The marketing field in general has shown interest in GIS, this was reflected in the absorption of the GeoBusiness Association into the American Marketing Association. GIS has been seen as being a critical component of a marketing information system (Hess et al., 2004). There are significant obstacles to the more widespread use of GIS in fields such as marketing. In business disciplines such as marketing, operational applications of GIS are less important than decision support applications. However, this group of potential SDSS users has little background in spatial processing and is inexperienced in the use of any type of DSS technology. Consequently, this category of users is not accustomed to the restrictions on model realism and the interface limitations that many users of DSS have been willing to put up with in the past. Such users will therefore require systems that are straightforward to use and which do not require the users to accommodate themselves to artificial restrictions on the problem representation. While the availability of user-friendly systems and interfaces incorporating spatial visualization will make modeling techniques in this field more accessible, potential users must gain experience with GIS-based systems in order to put them to effective use. GIS is therefore becoming more common, but is still far from universal, in education in business schools. GIS, which has been seen as the preserve of geographers and computer scientists, needs to also become the concern of managers (Reeve & Petch, 1999).

Service

Within Porter's value chain model, *service* refers to customer related activities other than direct sales and product delivery. This would include after sales service and support. With the routine high standards in modern manufacture and the outsourcing of logistics, service is often one area where companies can try to achieve a competitive advantage. There is increasing interest in the service dimension; this is reflected by the growth of IT systems such as customer relationship management (CRM) systems. A recent book, *The Support Economy*, (Zuboff & Maxmin, 2002) argues for the role of customized customer support. One element of "knowing your customer" is that customer's geographic location and good service requires an approach tailored to that location.

GIS-based techniques have an important role to play in customer service. Call centers will often use a customer's telephone number to identify where they are calling from, thereby providing a service appropriate to that customer. Spatial databases can be used to identify the nearest shop or repair center to a customer. Utilities can identify whether a customer is sufficiently close to a cable network or telephone exchange to avail of an improved service. Some service activities may overlap with sales or logistics, one example would be the use of GIS to improve both product dispatch and technician routing for Sears (Weigel & Cao, 1999).

New Areas for SDSS Use in Business

One area of growing importance for SDSS application is businesses where the importance of both spatial data and modeling is somewhat neglected at present, in sectors where decision-makers are less accustomed to using maps. Groups such as the insurance sector have been accustomed to using statistical and actuarial models, but have not tended to use information on the location of their customers. As insurance risks are often strongly spatially correlated, this sector needs to make more use of spatial techniques in the future (Morton, 2002). Software vendors are aware of the market for GIS related risk management software and are moving to provide solutions for this market (Francica, 2003).

Other recent developments in business, such as the growth of electronic commerce (Kalakota & Whinston, 1997), also provide opportunities for the use of GIS. Although the Internet is available throughout the world, the location of customers is of importance in the services offered in many electronic commerce applications. Many consumer electronic commerce applications offer goods that must be delivered to the customer. This mode of doing business requires a sophisticated delivery operation, and GIS techniques have an important role to play in the management of this function. In this environment, it can be argued that the move to electronic commerce will increase, rather than decrease, the importance of GIS.

Mobile computing and telecommunications is an emerging area of IT application that is of increasing interest to business (Mennecke & Strader, 2003). GIS is widely used by operational activities by mobile service providers for modeling service levels and locating signal masts. Mobile services can be largely distinguished from fixed Internet services by the presence of a locational element (MacKintosh, Keen, & Heikkonen, 2001). Mobile services can be divided into mobile commerce and Location Based Services (LBS) (Mitchell & Whitmore, 2003). LBS applications require the integration of wireless technology with GIS applications. Future developments will enhance the capabilities of these devices and we are likely to see the integration of mobile data devices and spatial technologies such as Global Positioning Systems (GPS). This will allow the location of the mobile user to be easily identified and will therefore provide the basis for a service customized to that location. This allows for the growth in situation-dependent services (Figge, 2004) directed at a particular customer in a particular location. For businesses in the mobile services sphere, spatial technologies are core to their business model and there are many opportunities for businesses that can successfully exploit this technology.

Conclusions

Trends in Business GIS

Business use of GIS covers a spectrum of GIS applications. The use of GIS applications is still somewhat fragmented and there is a need for further integration with other forms of IT. The trend in IT applications has been for initial operational use in specialized situations, followed by a more general information providing use, followed in turn by sophisticated specialist decision-making and executive management applications. The Nolan stages of growth model (Nolan, 1973, 1979) is a model of computing growth in an organization based on the organization's level of expenditure on IT. The more recent 1979 version of the model comprises the steps of *initiation, expansion/contagion, formalization/control, integration, data administration,* and *maturity*. Chan (1998) identifies a number of GIS researchers who have used the Nolan model and suggests that this research shows a common trend in public organizations towards GIS becoming an integral part of the overall corporate information system.

The availability of moderately priced data, and software for working with that data, means that the necessary conditions for *contagion* already exist. One significant difference in GIS, when compared to other forms of IT, is that applications in different parts of an organization may be concerned with the same geographic area. Geographic location provides a common feature facilitating this data integration. Business users are generally already reasonably sophisticated users of IT and there will be pressure for GIS to integrate in this infrastructure. The *data administration* phase of the Nolan model could be problematic for business, owing to the complexity of GIS data. One recent published example of a GIS failure in business noted difficulties with integrating diverse data sources (Birks, Nasirin, & Zailini, 2003). In general there are significant difficulties building a comprehensive spatial database in the first instance, but subsequently it is often easier to maintain this database. The GIS community generally is attempting to integrate multiple spatial data sources and new metadata standards and other initiatives aim to facilitate this (Goodchild & Zhou, 2003). With data organization made easier, business users will be able to take advantage of better-integrated GIS data to extend the areas of business where GIS is used. With concerns about data integrity and relationships resolved, business use of GIS can move to *maturity* in the Nolan model.

In IT adoption generally, the culture of organizations is an important factor in the adoption of technology. GIS applications are relevant to a wide range of sectors, from engineering related applications where technical solutions are readily accepted, to marketing departments where there is less tradition of using IT. GIS faces particular problems as most people in business have little training in spatial techniques and may consequently be slower to make full use of the technology. Where top managers have little appreciation of the technology, they are unlikely to be sufficiently enthusiastic in supporting it. The GIS failure discussed above (Birks et al., 2003) also resulted in part from a lack of senior management interest in the project. These researchers noted that lessons learned in different environments, such as the application of GIS in local government, could also have implications for the successful business use of GIS.

Summary

GIS represents a sophisticated information technology application that has grown in parallel with traditional business IT. As GIS techniques have come to focus on decision support, they have increasing potential for wider use in business, a potential that has yet to be fully realized. GIS has an important role to play in a variety of decision-making systems in specific functional areas, but GIS also needs to be incorporated in enterprise wide systems. Newer technologies such as e-commerce and location based services have an intrinsic spatial element and the spread of these applications will serve to further increase the importance of GIS.

References

Armstrong, M. P., Densham, P. J., & Rushton, G. (1986). Architecture for a microcomputer based spatial decision support system. Paper presented at the *Second International Symposium on Spatial Data Handling.*

Bell, P. C. (1994). Visualization and optimization: The future lies together. *ORSA Journal on Computing, 6*(3), 258-260.

Birks, D. F., Nasirin, S., & Zailini, S. H. M. (2003). Factors influencing GIS project implementation failure in the UK retailing industry. *International Journal of Information Management, 23*, 73-82.

Blaser, A. D., Sester, M., & Egenhofer, M. J. (2000). Visualization in an early stage of the problem-solving process in GIS. *Computers & Geosciences, 26*(1), 57-66.

Bodin, L. D., Golden, B. L., Assad, A., A., & Ball, M. O. (1983). Routing and scheduling of vehicles and crews: The state of the art. *Computers and Operations Research, 10*(2), 67-211.

Buttenfield, B. P., & Mackaness, W. A. (1991). Visualization. In D. J. Maguire, M. F. Goodchild, & D. W. Rhind (Eds.), *Geographical Information Systems, Volume 1: Principles* (Vol. 1, pp. 427-443). Harlow, Essex, UK: Longman Scientific & Technical.

Camm, J. D., Chorman, T. E., Dill, F. A., Evans, J. R., Sweeney, D. J., & Wegryn, G. W. (1997). Blending OR/MS, judgment and GIS: Restructuring P&G's supply chain. *Interfaces, 27*(1), 128-142.

Chan, T. O. (1998). *The Dynamics of Diffusion of Corporate GIS.* Unpublished PhD, The University of Melbourne, Melbourne, Australia.

Church, R. L. (2002). Geographical information systems and location science. *Computers & Operations Research, 29*(6), 541-562.

Coppock, J. T., & Rhind, D. W. (1991). The history of GIS. In D. J. Maguire, M. F. Goodchild, & D. W. Rhind (Eds.), *Geographical Information Systems, Volume 1: Principles* (Vol. 1, pp. 21-43). Harlow, Essex, UK: Longman Scientific & Technical.

Densham, P. J. (1991). Spatial decision support systems. In D. J. Maguire, M. F. Goodchild, & D. W. Rhind (Eds.), *Geographical Information Systems, Volume 1: Principles* (Vol. 1, pp. 403-412). Harlow, Essex, UK: Longman Scientific & Technical.

DeSanctis, G. (1984). Computer graphics as decision aids: Directions for research. *Decision Sciences, 15*(4), 463-487.

Eom, S., Lee, S., & Kim, J. (1993). The intellectual structure of Decision Support Systems (1971-1989). *Decision Support Systems, 10*(1), 19-35.

Figge, S. (2004). Situation-dependent services—a challenge for mobile network operators. *Journal of Business Research, forthcoming.*

Francica, J. (2003, October 22). MapInfo looks to take the risk out of insurance policy underwriting. *Directions Magazine.*

Fung, D. S., & Remsen, A. P. (1997). Geographic Information Systems technology for business applications. *Journal of Applied Business Research, 13*(3), 17-23.

Gardner, J., Griffith, R. J., Harness, H. D., & Larcom, T. A. (1838). Irish Railway Commission. (Atlas to accompany 2nd Report of the Railway Commissioners). London.

Good, M. (1999). SAP Business Data Visualization with MapObjects. Paper presented at the *1999 ESRI User Conference*, Redlands, California, USA.

Goodchild, M., & Zhou, J. (2003). Finding geographic information: Collection-level metadata. *Geoinformatica, 7*(2), 95-112.

Grace, B. F. (1977). Training users of a prototype DSS. *Data Base, 8*(3), 30-36.

Grimshaw, D. J. (2000). *Bringing Geographical Information Systems into Business* (2nd ed.). New York: Wiley.

Hearnshaw, H., & Medyckyj-Scott, D. (1993). The way forward for human factors in GIS. In D. Medyckyj-Scott & H. Hearnshaw (Eds.), *Human factors in Geographical Information Systems* (pp. 235-243). London: Belhaven Press.

Hess, R. L., Rubin, R. S., & West, L. A. (2004). Geographic information systems as a marketing information system technology. *Decision Support Systems, forthcoming.*

Ives, B. (1982). Graphical user interfaces for business information systems. *MIS Quarterly, 6*(5), 15-47.

Kalakota, R., & Whinston, A. B. (1997). *Electronic commerce: A manager's guide.* Reading, MA: Addison-Wesley.

Keenan, P. (1996). Using a GIS as a DSS Generator. In J. Darzentas, J. S. Darzentas, & T. Spyrou (Eds.), *Perspectives on DSS* (pp. 33-40). Greece: University of the Aegean.

Keenan, P. B. (1998). When the question is 'Where'? Integrating geographic information systems and management science. *OR Insight, 11* (1), 23-28.

Keenan, P. B. (2003). Spatial decision support systems. In M. Mora, G. Forgionne, & J. N. D. Gupta (eds.), *Decision Making Support Systems: Achievements and Challenges for the New Decade* (pp. 28-39). Hershey, PA: Idea Group Publishing.

Landis, J. D. (1993). GIS capabilities, use and organizational issues. In G. H. Castle (Ed.), *Profiting from a Geographic Information System* (pp. 23-53). Fort Collins, CO: GIS World Books.

MacKintosh, R., Keen, P. G. W., & Heikkonen, M. (2001). *The freedom economy: Gaining the mCommerce edge in the era of the Wireless Internet.* McGraw Hill.

Madon, S., & Sahay, S. (1997). Managing natural resources using GIS: Experiences in India. *Information & Management, 32*(1), 45-53.

Maguire, D. J. (1991). An overview and definition of GIS. In D. J. Maguire, M. F. Goodchild, & D. W. Rhind (Eds.), *Geographical Information Systems, Volume 1: Principles* (Vol. 1, pp. 9-20). Harlow, Essex, UK: Longman Scientific & Technical.

Mennecke, B. E. (1997). Understanding the role of geographic information technologies in business: Applications and research directions. *Journal of Geographic Information and Decision Analysis, 1*(1), 44-68.

Mennecke, B. E., & Strader, T. J. (2003). *Mobile commerce: Technology, theory, and applications.* Hershey, PA: Idea Group Publishing.

Mitchell, K., & Whitmore, M. (2003). Location based services: Locating the money. In B. E. Mennecke & T. J. Strader (Eds.), *Mobile commerce: Technology, theory, and applications* (pp. 51-66). Hershey, PA: Idea Group Publishing.

Morton, K. (2002). Mapping out risks. *Canadian Insurance, 107*(12), 8-10.

Nasirin, S., & Birks, D. F. (2003). DSS implementation in the UK retail organisations: A GIS perspective. *Information & Management, 40*(4), 325-336.

Nolan, R. L. (1973). Managing the computer resources: A stage hypothesis. *Communications of the ACM, 16*(7), 399-405.

Nolan, R. L. (1979). Managing the crises in data processing. *Harvard Business Review, 57*(2), 115-126.

Reeve, D. E., & Petch, J. R. (1999). *GIS, organisations and people: A socio-technical approach.* London: Taylor & Francis.

Reid, H. G. (1993). Retail trade. In G. H. Castle (Ed.), *Profiting from a Geographic Information System* (pp. 131-151). Fort Collins, CO: GIS World Books.

Sahay, S., & Walsham, G. (1996). Implementation of GIS in Ind Oa:organizational issues and implications. *International Journal of Geographical Information Systems, 10*(4), 385-404.

Snellen, I. (2001). ICTs, bureaucracies, and the future of democracy. *Communications of the ACM, 44*(1), 45-48.

Speier, C., & Morris, M. G. (2003). The influence of query interface design on decision-making performance. *MIS Quarterly, 27*(3), 397-423.

Sprague, R. (1980). A framework for the development of Decision Support Systems. *MIS Quarterly, 4*(1), 1-26.

Tegarden, D. P. (1999). Business information visualization. *Communications of the Association for Information Systems, 1*(4).

Walsham, G., & Sahay, S. (1999). GIS for district-level administration in India: Problems and opportunities. *MIS Quarterly, 23*(1), 39-65.

Weigel, D., & Cao, B. (1999). Applying GIS and OR techniques to solve Sears technician dispatching and home delivery problems. *Interfaces, 29*(1), 112-130.

Winchester, S. (2002). *The map that changed the world: A tale of rocks, ruin and redemption.* Penguin Books.

Zuboff, S. (1988). *In the age of the smart machine.* New York: Basic Books.

Zuboff, S., & Maxmin, J. (2002). *The support economy: Why corporations are failing individuals and the next episode of capitalism.* New York: Viking.

Chapter II

GIS and Decision-Making in Business: A Literature Review

Esperanza Huerta, Instituto Technológico Autónomo de México, Mexico

Celene Navarrete, Claremont Graduate University, USA

Terry Ryan, Claremont Graduate University, USA

Abstract

This chapter synthesizes empirical research from multiple disciplines about the use of GIS for decision-making in business settings. Todd & Benbasat's model (2000) was used as a theoretical framework to identify the variables that have been studied on decision-making at the individual and collaborative level. An extensive literature review in the fields of Information Science, GIS and Decision Science from 1990 to 2002 was conducted with a total of nine studies identified in six journals and two conferences. The scarcity of published research suggests that the impact of GIS on the decision-making process has not been extensively investigated. Moreover, researchers have paid more attention to the study of GIS to support individual decision makers. The effects of variables like desired effort *and* decision strategy *remain unexplored. More empirical work is needed to understand the impact of DSS capabilities, decision maker, task, and decision strategy on decision performance.*

Introduction

Geographic information systems (GIS) have been defined in several ways. For purposes of this chapter, "a GIS is a computer-based information system that provides tools to manage, analyze, and display attribute[s] and spatial data in an integrated environment" (Mennecke et al., 2000, p. 602). Spatial decision support systems (SDSS) are GIS specifically designed to support the decision-making process by providing both geographical data and appropriate tools for analysis (Densham, 1991; Murphy, 1995). *Figure 1* shows how SDSS can be viewed as occupying a place at the intersection of GIS and decision support systems (DSS).

This chapter is about research concerning the use of GIS to support decision-making — that is to say, research on SDSS. In particular, the chapter reviews empirical research to identify what has been learned and what areas remain unexplored.

Nowadays people are relying more on information technologies to make significant decisions (Todd & Benbasat, 2000). It is, therefore, important to understand the factors affecting the decision-making process. Such understanding is key in GIS for at least two reasons. First, decision makers might not be familiar with the geographic and cartographic principles essential to these systems (Mennecke & Crossland, 1996; Murphy, 1995; West, 2000). Second, one of the most common reasons to adopt GIS is the general assumption that the use of the system leads to better decisions (Mennecke & Crossland, 1996; Murphy, 1995; Todd & Benbasat, 2000). Therefore, it is important to determine under what circumstances the use of GIS improves the decision-making process (Mark, 1999; Mennecke, 1997; Mennecke & Crossland, 1996; University Consortium for Geographic Information Science [UCGIS], 1996).

The role of GIS in decision-making has been promoted as an area with great potential for study (Keenan, 1997; Mark et al., 1999; Mennecke & Crossland, 1996; UCGIS, 1996). The University Consortium of Geographic Information Systems (UCGIS), an organization of prominent researchers from academic and research institutions in the U.S., has identified geographic information cognition as an important research priority for GIS. Despite such recognition, the impact of GIS on the decision-making process has yet to be extensively investigated (Mennecke et al., 2000).

Figure 1. Relationship between GIS and DSS

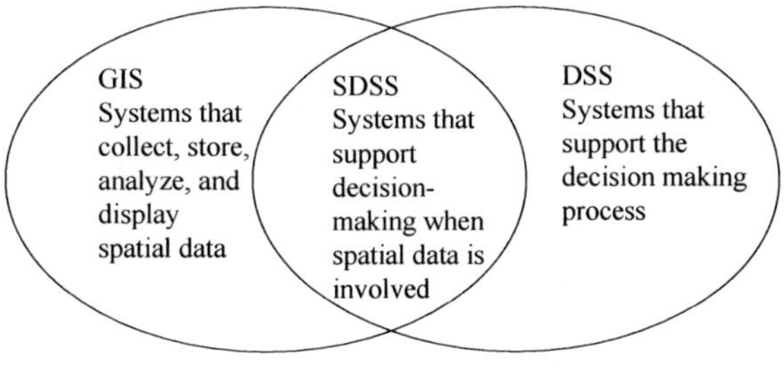

Decision making can be studied at two different levels of analysis: the person and the group (Todd & Benbasat, 2000). When a group makes decisions, the process is called collaborative decision-making. Our literature review aimed to include empirical research at both levels of analysis. However, we did not find any empirical research investigating collaborative decision-making in GIS. Therefore, this chapter concerns only studies investigating GIS and decision-making at the individual level. Todd & Benbasat's (2000) model identifies variables that affect decision performance at the individual level. Since SDSS are a particular type of DSS, we used this model as a theoretical framework to identify the variables that have been studied and to define areas for future research.

The chapter is organized as follows. The second section explains the theoretical framework that guides the chapter. The third section describes the methodology to identify previous research and presents a summary of the publications. The fourth section discusses findings from empirical studies, identifies the areas that have been studied, and suggests areas for future research. Finally, the fifth section presents conclusions.

Theoretical Framework

A theoretical framework is used, in reviewing the literature, to identify the variables that have been studied in GIS and decision-making research. This section presents a brief description of the framework and the variables in it.

According to Todd & Benbasat (2000), findings from studies investigating the relationship between the use of DSS and decision performance are equivocal. They argue that one of the reasons for these inconclusive findings is that there is no direct relationship between the use of DSS and decision outcomes, with multiple factors mediating and moderating the relationship. Based on previous research, they developed a comprehensive model of the factors influencing the impact of information technology (IT) on decision performance. In the first part of the model, three variables affect the decision strategy chosen by the decision maker: DSS capabilities, decision task, and decision maker. The "DSS capabilities" variable refers to the different types of support offered by a DSS. They suggest using Zachary's taxonomy (1988) to classify the types of support.

Based on general difficulties people face when making a decision, Zachary (1988) identifies six DSS support functions: process modeling, value modeling, information management, automated and/or semi-automated analysis and reasoning, representational aids, and judgment refining/amplification. Process modeling refers to simulating real world processes. Value modeling combines and makes trade-offs among competing decision criteria. Information management extends the ability to access information. Automated and/or semi-automated analysis and reasoning refers to tools to facilitate analytical reasoning. Representational aids makes data available in terms of the user's mental representation of the problem. Finally, judgment refining/amplification removes systematic inconsistencies in human judgment. Zachary (1988) comments that a DSS does not necessarily need to provide all six types of support. The types of support needed are based on the particular decision problem.

The "decision maker" variable points out that users' individual differences in cognitive capability must be taken into account. Cognitive capabilities might have an impact on how a DSS is used. In addition, individual differences in cognitive capabilities might help to inform DSS providers about "the type of DSS an individual user is likely to benefit from" (Todd & Benbasat, 2000, p. 7). Research in GIS has examined how users' spatial skills influence decision performance (Swink & Speier, 1999).

The "decision task" variable identifies the different types of tasks to be solved by the user. Decision tasks can be broadly classified into structured and unstructured problems. The effect that a DSS has on decision performance depends on the match between the task that needs to be solved and the type of tasks the system can support.

In the second part of the model, decision strategy influences the decision outcome. Decision strategy is the set of steps required to solve a problem. The implementation of each strategy is associated with time, effort, and resources. DSS capabilities, decision task, and decision maker have an influence on the decision strategy adopted by the user. However, this influence is moderated by the desired effort and accuracy. In general, it is desirable to reach an accurate solution with less effort. By far, effort is the most important factor influencing strategy selection.

Decision outcome is measured in terms of its quality. In short, Todd & Benbasat's model (2000) makes clear that an understanding of how the use of DSS influences decision performance must consider multiple intervening variables.

Methodology

An extensive literature review was performed to identify empirical research investigating the relationship between GIS and decision-making. The following three criteria were used to select relevant research. First, the study had to be an empirical contribution to the body of knowledge concerning spatial decision making in business settings. The unit of analysis could be both at the individual and the group level. Second, the study had to be published in a peer reviewed journal or conference proceedings in the fields of IS, GIS or decision science. Unpublished dissertations, research in progress, and book chapters were not included. Third, the study had to be published between 1990 and 2002. Past literature reviews in this area found no empirical research published before 1990 (Mennecke, 1997; Mennecke & Crossland, 1996).

To identify the literature, we performed a thorough search in three steps. In the first step, *ABI/Inform Global* was used to locate research articles in the disciplines of IS and decision science. This database contains citations from over 1,200 international periodicals in the areas of business and management. In the second step, we identified the journals within the set of top IS journals (Mylonopoulos & Theoharakis, 2001; Whitman et al., 1999) that are not included in *ABI/Inform* (e.g., *Communications of the AIS*). Then, we searched for relevant literature in the journals identified. In the third step, we searched

Table 1. Publications by Year

Year	Study	Journal/Conference
1993	1	HICSS
1994	1	HICSS
1995	1	*Decision Support Systems*
1997	2	*Decision Science,* AMCIS
1998	1	*Information Systems Research*
1999	1	*Decision Science*
2000	2	*Journal of End User Computing, MIS Quarterly*
Total	9	

for relevant literature in specialized journals and conferences concerning GIS. The Appendix lists the conferences and journals reviewed.

Table 1 lists the nine empirical studies that met our selection criteria. The small number of published empirical research indicates that little has been done in the area of spatial decision-making. Research in the last twelve years has been published in six journals and two conferences. Only *Decision Science* published more than one article. We are not aware of any publication from 1990 to 1992, nor in 1996, 2001, or 2002.

Research from 1990 to 2002

This section discusses, first, the research methodology and the main findings of the studies reviewed. Then it identifies and analyses the variables, from Todd & Benbasat's (2000) model, which have been investigated by the studies. Finally, based on this analysis, the Todd & Benbasat model is adapted to the context of GIS and decision-making. *Table 2* presents the research methodology and the main findings of the studies reviewed. It is organized chronologically rather than alphabetically to show how studies in this area have evolved. It is important to note that the column discussing main findings does not include all the findings for a given study, but rather the results that were judged most important for understanding GIS and decision-making. Crossland et al. (1993) presented in HICSS is not included in *Table 2* because it was published in a journal in 2000. The latter reference and the analysis are included in this table.

In terms of the methodology used, all the empirical studies are laboratory experiments. Laboratory experiments are the best approach for establishing causality. This is congruent with the goals of the studies, which aim to identify the factors affecting decision performance. However, laboratory experiments are limited in terms of generalizability. Research in the area will benefit from the use of multiple methodologies. For instance, field experiments offer less control but more generalizability.

In terms of the participants, in most studies participants were students, with only two studies having professionals as participants. Mennecke et al. (1997) replicated the Crossland et al. (1995) study, using professionals as participants. Later, Mennecke et al. (2000) compared the decision performance of experts and students. The validity of the results from studies having students as surrogates for professionals has been much argued. The only study that compared results from experts and students demonstrated that results did differ. Therefore, findings from the studies with students as participants should be interpreted carefully.

In terms of the theories used, two theoretical frameworks prevail: cognitive fit theory (CFT) and image theory (IT). Both theories have been used to explain performance in using GIS for spatial tasks. CFT describes the impact of graphical representations in the decision-making process. According to this theory, problem representation (graphical or tabular), the nature of the task, and the way the problem is represented in human memory influence the problem solution (Vessey, 1991). Although CFT was originally developed to study tasks that involved information in the form of graphs and tables, it has explicitly been expanded to include spatial tasks (Dennis & Carte, 1998). Similarly, IT explains how efficiency in the graphical constructions influences information assimilation and problem solving (Crossland et al., 2000, 1995). It classifies graphical constructions as "images" and "figurations" (multiple images) and measures their efficiency according to their observation time. The shorter the observation time needed for a specific problem, the more efficient the construction is (Crossland et al., 2000). Thus, the nature of the graphical representation can reduce the time needed for problem solution, and enhance decision-making.

It is important to note that cognitive fit theory and image theory are not, for the most part, competing theories. Both predict similar results in most cases.

In terms of decision performance, all studies assessed participants' performance in terms of the time elapsed to reach a solution and some measure of quality (i.e., percentage of error).

Based on the empirical studies we have learned among other things that:

- GIS are more useful for solving complex tasks (Crossland et al., 1995; Mennecke et al., 1997).
- GIS are more useful for solving problems involving geographical situations of adjacency (Smelcer & Carmel, 1997).
- GIS users perform better than users using paper maps (Crossland et al., 1995; Dennis & Carte, 1998; Mennecke et al., 1997, 2000; Smelcer & Carmel, 1997).
- Data dis-aggregation helps users when solving problems with highly dispersed data (Swink & Speier, 1999).
- Users with high visual skills perform better for large problems and for low data dispersion problems (Swink & Speier, 1999).

Table 3 identifies the variables from Todd & Benbasat's (2000) model that have been investigated by the studies presented in *Table 2*. Following *Table 3*, each variable is analyzed in the context of GIS decision-making.

Table 2. Summary of Empirical Studies Investigating the Relationship between GIS and Decision Making

Author and year	Research question	Variables and measures	Theoretical framework	Main findings
Crossland & Wynne (1994)	Do users using GIS perform better than users using paper maps?	DV Performance (elapsed time and accuracy) IV Task complexity (3 levels) IV Information presentation (paper maps and tabular data / GIS)	Image theory	GIS users performed better that users using paper maps for all 3 levels of task complexity
Crossland et al. (1995)	Do users using GIS perform better than users using paper maps?	DV Performance (elapsed time and accuracy) IV Task complexity (2 levels) IV Information presentation (paper maps and tabular data / GIS)	Image theory	GIS users performed better that users using paper maps for both levels of task complexity
Smelcer & Carmel (1997)	Do users using GIS perform better than users using tables?	DV Performance (elapsed time, accuracy measured but not statistically analyzed) IV Information presentation (GIS/tables) IV Task complexity (3 levels) IV geographic relationship (proximity/adjacency/containment) CV spatial skills (VZ-2 spatial visualization test) MV knowledge states (combination of alternative solution enumeration and problem-solving heuristics)	Cognitive fit theory Proximity compatibility principle Knowledge states	GIS users performed better than users using tables Maps keep time elapsed low for proximity and adjacency tasks Knowledge states explain performance times, maps keep knowledge states low

Note: DV = dependent variable; IV = independent variable; BV = blocking variable; CV = control variable

Table 2. (continued)

Author and year	Research question	Variables and measures	Theoretical framework	Main findings
Mennecke et al. (1997)	Do users using GIS perform better than users using paper maps?	DV Performance (elapsed time and accuracy) IV Task complexity (2 levels) IV Information presentation (paper maps and tabular data / GIS)	Image theory	GIS users performed better that users using paper maps for both levels of task complexity
Dennis & Carte (1998)	Do users using GIS perform better than users using tables?	DV Performance (elapsed time and accuracy) IV Information presentation (GIS / tables) IV geographic relationship (adjacent / nonadjacent) MV Decision process (perceptual / analytical)	Cognitive fit theory	GIS users performed better than users using paper maps for adjacent tasks GIS users spent less time but were less accurate than users using paper maps for nonadjacent tasks
Swink & Speier (1999)	Do problem size, data aggregation and dispersion affect performance when using GIS?	DV Performance (elapsed time and accuracy) IV Task complexity (2 levels) IV Data aggregation (4 levels) IV Data dispersion (3 levels) CV spatial skills (spatial orientation S-1 test)	Complexity theory Knowledge states Information load theory Flow dominance	Data dis-aggregation helps users when solving problems with highly dispersed data Users with high visual skills performed better for large problems and for low data dispersion problems

Note: DV = dependent variable; IV = independent variable; BV = blocking variable; CV = control variable

Table 2. (continued)

Author and year	Research question	Variables and measures	Theoretical framework	Main findings
Crossland et al. (2000)	Do users' individual characteristics affect their performance when solving spatial tasks?	DV Performance (elapsed time and accuracy) IV Task complexity (2 levels) IV Information presentation (paper maps and tabular data / GIS) BV Need for cognition BV Field dependency	Image theory	Non-field dependent users perform better than field dependent users in terms of time, but there is no difference in terms of solution accuracy Contrary to expectations, users with low need for cognition perform better than users with high need for cognition
Mennecke et al. (2000)	Do users using GIS perform better than users using paper maps?	IV Performance (elapsed time and accuracy) IV Information presentation (paper maps and tabular data/GIS) IV Task complexity (3 levels) BV Type of user (professionals/students) CV Need for cognition	Image theory Cognitive fit theory	GIS users spent less time that users using paper maps for more complex tasks No difference in accuracy for professionals using GIS and those using paper maps

Note: DV = dependent variable; IV = independent variable; BV = blocking variable; CV = control variable

Table 3. Variables from Todd and Benbasat's Model (2000) Investigated in the Literature

Article	DSS capabilities	Decision maker	Decision task	Decision strategy	Desired effort expenditure	Desired accuracy	Decision performance
Crossland & Wynne (1994)	X		X				X
Crossland et al. (1995)	X		X				X
Smelcer & Carmel (1997)	X	X	X				X
Mennecke et al. (1997)	X		X				X
Dennis & Carte (1998)	X			X			X
Swink & Speier (1999)	X	X	X				X
Crossland et al. (2000)	X	X	X				X
Mennecke et al. (2000)	X	X	X				X

DSS Capabilities:

DSS capabilities can be identified using Zachary's (1988) taxonomy. According to this taxonomy, all the studies but one investigated the "information management" capability. That is, the studies compared whether decision performance from users using a GIS, which supports information management, yield different results from those using maps or tables. Only Swink & Speier's (1999) study investigated the "representational aids" capability. Swink & Speier (1999) manipulated data aggregation and data dispersion. Then, they analyzed the effects of this manipulation on decision performance.

Three DSS capabilities can be explored further. First, past research has identified several biases people have when dealing with geographic information (West, 2000). A "judgment refining/amplification" capability can be used to remove those biases. Second, SDSS have the capability to perform simulations (Densham, 1991; Murphy, 1995). The ability to perform simulations is a "process modeling" capability. Third, complex models involve multiple decision criteria. GIS can include a "value modeling" capability to deal with competing goals.

Decision Maker:

Decision makers' cognitive styles influence decision strategy. Four studies measured some type of user's cognitive style, but the definition and operationalization of the construct were different in all cases. Of the four studies, two measured cognitive style variables not specifically related to spatial skills. The other two studies measured spatial skills, but with different constructs. Smelcer & Carmel (1997) used the VZ-2 spatial visualization standard test. This test measures the ability to mentally manipulate figures. In their study, the spatial skill variable did not show a significant effect on decision performance. Smelcer & Carmel (1997) considered that the construct they used was not measuring the spatial visualization ability required to interpret maps. Swink & Speier (1999) agreed on the inadequacy of the test. They used, therefore, a different construct in their study: the S-1 spatial ability standardized test. Results from their study showed a significant relationship between the participants' spatial ability and decision performance (in terms of quality). Based on their results, Swink & Speier (1999) encourage further research in this area. However, to date no published research has investigated the effect of spatial skills on decision performance.

Decision Task:

All but one of the studies investigated the relationship between decision task and decision performance. Decision task is manipulated in terms of task complexity. The number of levels of task complexity manipulated varies from two to three; however, complexity is consistently manipulated in terms of the number of variables involved in the problem. The higher the number of variables involved the more complex the problem is.

In terms of the type of task used, all the studies used structured problems. Structured problems have the advantage of measuring decision accuracy objectively. However, the main goal of a DSS is to support users in solving ill-structured problems. Refocusing research on unstructured problems will help to tap the real potential of GIS. Therefore, future research should investigate the effect of GIS on decision performance when solving unstructured problems.

Decision Strategy:

Only Dennis & Carte (1998) analyzed the impact of decision strategy on decision performance. They used the Cognitive Fit Theory (CFT) as their theoretical framework. The CFT distinguishes two types of decision processing: analytical and perceptual. As predicted by the theory, they found a significant relationship between information presentation and decision process. They suggest further research exploring different forms of information presentation.

Desired Effort and Accuracy:

None of the studies directly investigated the moderating effect of desired effort expenditure and desired accuracy. Dennis & Carte (1998) mentioned the effect of effort and desired accuracy on decision performance. They found that using maps to present geographic data, when the problem does not need to understand relationships among data, leads to lower accuracy. They argue that users traded-off accuracy for time. However, they did not measure the desired effort and accuracy. Further research in this area is needed.

Figure 2. Factors Affecting Decision Performance in GIS

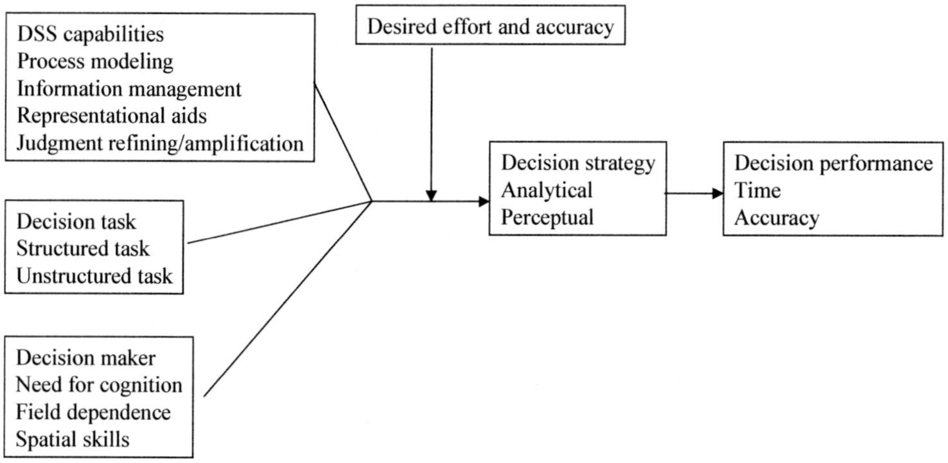

Decision Performance:

As it is reasonable to expect, decision performance is the central construct in all studies. The time elapsed to reach a solution, as well as the quality of the solution, are the constructs consistently used for measuring decision performance. As mentioned earlier, structured tasks were used in all studies leading to objective measures of decision quality. New measures of decision quality must be developed for unstructured tasks.

Based on the limited number of studies conducted during the last 12 years as well as the small number of factors influencing decision performance investigated, it is clear that this area has still a great potential for study. *Figure 2* adapts Todd & Benbasat's (2000) model to show the variables of interest in GIS.

The variables in the model that remain unexplored are desired effort and decision strategy. However, most of the other variables are not completely understood yet. For instance, DSS capabilities have dealt mainly with information management. Future research can explore other types of capabilities such as judgment refining/amplification, process modeling, and value modeling. In terms of the decision maker cognitive abilities, there is need for further research to identify the spatial skills required to make the most of a GIS. Similarly, decision task can be further explored. Research is needed to investigate the use of GIS when dealing with unstructured problems. Finally, the effect of decision strategy on decision performance will benefit from future research. Operationalizations and constructs are needed in this area to be able to measure the type of strategy employed by the user.

Conclusions

This chapter reviewed relevant work on spatial decision-making from 1990 to 2002. Nine empirical studies published in journals and conference proceedings of the IS and decision sciences fields were reviewed. The examination of the literature indicated that researchers have paid more attention to the study of GIS supporting individual decision makers than to the support of collaborative decisions with GIS. Empirical studies investigating the use of GIS to support decision-making in groups were not found.

Todd & Benbasat's (2000) model was used to identify the variables affecting decision performance investigated in the literature. This model was adapted to decision making using GIS according to the analysis of variables studied in previous research. However, more empirical work is needed to understand the impact of DSS capabilities, types of cognitive capabilities, decision strategies, and levels of desired effort and accuracy on decision performance. In general, knowledge in all these areas is required to better understand the influence of GIS technology on decision-making. A better understanding will lead to the creation of more efficient technologies. Research in these areas will benefit from theoretical frameworks specifically designed for GIS, such as Jarupathirun & Zahedi's (2001, 2003) conceptual model for SDSS utilization and spatial decision performance.

Finally, it is important to mention that using different research methodologies can enrich the study of spatial decision support systems. All studies used experiments as the research approach. Experiments are of great value in many cases, but they also have problems of ecological validity[1]. In-depth qualitative studies, for instance, could reinforce theory and provide new insights on the mechanisms through which GIS assists decision-making.

References

Crossland, M. D., & Wynne, B. E. (1994). Measuring and testing the effectiveness of a spatial decision support system. *Proceedings of the 27th Annual Hawaii International Conference on System Sciences*, (Vol. 4, pp. 542-551).

Crossland, M. D., Herschel, R. T., Perkins, W. C., & Scudder, J. N. (2000). The impact of task and cognitive style on decision-making effectiveness using a geographic information system. *Journal of End User Computing, 12* (1), 14-23.

Crossland, M. D., Scudder, J. N., Herschel, R. T., & Wynne, B. E. (1993). Measuring the relationships of task and cognitive style factors and their effects on individual decision-making effectiveness using a geographic information system. *Proceedings of the 26th Annual Hawaii International Conference on System Sciences*, (Vol. 4, pp. 575-584).

Crossland, M. D., Wynne, B. E., & Perkins, W. C. (1995). Spatial decision support systems: An overview of technology and a test of efficacy. *Decision Support Systems, 14*, 219-235.

Dennis, A. R., & Carte, T. A. (1998). Using geographical information systems for decision making: Extending cognitive fit theory to map-based presentations. *Information Systems Research, 9*(2), 194-203.

Densham, P. J. (1991). Spatial decision support systems. In D. J. Maguire, M. F. Goodchild, & D. W. Rhind (Eds.), *Geographical information systems* (Vol. 1, pp. 403-488). London: Longman Scientific & Technical.

Jarupathirun, S., & Zahedi, F. M. (2001). A theoretical framework for GIS-based spatial decision support systems: Utilization and performance evaluation. Paper presented at the *Americas Conference on Information Systems (AMCIS)*.

Keenan, P. B. (1997). Geographic information systems: Their contribution to the IS mainstream. Paper presented at the *Americas Conference on Information Systems (AMCIS)*.

Mark, D. M. (1999). Geographic information science: Critical issues in an emerging cross-disciplinary research domain. Retrieved May 1, 2003: *http://www.geog.buffalo.edu/ncgia/workshopreport.html*.

Mark, D. M., Freksa, C., Hirtle, S. C., Lloyd, R., & Tversky, B. (1999). Cognitive models of geographical space. *International Journal of Geographical Information Science, 13* (8), 747-774.

Mennecke, B. E. (1997). Understanding the role of geographic information technologies in business: Applications and research directions. *Journal of Geographic Information and Decision Analysis, 1* (1), 45-69.

Mennecke, B. E., & Crossland, M. D. (1996). Geographic information systems: Applications and research opportunities for information systems researchers. *Proceedings of the 29th Annual Hawaii International Conference on System Sciences*, (Vol. 3, pp. 482-491).

Mennecke, B. E., Crossland, M. D., & Killingsworth, B. L. (1997, August 15-17). An experimental examination of spatial decision support system effectiveness: The roles of task complexity and technology. Paper presented at the *Americas Conference on Information Systems (AMCIS)*, Indianapolis, IN.

Mennecke, B. E., Crossland, M. D., & Killingsworth, B. L. (2000). Is a map more than a picture? The role of SDSS technology, subject characteristics, and problem complexity on map reading and problem solving. *MIS Quarterly, 24* (4), 601-629.

Murphy, L. D. (1995, January). Geographic information systems: Are they decision support systems? Paper presented at the *28th Hawaii International Conference on System Sciences*.

Mylonopoulos, N., & Theoharakis, V. (2001). On-site: Global perceptions of IS journals. *Communications of the ACM, 44* (9), 29-33.

Newman, L. W. (2003). *Social research methods* (5th ed.). Boston, MA: Allyn and Bacon.

Smelcer, J. B., & Carmel, E. (1997). The effectiveness of different representations for managerial problem solving: Comparing tables and maps. *Decision Sciences, 28* (2), 391-420.

Suprasith, Jarupathirun, S., & Zahedi, F. M. (2003). The value of GIS information in improving organizational decision making. In J. B. Pick (Ed.), *GIS In Business*. Hershey, PA: Idea Group Publishing.

Swink, M., & Speier, C. (1999). Presenting geographic information: Effects of data aggregation, dispersion, and users' spatial orientation. *Decision Sciences, 30* (1), 169-195.

Todd, P., & Benbasat, I. (2000). The impact of information technology on decision making: A cognitive perspective. In R. W. Zmud (Ed.), *Framing the Domains of IT Management* (pp. 1-14). Cincinnati, OH: Pinnaflex Education Resources.

University Consortium for Geographic Information Science. (1996). *Research priorities for geographic information science*. Retrieved May 1, 2003: *http://www.ncgia.ucsb.edu/other/ucgis/CAGIS.html.*

Vessey, I. (1991). Cognitive fit: Theory-based analyses of the graphs versus tables literature. *Decision Sciences, 22* (2), 219-241.

West, L. A. J. (2000). Designing end-user geographic information systems. *Journal of End User Computing, 12* (3), 14-22.

Whitman, M., Hendrickson, A., & Townsend, A. (1999). Academic rewards for teaching, research and service: Data and discourse. *Information Systems Research, 10* (2), 99-109.

Zachary, W. W. (1988). Decision support systems: Designing to extend the cognitive limits. In M. Helander (Ed.), *Handbook of Human-Computer Interaction*. Elsevier Science Publishers.

Endnote

1 *Ecological validity*: a way to demonstrate the authenticity and trustworthiness of a field research study by showing that the researcher's description on the field site matches those of the members from the site and that the researcher was not a major disturbance (Newman, 2003).

Appendix

Journals and Conferences Included in the Literature Review

Journals in Information Science:

Communications of the ACM

Communications of the AIS

Decision Sciences

Decision Support Systems

European Journal of Information Systems

IBM Systems Journal

IEEE Transactions

Information Systems Journal

Information Systems Research

Journal of the Association for Information Systems

Journal of End User Computing

Journal of Management Information Systems

MIS Quarterly

Conference Proceedings:

Hawaii International Conference on System Sciences (*HICSS)*

Americas Conference on Information Systems (AMCIS).

IEEE Conference Proceedings

Journals in Geographic Information Systems:

Geoinformatica

International Journal of Geographical Information Science (formerly International Journal of Geographical Information Systems)

Journal of Geographic Information and Decision Analysis (GIDA)

Transactions in GIS

Chapter III

Techniques and Methods of GIS for Business

Richard P. Greene, Northern Illinois University, USA

John C. Stager, Claremont Graduate University, USA

Abstract

This chapter reviews some standard techniques and methods of geographic information systems for business applications. Characteristics of spatial databases are first reviewed and discussed. Methods of displaying spatial data are compared and contrasted and GIS overlay procedures are described. Two case studies showcase many of the techniques introduced. The first case study illustrates the use of GIS for analyzing an urban labor market while the second demonstrates the integration of modeling functions into a GIS with an application of the gravity model.

Introduction

Managing a company requires a multitude of decisions. The decision makers historically rely on statistical analysis of their customer sales records to make future decisions. They look at charts to track a product's sales trends. They look at store sales records to see which stores are doing well and not so well. The company's database contains lots of other information that often is not even recognized. Almost every database table contains some location information like an address, phone number, or zip code. Most companies use these attributes to keep contact lists, print mailing labels, and send billing statements and advertisements.

If a decision maker is solely looking at statistical charts and tables they are sadly missing out on a gold mine of geographic information, on which a type of geographic analysis called spatial analysis can be performed to yield trade area information, new customers, and competitor area analysis.

This chapter discusses selected techniques and methods of geographic information systems (GIS), with a focus on their applications to business. First, the standard techniques available within GIS software packages are presented. Standard GIS techniques include buffer delineations, overlay analyses, and geo-coding, all of which underlie many GIS applications already in wide use by businesses. Secondly, GIS allows for advanced spatial analyses and interpretation of spatial data. Simply stated, spatial analysis manipulates geographic coordinates and associated attribute data for the purposes of solving a spatial problem. Spatial analyses that are especially relevant for businesses are illustrated. Two example applications illustrate the range of analyses that GIS can provide for decision support in business. The first example illustrates how GIS can assist in the spatial analysis of an urban labor market's industrial specialization. The second example illustrates the use of the gravity model by businesses for determining the spatial extent of their market areas. These examples can be replicated with any commercial GIS software.

The use of GIS for business applications is growing immensely. There are many examples of excellent uses for a GIS from a business point-of-view and the literature on the topic has grown in recent years (see Thrall, 2002; Boyles, 2002; and Grimshaw, 2000, for an overview). It is the purpose of this chapter to not only highlight GIS techniques relevant for spatial analysis in business, but more importantly, showcase their use in a couple of case studies.

Spatial Databases

Companies consider data to be a company asset. Spatial data, falling within this definition of data, would also be an asset. However, spatial data is often not utilized to the extent that it could be to fully leverage its value. Many existing attributes of company databases are spatial, including addresses, zip codes, and telephone area codes. These are not typically thought of as spatial attributes because they are not specified as latitude and longitude coordinates. For example, an address can be used regularly for mailing statements by a bank. The bank may want to put in a new branch and will base their decision on surveys or guesses to determine the location. Using existing transaction data by existing branches, the bank could plot, using a GIS, this transaction data by geocoding the customer home address and at which branch the transaction was made. Using the GIS the bank could then perform analyses, such as a what-if analysis, based on the distance from all of the homes to the branch used. This analysis could determine possible new locations for a branch based on reducing the distance from customers to the branch they use. This is a very simple example, but it does illustrate the possible use of existing spatial data that is not generally thought of as spatial.

Typically, what makes a database spatial is the connection of the data to a geographically referenced coordinate system. A geographical coordinate system precisely locates features on the earth in terms of an X and Y coordinate position. Latitude and longitude are the most frequently used reference coordinates. Both are measured as angles from the center of the earth as a point to a point on the surface of the earth. Many GIS databases are geographically referenced with transformed coordinates from a different map projection and associated coordinate system and are typically referred to as X and Y coordinates.

Commercially available Data Base Management Systems (DBMSs) have, directly or through extensions, implemented support for these spatial data. Three examples of this are the Oracle Database 9.X Server with Oracle Spatial 9.X spatial database; Informix's spatial data-blades including 2D, 3D, and Geodetic; and ESRI's Spatial Data Engine (SDE). However, it is possible to store spatial data in a traditional DBMS, especially given that spatial data are often readily stored in most databases (e.g., address information). It is also possible to store other spatial data (e.g., base maps, overlays) in traditional DBMS's. This can be accomplished using Binary Large Object (BLOB), which is a collection of bits that can contain anything (e.g., video, text, music, raster data, vector data, and any digital data). Moreover, a BLOB can store a shape file (.SHP) in a record. The shape file has become an industry standard for storing some types of spatial data.

Spatial data requires a unique set of operations to manipulate the X and Y coordinates stored in the DBMS. For example, a spatial query may be a SQL Select statement with a where clause of where Street_1 *intersects* Street_2. The GIS intersect operation is yet another operation that involves two polygons that overlap in the X and Y coordinate space. These unique operations are not generic to SQL and the database world. That is, a GIS intersect is not as one would have in set processing (e.g., an intersection of two sets), but rather a spatial intersection (e.g., does the trade area of Sears intersect with the trade area of Kohl's?). Other familiar database operations are also different in a GIS context. Joins, for example, need to be oriented toward spatial data. Normally, we join on like fields and like values. In the spatial world, we will need to support a join that joins a point (e.g., a store location) to a polygon (e.g., a store's trade area) based solely on its spatial attributes that may not be a textual representation of a spatial attribute.

Finally, DBMS indexing needs to allow for retrieving and displaying items in the database without the necessity of doing a table, in the case of a relational database, scan (i.e., sequentially processing the entire data table). The advantage of indexing is for speed and efficiency. Indexing in a DBMS is similar to the indexing of a book. If one were reading a book with no index, the only method for finding a specific item in the book would be to read the entire book in sequence from beginning to end. With an index that is arranged alphabetically, one simply looks up the term and turns to the reference page number. Similarly, an index can be built for a DBMS, which improves the efficiency of various queries and operations.

Most enterprises have already implemented rudimentary spatial data into their operations. They have ever since they placed the first address into a database. Do enterprises have to switch to a DBMS that fully supports spatial data? They do if that is a requirement or is wholly or partially a core competency of the firm. However, in utilizing spatial data to increase productivity, provide for better customer service, or increase sales, it may be fine to use what is available and add what is possible within the current DBMS.

One technique that can be used to transform an address of a customer into spatial data represented as an X and Y location is to perform geocoding or what is often referred to as address matching. Address matching is a process that compares two addresses to determine whether they are the same. To match addresses, the GIS software examines the components of addresses in both the database file where addresses are maintained and the attribute table associated with a GIS layer of roads. The U.S. Census Bureau's digital street maps are commonly used for this purpose, as their attribute tables have four street address numbers ranging from low to high for each side of a street segment. The range indicates the possible numbers that could fall within a particular block, and the numbers are divided into even numbers on one side of the street and odd numbers on the other. The address components for this type of street are typically represented as:

Left_from	*Left_to*	*Right_from*	*Right_to*	*Street_name*	*Type*
201	*299*	*200*	*298*	*SUNSET*	*ST*

GIS software tools can then take a table of addresses and interpolate a point for each address based on these address ranges. For example, a customer's home address could be geocoded into latitude and longitude coordinates and then stored in the customer database as an attribute of the customer, as is the address. When a delivery is scheduled to that customer, the latitude and longitude could be placed into a shipping file for all the shipments that must be made on a given day. Using this file, the shipments can be divided into the number of available trucks. Then using the list of shipments on a given truck, the latitude and longitude could be used by a GIS to route the truck on the most efficient path for all of the deliveries.

One other example would be to keep spatial data on the company assets used by customers of an electric utility. Then a problem occurs — let us say that someone accidentally knocked down a utility pole. Based on the connections that exist to customers that somehow relate to that pole (e.g., electric transmission and delivery circuits), the utility could know the extent of an outage and take proactive action to notify affected customers about the problem and the estimated time that the problem will be resolved. The spatial data processing of a geometric network of electric lines and the addresses represented as points would be employed to generate the list of affected customers.

Querying Attributes and Spatial Display of Data

The benefit for a company that stores its customer data in a GIS is that it allows them to visualize the spatial patterns of those customers. Consider the case of a company that has a large database filled with tables of all aspects of its customers collected over the years. The value in the database is not its large size, but rather in its ability to answer questions. You can make queries on the database that result in a smaller subset of data. For example you might make a query such as, "show me last month's customers that spent over $200." The value in your database is the ability to structure these queries, as well as the methods available for displaying the results of those queries. For instance, if the above query yields 20 customers, then one may choose to display them in a list, sort them, and sum their total spending. Alternatively, one could display the results in a chart. It would also be useful to display each customer on a map, which is made possible when the data are stored in a GIS. Such a customer map might include the store locations in order to visualize patterns such as clustering around a certain store or zip code. The observations may lead to a decision to mail advertisements to nearby zip codes that appear underserved.

Some GIS support a standardized language to perform such queries. It is called the Structured Query Language or SQL (some pronounce SQL as sequel). A SQL statement is made up of three parts: a field name, an operator, and a value. Such statements can also be connected with other statements with connectors. For instance, from the sales example above, if you wanted to visualize customers originating from high income zip codes the statement may be written as SELECT SALES_AMOUNT>200 AND ZIP_CODE = HIGH INCOME. The GIS would then highlight all of the customers represented as dots that met the above criteria.

Following a set of queries, a company will wish to present the results on a map for decision-making purposes, such as convincing a decision maker to take a certain action. Many businesses in the days before GIS would place a map of an area on their wall and push colored pins into the map to signify locations of importance for strategic decision-making. Similarly, when traveling, a business person would take a highlighting marker and mark the route to be taken or record the route that was taken. Today, a critical component included in GIS maps is a legend consisting of appropriate symbols, colors, and classifications used for drawing the map layers shown on the map. These legends vary in type ranging from one color or one symbol displays to the display of many colors and symbols. For instance, a metropolitan map illustrating disposable income patterns may vary the symbol size for zip code points to denote levels of income originating from an income attribute contained in the zip code attribute table. The latter technique, referred to as a graduated symbol map, is illustrated in the first case study at the end of the chapter.

Lines are another way that features are represented on maps, and similar to areas and points they can be colored or symbolized based on an attribute contained in a table. For instance, if a company has captured the flow of its customers with an origin, such as a zip code location, and a destination, such as a store location, then the business may decide to illustrate the relationship by lines connecting the zip codes to the store and varying the width based on the volume originating by zip code. A flow map of this nature will quickly reveal the directions in which the store is attracting customers as well as the location of underserved areas.

Areas referred to as polygons are also used to communicate information other than just its location and relative size. For instance, the same metropolitan map illustrating disposable income patterns above, but this time census blocks rather than zip code points, may use different colors to denote levels of income originating from an income attribute contained in a census block attribute table. This latter example is referred to as a graduated color map that is created by picking a color ramp (a spectrum of color that allows a distinct color for each represented value). For our example we would pick a red color ramp that shows a spectrum of red from a white to a pink to a bright red to a dark red. We would equate the dark red with higher disposable incomes. The way we equate the income to a specific color is done using a mapping classification technique discussed in the next section.

Another symbolization technique is a dot density map. This technique is performed by equating the income to a number of dots that are shown within the census blocks of the metropolitan area. Let us say that each dot represents $100 of disposable income. If the disposable income of a specific census block were $2,000 then there would be 20 dots within the boundary of that block on the map. There is only one potential problem with this technique, and that is that the placement of the dots is random within the block.

Mapping Classification Methods

Just as in preparing a graph of sales, mapping will often require a user to first classify the data into classes in order to simplify the display. Thus, when you use the symbolization techniques described above for map layers, you decide how many classes each needs and you decide how to break the data into classes. Each class has a beginning and an ending value, you can pick these values through your own criteria, or you can use an established classification method. Each method uses a math equation to calculate the range of each class, and some of the more common ones employed in GIS software are described.

Natural Breaks

Natural breaks are said to occur within the data when there are large jumps between values of observations in the dataset. The natural breaks method then looks for obvious breaks or gaps in the data to establish classes. If one were to request three classes to be generated from the data, the GIS would attempt to find three areas that are separated by a gap in the clusters of values.

Equal Interval

In the equal interval classification the lowest value is subtracted from the highest value to compute a range. Then the range is divided by the number of classes that are desired. The resulting number is then added to the low to get the upper range for the low class. The resulting number is then added to that, to get the upper range for the second class, and then repeated for additional classes. An example of the equal interval classification is presented in the first case study at the end of the chapter.

Equal Area

An equal area classification sets class boundaries so as to include an equal proportion of a map area into each established class. Thus, the map will appear balanced in that each class will represent approximately the same area in extent. In business this may be of use if a company wishes to map a product that it wishes to distribute equally over a trading area.

Standard Deviation

Many GIS software packages have introduced the standard deviation or another measure of dispersion for establishing a classification. A standard deviation is basically the average difference of the set of values from the mean of the set of values. To calculate the standard deviation one calculates the mean of the values, which is the total of the values divided by the number of values. Next a sum of the difference of each of the values is computed, then squared, divided by the number of items; finally the square root of the result is computed. The formula to compute a standard deviation is:

$$s = \sqrt{\frac{\sum_{i=1}^{n}(X_i - \bar{X})^2}{n}}$$

where s is the standard deviation, n is the number of items in the list, X is the value of an item, $\bar{X}$ is the mean of the items in the list. For classification breaks, a user decides on the number of standard deviations, for instance two will result in four classes, two above the mean and two below the mean. This is an effective method for showing extremes in the data: as in the case of sales, a business can quickly visualize extreme low and high sales volume areas.

Table Joins, Buffer, and Overlays

A number of advanced database operations are available in GIS. Although not unique to GIS, table joins are a feature of all relational databases of which GIS is included. Joining tables is the technique for using data from multiple sources in your analysis. For example, you have a table of data of all of your customers' transactions (e.g., sales) for a particular time that also contained a customer identification number. You also have a table of data about your customers that contains the same customer identification number. The second table has the customers' address but nothing about their volume of sales. By joining the two tables on the same characteristic, you could obtain the total sales for a customer and plot it on a map at their address or even aggregate the sales for a given area and plot that data. The join is not a permanent change to your data but exists only as long as the analysis process takes to perform. It can be saved and performed each time it is needed, thereby ensuring a fresh look at the data with the data chosen by the analyst (e.g., last month's sales data, all sales).

Table joins are also useful for dealing with legacy systems. Legacy systems are older systems that were written in languages that are not actively used in software development today. They may have also used methodologies and techniques that are also not generally used in today's development environment. These systems are often not replaced because they may be large and contain massive amounts of an enterprise's data. The development effort to replace the system is large and overshadowed only by the effort required to convert the data to the new system. Legacy systems may contain addresses, latitude and longitude, and other spatial information. Through the join process, these data can be utilized in a modern GIS system.

A number of map overlay operations are available in a GIS including a union procedure. The union of two layers in a GIS is done to combine an input layer (base map) and an overlay layer to produce a third layer. This new layer contains the attributes of both component layers and the total extent of both layers. What this means is that each polygon of the resultant layer has the attributes of its constitute polygons and the extent (boundary) of the resultant set of polygons is the total of the two components (*Figure 1*).

In this example, one input map contained census blocks and the other map contained three store trade areas with the resulting map showing the combination. The new union layer also contains the attributes contained in the input map layers. The advantages of this technique for businesses are numerous, including their ability to query multiple map layers. In the above example, a business could use this union procedure to summarize the demographic variables contained in the census block attribute table by the trade area definition.

Another common GIS operation, but unique to GIS, is a buffer. A buffer is an area around a point, line, or area defined by a radius distance. Consider a point of interest on a map. You are interested in the area within one mile of that point. With a GIS, you can place a buffer around that point with a radius of one mile. Essentially, the display on the map

Figure 1. A GIS Union Procedure

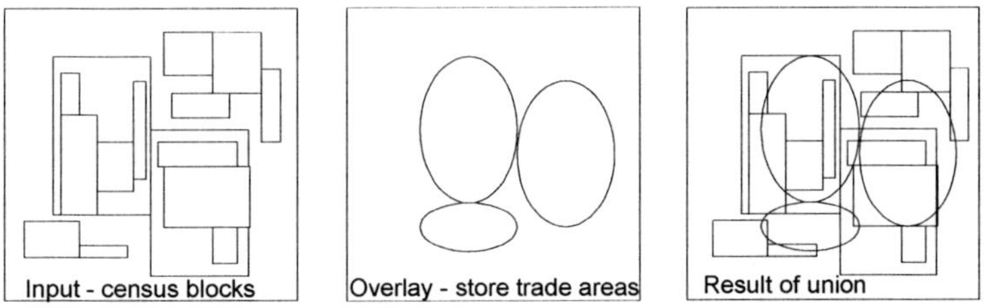

would show the point and a circle with the point as the center of the circle. The circle has a radius of one-mile. The same is done with lines and areas (polygons). However, the method used for those two features is not as simple as a circle with a radius of the desired buffer size. For a line the buffer is displayed as an area around the entire feature at the specified distance. For an area, the edge of the area (polygon) is used for the starting point of the distance of the buffer. For an example of using this technique for a point, imagine that you are responsible for a new store being opened. Your marketing information shows that 85% of your sales will be from people living within a one-half mile radius from the store location. Using our GIS you construct a one-half mile buffer around the store location. All of the addresses that fall within the area created will be targeted for a direct mail campaign.

An example of a business using a buffer around a line might be for gauging competition along a lengthy commercial strip. Consider a business that wants to build a new store in an area that it knows has good market potential. It has been able to get access to zip code level customer sales generated at competing stores along the commercial strip. So the business decides to buffer the commercial road by a few miles and union the resulting buffer polygon with the zip code polygons. Now the business can examine customer gaps within the buffer region in order to assist in the decision on where to place the new store.

Finally, for an example of a buffer around an area (polygon), imagine a new city ordinance that requires that there be no liquor stores within 1,000 feet of a schoolyard. A business wishing to open a liquor store in this context might use the GIS to find all of the local schoolyards and place a buffer at a 1,000-foot distance around them. Any location outside of these identified buffers should be acceptable to the city on the ordinance requirement.

The following two sections are case studies that highlight many of the GIS procedures already discussed. The case studies are also business applications, which should stimulate additional ideas for GIS project development.

GIS Analysis of the Industrial Specialization of an Urban Labor Market

Businesses need to know the availability of types of workers appropriate to their workplaces. This information is useful as one factor in deciding on the siting of facilities, and for existing facilities in determining the costs of hiring, given job market abundances or scarcities. Urban labor market spatial analysis can pinpoint the intensity distributions of labor force for industrial specializations of workers in an urban area.

The location of business activities within an urban labor market may at first glance appear widely dispersed. However, if one disaggregates the economic activity by economic sector with mapping functions in a GIS, there appears to be much spatial logic to the pattern (*Figure 2*). Using graduated symbols based on an equal interval classification of manufacturing and professional employment location-quotients this map shows a

Figure 2. Location Quotients for Manufacturing and Professional Employment

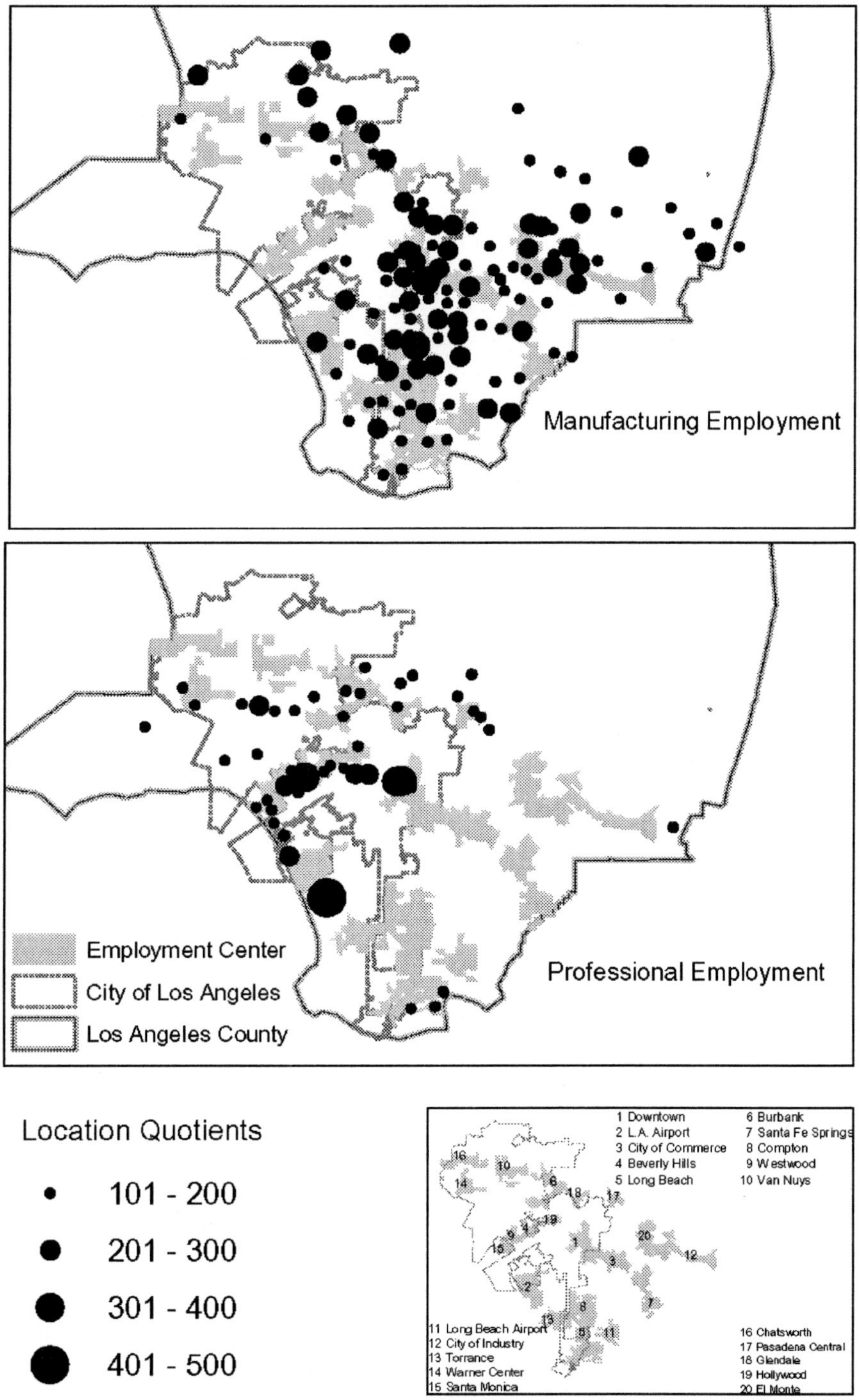

geographic difference between the two industry types. For the Los Angeles County labor market, manufacturing activities appear highly clustered in the center while professional services appear less centralized, but the amount of clustering of the two industries and reasons are quite different. The GIS in this example consists of a zip code point map layer and a job center polygon map layer with industrial sector attributes for zip codes drawn from the 1997 Economic Censuses.

Starting with its 1997 Economic Census, the U.S. Census Bureau replaced its long-established Standard Industrial Classification (SIC) with the North American Industry Classification (NAICS). The NAICS recognized 361 industries not previously identified separately by the SIC system. The zip codes in this example were represented as points with a latitude and longitude coordinate located inside the original zip code area. NAICS industry types were extracted from the 1997 Economic Census at the zip code level and, using a GIS join operation, the data were related to the zip code map by the unique zip code field of each database. The job center map layer is a reference map of the principal job centers in Los Angeles County.

Location Quotient

A *location quotient* expresses the share of employment in a given industry in a specified sub-area as a percentage of the share of employment in the same industry within the larger area.

Consider the following example where Zip Code A has 250 workers in the bottling industry out of a total labor force of 1,000, while in the county containing Zip Code A, there are 50,000 workers in the bottling industry and a one million labor force total. The location quotient is calculated as follows:

$$LQ = (E_{ZIP}/LF_{ZIP}) / (E_{COUNTY}/LF_{COUNTY}) * 100 = (250/1,000)/(50,000/1,000,000) * 100 = 500$$

Zip Code A is then said to be specialized in the bottling industry because the location quotient is greater than 100. In the Los Angeles County GIS, the sub-areas are zip codes while the larger area is Los Angeles County.

$$LQ = (E_{ij}/E_j)/(E_i/E_t) * 100.$$

Where E_{ij} = Employment in sub-area j in sector i;

E_j = Total employment in sub-area j;

E_i = County employment in sector i;

E_t = Total County employment.

A location quotient greater than 100 in a zone indicates more specialization in the industry category, and less than 100 indicates that the zip Code is less specialized in that category.

Spatial Patterns of Manufacturing and Professional Services

To illustrate how a GIS database can elucidate specialization patterns identified by the location quotient, we will examine differences between the geographic distribution of manufacturing and professional services. The professional services map indicates that a heavy concentration of professional services has formed in an arch-like pattern both within and outside the City of Los Angeles (*Figure 2*). Starting with the end containing the largest concentration of services to the other, the arch-like pattern runs from downtown Los Angeles, through Hollywood, into Beverly Hills, Century City, and West Hollywood, toward Westwood — West Los Angeles and ends in Santa Monica. The second largest concentration of services within the Professional group is occurring between Beverly Hills — Century City — West Hollywood and Westwood — West Los Angeles. Wilshire Boulevard is the major road connecting these centers. To the north of this same formation is a lesser concentration of services than the first, but one that should be indicated as a heavier concentration of Professional Services than typical. This concentration forms a linear pattern.

There are only minor concentrations of Professional Services in southern Los Angeles. While the map on Manufacturing also shows heavier concentrations, including in downtown Los Angeles, it does not show that these activities string together to create any solid formation of manufacturing over significant distance, like Professional Services (*Figure 2*). The area northwest of downtown Los Angeles shows a heavier grouping of manufacturing than do areas south and west of the downtown area. Heavier manufacturing areas exist further south. A larger, heavier area of concentration exists outside the city, south and east of the city limits.

Professional services and manufacturing complement each other. Where one dominates, the other has small concentrations, with the exception being downtown Los Angeles itself. Here both services exist even though it looks as if they do in different sections of the downtown area — Professional services appear higher in the northern end of the downtown area while manufacturing dominates more toward the center and in the southeast portion of the downtown area.

Confirming these patterns, Forstall & Greene (1999) found that manufacturing is relatively important in the Commerce-Vernon, Compton, Los Angeles Airport, and Santa Fe Springs employment concentrations and relatively unimportant in the Beverly Hills and Westwood concentrations. Scott (2002) noted that the garment industry, a category of manufacturing, in Los Angeles appears in the form of a dense agglomeration of firms near the center. Meanwhile, professional services bulk largest in Downtown, Beverly Hills, and Westwood. Professional services and manufacturing are the two highest job concentrations in Downtown.

These observations on the labor market differences between professional and manufacturing industries are relevant for some business decision-making as they reveal different requirements by various industries. Perhaps a manufacturing firm considering whether or not to relocate within the county would want to conduct such a GIS analysis. A concentration of manufacturing workers, as revealed in the location quotient map, may be a good indication of a plentiful labor supply from which to draw its workers for the firm.

Trade Area Analysis

In retail businesses and retail-oriented nonprofits, it is useful to estimate, for a market region, the probability that a customer will decide to visit a particular facility, given the presence of competing facilities. This section explains the gravity model method for doing this. It then illustrates a simple case application of the model by estimating the probability of consumers in southeastern Wisconsin and northeastern Illinois visiting one of three major opera houses in the region. Contours are mapped for the probabilities of visiting the Oscar Meyer Opera House in Madison, Wisconsin.

In addition to descriptive mapping techniques, advanced spatial analysis methods can be employed for prediction of spatial interaction. The gravity model is just one example of such methods, a model that has been applied to a number of business location analysis problems. Although the application of the gravity model had been in use for migration studies prior to the 1930s, Reilly (1931) was the first to apply the model to the study of retail trade. The general gravity model holds that any two bodies attract each other with a force proportional to the product of their masses and inversely proportional to the square of the distance between them. Reilly then interpreted this for retail as meaning that two cities would attract consumers from some smaller, intermediate city in direct proportion to their population sizes and in inverse proportion to the square of their distances from the intermediate city. Reilly later went on to develop a technique, now referred to as *Reilly's law of retail gravitation*, which allowed one to determine trade area boundaries around cities based solely on population and distance measures.

Huff (1964) offered an alternative model referred to as the probabilistic model of consumer spatial behavior. Recently, Huff (2003) revisited his earlier formulation in the context of its applicability in GIS:

The Huff Model has endured the test of time — more than 40 years. Its widespread use by business and government analysts, as well as academicians, throughout the world is remarkable. With the development of GIS, the model has received even more attention (p. 34).

Huff's model results computed in a GIS are often depicted as probability contours and interpreted, for instance, that a person from neighborhood X has a probability of 0.9 of going to a nearby supermarket for groceries, while a person from neighborhood Y has a probability of 0.2 of going to the same supermarket. The person from neighborhood Y is also assigned a probability of visiting competing supermarket destinations including the one in her own neighborhood.

Huff first formulated his ideas by establishing an attraction index to be associated with a given retail facility:

$$v(j) = A_j^{\gamma} D_{ij}^{-\lambda} \qquad (1)$$

where A_j is an attraction index associated with a particular retail facility j; D_{ij} is the accessibility of a retail facility j to a consumer located at i; and γ and λ are empirically derived parameters. The quotient derived by dividing A^{γ}_j by D^{λ}_{ij} is regarded as the perceived utility of retail facility j by a consumer located at i (Huff & Black 1997, p. 84).

The probability contours of the probability that a consumer located at i will choose to shop at retail facility j, is determined as follows:

$$P_{ij} = \frac{A_j^{\gamma} D_{ij}^{-\lambda}}{\sum_{j=1}^{n} A_j^{\gamma} D_{ij}^{-\lambda}} \qquad \textbf{(2)}$$

Today, this probabilistic model of consumer spatial behavior appears in many commercial GIS software packages and, if not available, it is feasible to employ it directly in a standard GIS system as distance calculations are derived easily with X and Y coordinates.

To illustrate why GIS lends itself well to computing the gravity model, consider the application of Huff's probability model for estimating the attendance of three major opera houses in the Chicago, Milwaukee, and Madison metropolitan triangle region. In this example we wish to create probability contours depicting the probability of an area sending customers to an opera house based on its attraction measure. The seating capacity of the three principal opera houses within the region was used as the attraction index: the Lyric in Chicago with 3,563 seats, the Skylight in Milwaukee with 358 seats, and the Oscar Mayer in Madison with 2,200 seats. A more sophisticated measure of attraction might consider the number and quality of performances, cost for a seat, and other attributes of an opera event. Census tract centroids (an internal x and y coordinate pair) were used for the origins of those potential customers attending an opera (*Figure 3*).

In this latter step, having the data compiled in a GIS is critical because the X and Y coordinates are intrinsic to the database design. Thus, GIS software algorithms can return those X and Y coordinates to a given map layer's associated attribute table. In this example case, the census tracts and opera house locations are registered to the UTM map projection and coordinate system, a Cartesian coordinate system that makes distance computations quite simple to perform in any GIS or computer spreadsheet package. Similarly, X and Y coordinates were generated for the opera house locations and returned to that layer's table (*Figure 4*).

The opera house map layer's attribute table was previously populated with the seating capacity of each opera house.

Once the X and Y coordinates were computed for each map layer's features, they were exported out to a spreadsheet where the gravity model probabilities would be computed for each census tract in the study region (*Figure 5*). A critical component of this spreadsheet design was the inclusion of a relate item, in this case a census tract FIPS code, that will be used to link the probability calculations back to the GIS census tract layer in order to map the 50 percent probability areas around each opera house. The next

Figure 3. Adding X and Y Coordinates from the Census Tract Layer to its Attribute Table

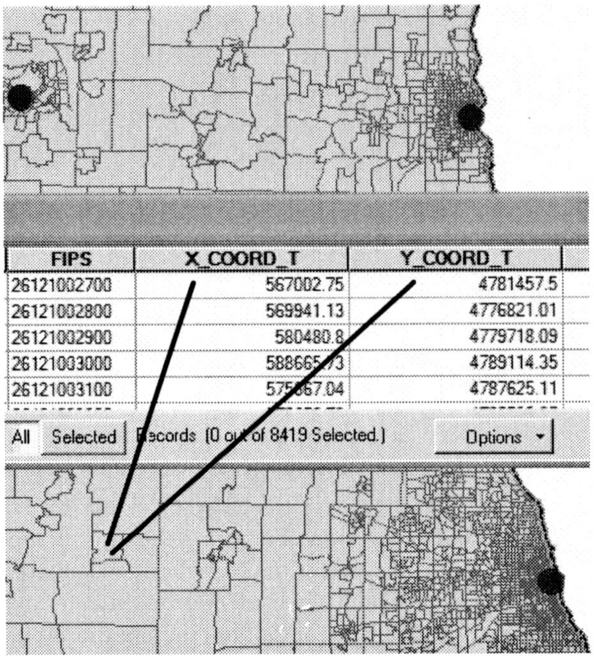

Figure 4. Adding X and Y Coordinates from the Opera House Layer to its Attribute Table

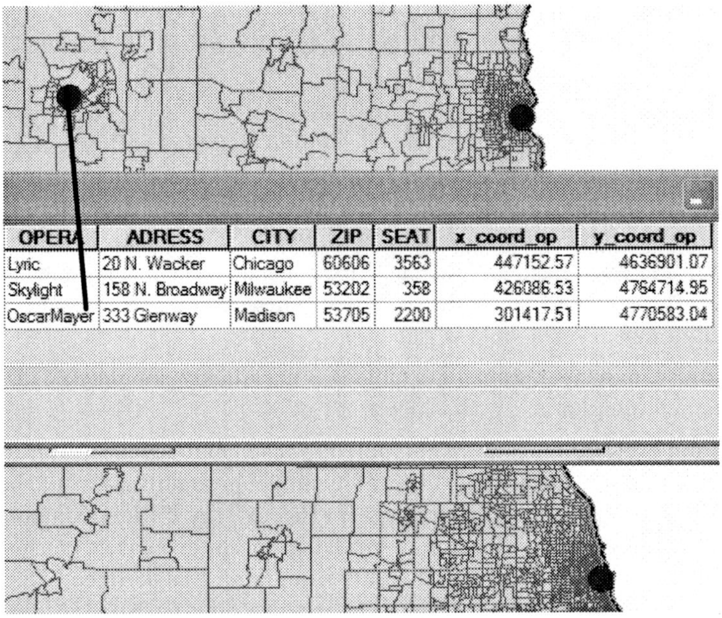

Figure 5. Table Exported with X and Y Coordinates and Relate Field to a Spreadsheet

Relate field

A	B	C	D	E	F	G
FIPS	x_coord_t	y_coord_t		seats	opera_x	opera_y
26121002700	567002.75	4781457.50	chicago	3563	447152.57	4636901.07
26121002800	569941.13	4776821.01	milwaukee	358	426086.53	4764714.95
26121002900	580480.80	4779718.09	madison	2200	301417.51	4770583.04
26121003000	588665.73	4789114.35				
26121003100	575667.04	4787625.11				
26121003200	572656.79	4790580.95				
26121003300	573239.26	4803647.64				
26121003400	563445.32	4796305.45				
26121003500	554418.96	4798035.87				
26121003600	564337.39	4801919.04				
26121003700	552269.04	4803116.22				

step was to compute the attraction index for each tract to each opera house as stated in equation one above (*Figure 6 a-c*).

The numerator in each equation across for the first census tract is simply the seating capacity of the given opera house and the denominator is the distance calculation with a conversion factor for meters to miles given that the reference units for the UTM coordinate system is meters. Once the utility or attraction was computed for the first census tract, the formula was copied to the remaining tracts. A final step was computing the probabilities by dividing the attraction by the sum of the three attractions (*Figure 7 a-c*). In all cases the three probabilities computed for each tract should sum to one.

Once the probabilities were computed they were linked back to the GIS with the relate-item (FIPS) that was contained in each table (*Figure 8*). Drawing upon additional mapping functions, typically available as software extensions, it was possible to draw the probability surfaces for a given opera house based on the probability field contained in the GIS attribute table (*Figure 9*). This is a map of the probability surface for the Oscar Mayer opera house in Madison. The advantage of this mapping technique is that it shows the full range of probabilities and highlights additional spatial features such orientation of market penetration.

Figure 6a. Computing the Attraction Index for Chicago's Lyric Opera House

fx =E$2/(SQRT((B2-F$2)^2+(C2-G$2)^2)/1609)^2

B	C	D	E	F	G	H	I	J
ord_t	y_coord_t		seats	opera_x	opera_y	Chicago	Milw_util	Madis_
102.75	4781457.50	chicago	3563	447152.57	4636901.07	0.2616	0.04602	0.081
141.13	4776821.01	milwaukee	358	426086.53	4764714.95	0.2662	0.04447	0.079
80.80	4779718.09	madison	2200	301417.51	4770583.04	0.2416	0.03852	0.073
65.73	4709114.35					0.2135	0.03429	0.069
67.04	4787625.11					0.2351	0.04047	0.075
56.79	4790580.95					0.2343	0.04184	0.077
39.26	4803647.64					0.2111	0.04	0.076
45.32	4796305.45					0.2369	0.04665	0.082

Figure 6b. Computing the Attraction Index for Milwaukee's Skylight Opera House

fx =E$3/(SQRT((B2-F$3)^2+(C2-G$3)^2)/1609)^2

B	C	D	E	F	G	H	I	J
ord_t	y_coord_t		seats	opera_x	opera_y	Chicago	Milw_util	Madis_
102.75	4781457.50	chicago	3563	447152.57	4636901.07	0.2616	0.04602	0.081
141.13	4776821.01	milwaukee	358	426086.53	4764714.95	0.2662	0.04447	0.079
80.80	4779718.09	madison	2200	301417.51	4770583.04	0.2416	0.03852	0.073
65.73	4709114.35					0.2135	0.03429	0.069
67.04	4787625.11					0.2351	0.04047	0.075
56.79	4790580.95					0.2343	0.04184	0.077
39.26	4803647.64					0.2111	0.04	0.076
45.32	4796305.45					0.2369	0.04665	0.082

Figure 6c. Computing the Attraction Index for Madison's Oscar Mayer Opera House

fx =E$4/(SQRT((B2-F$4)^2+(C2-G$4)^2)/1609)^2

B	C	D	E	F	G	H	I	J
ord_t	y_coord_t		seats	opera_x	opera_y	Chicago	Milw_util	Madis_
102.75	4781457.50	chicago	3563	447152.57	4636901.07	0.2616	0.04602	0.081
141.13	4776821.01	milwaukee	358	426086.53	4764714.95	0.2662	0.04447	0.079
80.80	4779718.09	madison	2200	301417.51	4770583.04	0.2416	0.03852	0.073
65.73	4709114.35					0.2135	0.03429	0.069
67.04	4787625.11					0.2351	0.04047	0.075
56.79	4790580.95					0.2343	0.04184	0.077
39.26	4803647.64					0.2111	0.04	0.076
45.32	4796305.45					0.2369	0.04665	0.082

Figure 7a. Computing the Probabilities for Visiting Chicago's Lyric Opera House

=H2/(H2+I2+J2)

	C	D	E	F	G	H	I	J	K	L	M
d_t	y_coord_t		seats	opera_x	opera_y	Chicago	Milw_util	Madis	prob_chic	prob_milw	prob_madi
2.75	4781457.50	chicago	3563	447152.57	4636901.07	0.2616	0.04602	0.081	0.673817	0.118647	0.207636
1.13	4776821.01	milwaukee	358	426086.53	4754714.95	0.2662	0.04447	0.079	0.683212	0.114149	0.202639
0.80	4779718.09	madison	2200	301417.51	4770583.04	0.2416	0.03852	0.073	0.684118	0.109047	0.206835
5.73	4789114.35					0.2135	0.03429	0.069	0.674545	0.10032	0.217136
7.04	4787625.11					0.2351	0.04047	0.075	0.669793	0.115305	0.214902
6.79	4790580.95					0.2343	0.04184	0.077	0.663484	0.118478	0.218037
9.26	4803647.64					0.2111	0.04	0.076	0.64541	0.122317	0.232273
5.32	4796305.45					0.2369	0.04665	0.082	0.647787	0.127564	0.224649

Figure 7b. Computing the Probabilities for Visiting Milwaukee's Skylight Opera House

=I2/(H2+I2+J2)

	C	D	E	F	G	H	I	J	K	L	M
d_t	y_coord_t		seats	opera_x	opera_y	Chicago	Milw_util	Madis	prob_chic	prob_milw	prob_madi
2.75	4781457.50	chicago	3563	447152.57	4636901.07	0.2616	0.04602	0.081	0.673817	0.118647	0.207636
1.13	4776821.01	milwaukee	358	426086.53	4754714.95	0.2662	0.04447	0.079	0.683212	0.114149	0.202639
0.80	4779718.09	madison	2200	301417.51	4770583.04	0.2416	0.03852	0.073	0.684118	0.109047	0.206835
5.73	4789114.35					0.2135	0.03429	0.069	0.674545	0.10032	0.217136
7.04	4787625.11					0.2351	0.04047	0.075	0.669793	0.115305	0.214902
6.79	4790580.95					0.2343	0.04184	0.077	0.663484	0.118478	0.218037
9.26	4803647.64					0.2111	0.04	0.076	0.64541	0.122317	0.232273
5.32	4796305.45					0.2369	0.04665	0.082	0.647787	0.127564	0.224649

Figure 7c. Computing the Probabilities of Visiting Madison's Oscar Mayer Opera House

=J2/(H2+I2+J2)

	C	D	E	F	G	H	I	J	K	L	M
d_t	y_coord_t		seats	opera_x	opera_y	Chicago	Milw_util	Madis	prob_chic	prob_milw	prob_madi
2.75	4781457.50	chicago	3563	447152.57	4636901.07	0.2616	0.04602	0.081	0.673817	0.118647	0.207636
1.13	4776821.01	milwaukee	358	426086.53	4754714.95	0.2662	0.04447	0.079	0.683212	0.114149	0.202639
0.80	4779718.09	madison	2200	301417.51	4770583.04	0.2416	0.03852	0.073	0.684118	0.109047	0.206835
5.73	4789114.35					0.2135	0.03429	0.069	0.674545	0.10032	0.217136
7.04	4787625.11					0.2351	0.04047	0.075	0.669793	0.115305	0.214902
6.79	4790580.95					0.2343	0.04184	0.077	0.663484	0.118478	0.218037
9.26	4803647.64					0.2111	0.04	0.076	0.64541	0.122317	0.232273
5.32	4796305.45					0.2369	0.04665	0.082	0.647787	0.127564	0.224649

Figure 8. Probabilities are Related Back to the GIS Attribute Table

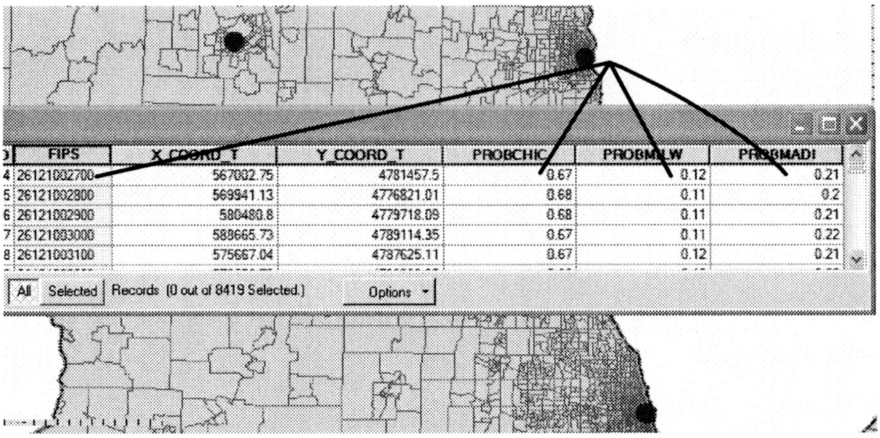

Figure 9. Probability Surface for the Oscar Mayer Opera House in Madison, WI

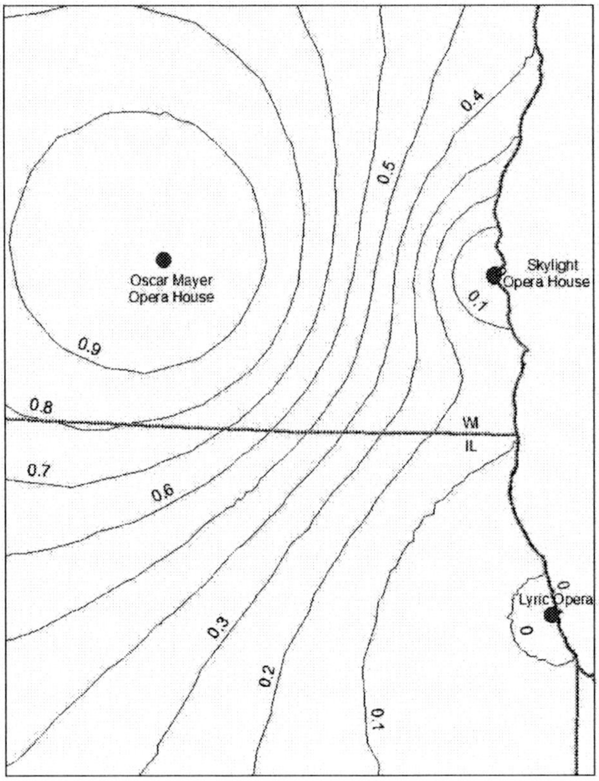

Conclusions

The GIS techniques and methods available for business applications are numerous. Many of the techniques and methods described in this chapter are not generic to business but are used in many different types of GIS applications. However, the two case studies have highlighted their use for decision support within a business setting. The application of GIS for analyzing an urban labor market's industrial specialization shows the benefits of linking economic census to a zip code map layer for understanding the spatial relationships of industry types. The spatial relationships revealed through graduated symbol mapping of locations quotients can yield better understanding of the diversity of the labor market and may even assist in a site location decision. The gravity model application introduced how predictive capabilities can be harnessed from a GIS database, including a better understanding of where customers originate. More sophisticated analyses of trade areas can be performed with additional information that could be included into the attraction measure, as well as other map layers including one with travel times. These and other improvements are made easy if time is invested early on in the spatial database design stage of the GIS project.

References

Boyles, D. (2002). *GIS means business (Vol. 2)*. Redlands: ESRI Press.

Forstall, R. L., & Greene, R.P. (1997). Defining job concentrations: The Los Angeles case. *Urban Geography, 18*, 705-739.

Grimshaw, D.J. (2000). *Bringing geographical information systems into business (2nd Ed.)*. New York: John Wiley & Sons.

Huff, D.L. (1964). Defining and estimating a trading area. *Journal of Marketing, 28*, 34-8.

Huff, D.L. (2003, October-December). Parameter estimation in the Huff Model. *ArcUser*, 34-36.

Reilly, W.J. (1931). *The law of retail gravitation*. New York: Knickerbocker Press.

Scott, A.J. (2002). Competitive dynamics of Southern California's clothing industry: The widening global connection and its local ramifications. *Urban Studies, 39*, 1287-1306.

Thrall, G.I. (2002). *Business geography and new real estate market analysis*. New York: Oxford University Press.

Chapter IV

Costs and Benefits of GIS in Business

James Pick, University of Redlands, USA

Abstract

This chapter examines the costs and benefits of geographic information systems (GIS). It focuses on the research questions of what components to include in GIS cost-benefit (C-B) analysis, what distinguishes GIS C-B analysis from non-spatial C-B analysis, what methods to use, and how to invest in GIS systems in order to obtain net payoffs over time. It categorizes costs and benefits of GISs. It considers the topics of systems analysis sub-steps in cost-benefit analysis, the feasibility decision, stakeholders and externalities, and the importance of timing and timeliness in investing in GIS and assessing payoffs. It examines the C-B aspects of a well-known GIS case study of Sears Roebuck's delivery system. The literature on the value of investment in IT and productivity paradox is analyzed for its relevance to GIS investment. The major findings are, first, that the costs and benefits of a GIS can be estimated through modifications of standard non-spatial IS methods. Second, the key factors that differentiate GIS cost-benefit analysis from that of non-spatial IS are more extensive analyses of the costs of data acquisition, need to pro-rate the GIS costs and benefits for tightly linked combinations of GIS and other systems and technologies, and need for improved techniques to estimate the costs and benefits of the GIS's visualization features.

Introduction

Geographical Information Systems (GISs) have become an important tool for government and business decision-making (Huxhold, 1991, 1995; Grimshaw, 2000; Tomlinson, 2003;

Clarke, 2003). One early definition of GIS is the following: "The purpose of a traditional GIS is first and foremost spatial analysis... Capabilities of analyses typically support decision-making for specific projects and/or limited geographic areas" (Exler, 1988). GIS is more than mere mapping, extending much further in its capabilities to analyze spatial information through such techniques as overlays, queries, modeling, statistical comparisons, and optimization (Mitchell, 1999; Clarke, 2003; Greene & Stager, 2003).

GIS performs spatial analysis based on data and boundary files stored in databases to support decision-making (Huxhold, 1991; Murphy, 1995; Jarupathirun & Zapedi, 2001, 2004). GISs provide mapping and analysis for marketing, transportation, logistics, resource exploration, siting, and other business sectors (Harder, 1997; Grimshaw, 2000; Boyles, 2002). GIS can take advantage of spatial factors to improve response times, decide more efficiently on locations, optimize movements of goods and services, market more effectively, and gain enhanced knowledge of routing, siting, and territories. GIS can be linked to the Internet to support virtual applications. The core of the GIS market, consisting of software, services, data, and hardware, is estimated to total $1.75 billion in 2003 (*Directions Magazine*, 2003). Projections indicate this GIS market grew by 8 percent during 2003, which occurs as the U.S. came out of recession (*Directions Magazine*, 2003).

Cost benefit analysis for GIS is done for three main reasons (King & Schrems, 1978).

1. Cost-benefit analysis assists in planning for an organization. Planning involves tradeoffs between competing demand for organizational investment and resources. C-B analysis can help in deciding between competing demands.
2. Cost-benefit analysis is useful in auditing. The organization may decide after a GIS system has been put into effect to perform a C-B analysis retrospectively, to assess what occurred.
3. To prepare and support participants in political decision-making (King & Schrems, 1978). This is less formal and more rapid than the planning uses in numbers 1 and 2. It is mostly done with less complete information, but is more commonplace than the other reasons (King & Schrems, 1978).

The objective of this chapter is to answer three research questions: (1) How can the costs and benefits of a GIS be estimated? (2) What differentiates cost-benefit analysis of GIS from standard cost-benefit analysis of information systems? and (3) What are the appropriate data and instruments to measure costs and benefits of GIS? In answering the questions, the chapter considers financial, technical, institutional, and integration costs and benefits. It also looks at the value of IT investment methods, to assess their relevance for GIS.

Several prior research studies have examined cost-benefit analysis of GIS and IS. Obermeyer (1999) summarized several methods to analyze costs and benefits of GIS-based systems. Tomlinson (2003) discussed the principles and rationale for cost-benefit analysis of GIS, emphasizing certain features that are distinctive about spatial analysis. Taking a wide view of costs and benefits, on the cost side he added to the usual items the broader categories of liability, software interfaces, and communications. On the benefit side, he added better timing of projects, richer information to users, improved organizational workflow, and more effective expenditures by the firm. However, often firms must wait over the lifetime of a project to receive the benefits (Tomlinson, 2003). It may take a GIS five or more years to reach the breakeven point where cumulative benefits start to exceed cumulative costs (Tomlinson, 2003). Huxhold & Levinsohn (1995) discussed the benefits of GIS from a project management perspective. There is also research that focuses on the investment in, and value derived from, IT for different units of analysis, such as the project, firm, and industry (Ahituv & Neumann, 1990).

A related line of literature has involved the information technology productivity paradox, which refers to studies of the investments in, and payoffs from information technology (IT). Some of this research has indicated low or no payoff (Brynjolfsson, 1993; Strassmann, 1997, 1999; Lucas, 1999). Several conceptual frameworks exist for the productivity paradox, including normative value, realistic value, and perceived value (Ahituv, 1980, 1989). Recent literature in this area has been more likely to conclude that there are net benefits from IT investment (Ragowsky et al., 2000; Deveraj & Kohli, 2002). One reason for a more optimistic picture is that methods and data collection have improved, so that benefits that might have been missed before can now be accounted for. In addition, studies are better able to judge the appropriate timeframes over which to measure the benefits, taking into account lagged effects, and to disaggregate the unit of analysis into smaller units (Brynjolfsson, 1993; Ragowsky et al., 2000; Deveraj & Kohli, 2002; Navarrete & Pick, 2002).

The specific topics covered in this chapter are: (1) Summary of costs and benefits, (2) Systems analysis sub-steps for cost-benefit analysis, (3) Comparison of costs and benefits, (4) Analysis steps in cost-benefit analysis, (5) Stakeholders and externalities, (6) The case of GIS costs and benefits at Sears Roebuck Delivery, (7) Value of IT investments: its relevance to GIS, and (8) Conclusion.

The methodology utilized in the present chapter is critical evaluation of whether or not non-spatial IS cost-benefit techniques can be applied to GIS, and the analysis of the Sears case example. The chapter is broad and draws on a variety of literature. It does not perform empirical analysis with real-world data to determine net positive benefits. However, the chapter may be useful as a framework to researchers conducting such empirical studies.

Setting the Context for Cost-Benefit Analysis

A GIS cost-benefit analysis must consider and clarify at the beginning its broad context, especially four factors (King & Schrems, 1978):

1. *Statement of purpose.* This statement indicates whether the C-B analysis is being utilized directly to make a decision, to provide background data for a decision, or to politically influence decision-making (King & Schrems, 1978). It is important to state the purpose, since the methods, quality of the data, and reporting of findings, among other things, differ by purpose.
2. *Time simultaneity.* The C-B analysis must indicate whether the C-B analysis is prospective to a future GIS system, for a current GIS system, or retrospective. All three are useful in certain circumstances.

3. *Scope*. The C-B analysis may be comprehensive in examining all possible costs and benefits. On the other hand, it may be severely limited in scope: for instance, it may only manipulate a single cost item, in a sensitivity analysis, to see if change in that item affects the benefit outcomes.
4. *Criterion*. The last contextual factor is the method that is used to compare the costs and benefits, after they have been compiled. This may be a quantitative index measure, graphical comparison, or involve a lengthier model. The criterion needs also to state whether or not the values of the costs and benefits are to take into account inflation and opportunity costs through present-value calculations.

Among the challenges in GIS cost-benefit analysis is that GIS usually has higher costs than for conventional information systems (ISs), due to its considerable data acquisition and data management. This is because a GIS is based on both attribute data and spatial data (Huxhold, 1991; Clarke, 2001). The extra time and effort, versus non-spatial IT, stems from the need to do the following:

- Gather boundary data and associated attributes
- Convert the data to digital form
- Design or configure topological data structures and link non-spatial with spatial data
- Maintain GIS boundary files and data

Estimates indicate data collection for GIS may constitute 65 to 80 percent of the total cost for conventional systems development/implementation (Huxhold & Levinson, 1995; Obermeyer, 1999; Tomlinson, 2003). Further, the attribute and digital boundary data need to be linked together. These linkage tasks add to costs, relative to non-spatial IS. Another difference has to do with GIS's feature of visualization (Jarupathirun & Zahedi, 2004). In this respect, GIS is comparable to a multimedia business application, which may have higher costs, due to its visual aspects. Because the benefits of visualization are predominantly intangible, GIS tends to have a higher proportion of intangible costs and benefits than a non-spatial IS. Another distinctive aspect of GIS is that it tends to be more linked or coupled with other software systems and technologies than is normally present in IS applications. Among the systems and technologies with which GIS is often interfaced are global positioning systems (GPS), remote sensing (Meeks & Dasgupta, 2003, 2004), and marketing information systems (Allaway, Murphy, & Berkowitz, 2005). As seen in *Figure 1*, because of this linking together of several types of systems and technologies, the cost and benefit calculation for any one of them may be more difficult.

Figure 1 demonstrates that the assessment of costs and benefits of a simple, uncoupled GIS system can be done by performing an assessment of that system's costs and benefits, followed by a cost-benefit comparison such as break-even analysis. However, the cost-benefit analysis becomes more complicated for a GIS that is closely linked with another system or technology, such as point-of-sale, remote sensing, or GPS. For linked systems, it is difficult to disaggregate the costs and benefits of the GIS from those of the other parts.

Figure 1. Comparison of Cost-Benefit for a Simple GIS System to a GIS System Closely Coupled with Another System or Technology

Cost-benefit analysis for a simple uncoupled GIS

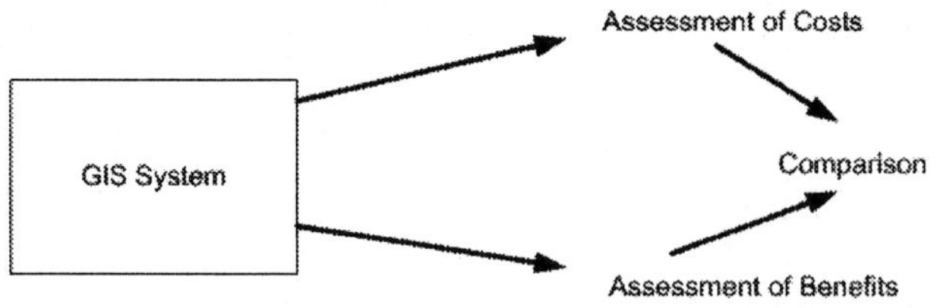

Cost-benefit analysis for a GIS system closely coupled with another system or technology

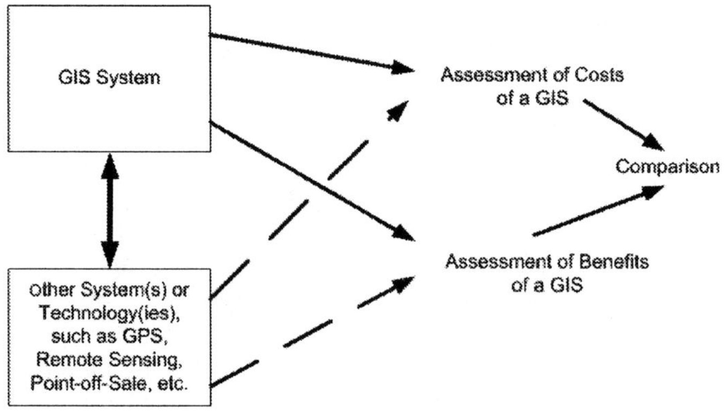

Summary of Costs and Benefits of a Geographic Information System

This section discusses the categories of costs and benefits of a geographic information system. Costs and benefits may be divided into tangible — i.e., able to be converted into monetary amounts — and intangible — i.e., not convertible to monetary values (King & Schrems, 1978). Costs for information systems including GIS are predominantly tangible, while benefits are a mixture of tangible and intangible. For instance, the cost of a GIS managerial employee can be estimated by the tangible value of his/her salary and job benefits. However, the most significant benefit of the employee is his/her effective leadership and decision-making, outcomes difficult if not impossible to convert to dollar amounts.

Costs

The costs of a GIS can be classified into the categories given in *Table 1*. These costs are all tangible and are possible to estimate. However, some problems occur in accurately accounting for the costs, and by the risk of exclusion of some of the full range of costs. It is important to avoid the many common errors prevalent in cost accounting (King & Schrems, 1978), such as not identifying hidden costs, counting costs twice, or omitting important costs. An example of hidden costs would be costs located at other places in the organization that are not being counted for the GIS. For instance, the development of a GIS system depends on the ideas of several top managers who spend significant time with the applications development team. However, the overhead of the time spent by the top managers with the team is not included in the costs. Omitted costs are ones that are not obvious, but are in fact dedicated to the project. For example, space, site, and utility costs are commonly omitted, but may be important, particularly in expensive locations, such as midtown Manhattan.

Benefits

Benefits of GIS and of ISs in general are more difficult to measure than costs (King & Schrems, 1978; Obermeyer, 2000; Tomlinson, 2003). The reason is that benefits often accrue in the form of a more informed, ready, efficient, and high-performance organization, which is hard to measure, since there are many beneficiaries, time lags, and intervening causes. A deeper problem is the benefit of information value may increase, but the value of information is difficult to measure (Ahituv, 1989). The difficulty stems

Table 1. Tangible Costs of a GIS

· Hardware
· Software
· Data collection
· Transformation of manual maps and data into digital format
· Maintenance costs for hardware and software
· Maintenance of data
· Supplies
· Design and construction of databases
· Hiring more staff
· Training present staff
· Outsourcing (e.g., GIS applications programming)
· Consulting
· Licensing
· Communications interfaces and networks
· Space, site, and utilities

Modified and expanded from Huxhold (1991), Obermeyer (1999), Tomlinson (2003)

from information value being dependent on its timeliness, users, and what decisions it is influencing. Another problem with measuring benefits has to do with quantity versus quality. If a GIS system leads to a higher quantity of maps being produced than the former manual system, the key related question is how does the quality of the GIS maps compare to that of the manual ones. Hopefully, the GIS-based maps would have higher quality, but that may not be so, in which case the output gains in quantity of maps would be offset by reduced quality.

In spite of the problems, benefits can be, and are calculated and utilized. Practical suggestions for evaluating benefits include the following.

1. Larger benefits may be disaggregated into smaller pieces, which may be more amenable to quantification. Consider an example. A truck transport company has identified the benefit of better control of its fleet by its GIS. However, the overall control cannot be made tangible. Nevertheless, if control is broken down into a set of small control items, such as local supervisory knowledge of trucks arriving in each city, supervisory knowledge of trucks loading in each city, etc., control for the small items is now more amenable to assignment of dollar values, although it is still not easy.
2. The C-B analysis can be restricted to only tangible costs and tangible benefits (King & Schrem, 1978). If that result indicates net benefit, then the intangible benefits may be regarded as an added plus. Another variation on this approach is to perform a break-even analysis for a future time point, just restricted to tangibles (King & Schrem, 1978). If costs exceed benefits, the difference of benefits minus costs can be compared against intangible benefits. They may be close enough to yield a compelling argument of justification.

Unit of Analysis

The difficulty in measuring the costs and benefits of GIS varies by the unit of analysis (Obermeyer, 1999; Tomlinson, 2003). Among the units of analysis for GIS are the following:

- Industry
- Company
- Department or division
- Project
- Individual

Most commonly, the unit of analysis is the company or project. For units with small scope, such as a single-user desktop project, costs and benefits may be more readily estimated. For them, there is limited integration with other systems and technologies; intervening factors are reduced, and the external environment is not as influential. On the other hand, if the unit of analysis is a corporate-wide GIS system, it may be challenging

Table 2. Benefits of GIS (Tangible and Intangible)

Tangible
- Salary and benefits lowering from reducing the workforce
- Cost reduction (through employees performing their tasks more efficiently)
- Cost avoidance in the future (projected greater workload per employee)
- Expansion of revenues (achieved through improved data quality, improved efficiency)
- Improved productivity
- Improved performance
- Higher value of assets

Intangible
- Improved decision-making (at issue, how is the tangible value of better decisions estimated. The data for decision-making may be faulty. GIS can only contribute so much if the data are faulty.)
- Effectiveness of managers and executives
- Reaching strategic objectives
- Environmental scanning
- Speed and timeliness of information
- Volume and quality of information
- Better capability to sell products (CDs, web services, manual maps)
- Improved collections of money
- Identification of missing revenue sources (e.g., in government, identification of properties not being taxed)
- Better operational efficiency and workflow
- Better utilization of assets
- Reduced error
- Reduced liability (e.g., GIS for security monitoring)
- External benefits (i.e., benefits to organizations other than the one implementing the GIS; an example would be benefits from a marketing GIS product to the customer buying the product)

Modified and expanded from Huxhold (1991), Obermeyer (1999), Stein & Nasib (1997), Deveraj & Kohli (2002), Tomlinson (2003)

to separate its costs and benefits from those of other systems inside the company, from inter-organizational systems, or from the outside environment. Furthermore, for enterprise-wide GIS, the attribute data are commonly shared with other company systems, such as marketing, or with the firm's enterprise resource planning (ERP) system. The shared aspect of the business data complicates the separation of GIS's costs and benefits.

At the firm level, the task of separating the benefits of GIS from other intervening factors is even more prevalent, since extraneous influences become more important, such as economic impacts, competitors, and government policy changes. An illustration of the challenges of correlating returns on IT investment at the firm level was seen in a longitudinal study at the firm level for the Mexican Banking industry from 1981 to 1992 (Navarrete & Pick, 2002). The economic cycle during this period was seen to be influential on the level of IT investment for a given year. Availability of funds for IT investment depended not only on bank cash flow, but also on the outside factors of economic and market conditions of interest rates, disposable incomes of customers, banking-industry competitive pressures, and pricing of hardware and software products.

Since GIS technology is fast moving, there may not be precedents for modeling or benchmarking some future costs or benefits. For instance, hand-held GIS products such as ArcPad have appeared in the last five years, and consist of a limited GIS software package, a keypad or other means to enter data, a small display screen, and cabling connectors for information import and export (ESRI, 2003). However, given that these hand-held devices are early in their product cycle, there may not be benchmarking studies available on their performance, reliability, and ease of use (Day & Schoemaker, 2000). As new products, their costs may be unstable. On a long-term basis, technological changes for them may be hard to evaluate, since some new technology products may fail, while others remain relatively stable, and yet others are redesigned rapidly (Day & Schoemaker, 2000; Doering & Parayre, 2000).

Systems Analysis: Sub-Steps in Cost-Benefit Analysis for GIS

In systems analysis and design, the analysis stage includes a feasibility study that contains cost-benefit analysis as a part of it (Ahituv & Neumann, 1990; Satzinger, Jackson, & Burd, 2002).

The following are the standard cost-benefit sub-steps:

1. Develop an overall plan for the C-B analysis.
2. Decide on the analyst or analyst team.
3. Determine the alternative C-B analyses to be conducted.
4. Determine all the material factors for costs and all the material factors for benefits.
5. For each tangible factor, decide how it will be measured.
6. Measure the costs and benefits.
7. Compare the cost-benefit results over the entire time period of the study. Include summary measures such as break-even point, etc., and criteria to evaluate alternatives.
8. Perform a comparative analysis of the alternatives from step 3.
9. Decide on what recommendations to make to management, based on these results. Present the findings to management (modified from Ahituv & Neumann, 1990; King & Schrems, 1978).

Before commencing C-B analysis, an analyst must be selected to conduct the analysis — a choice that is often critical for C-B's success or failure. There are a number of places to find the analyst (King & Schrems, 1978; Ahituv & Neumann, 1990; Satzinger, Jackson, & Byrd, 2002): (1) *Inside person.* This is someone within the organization who has the necessary financial management and technical skills, as well as experience and knowledge in GIS. (2) *Outside consultant.* Auditing, accounting, and GIS and IS consulting firms can provide such persons. (3) *Persons from other organization(s).* Someone from an affiliated organization may be loaned or made responsible to perform the C-B analysis. They may stem from government or corporate oversight, or through corporate alliances.

For GIS, the C-B substep steps differ in some respects. For Step 2 (develop an analyst team), GIS is influenced by its historical origination in the public sector, outside of the mainstream of business IT (Huxhold, 1991; Tomlinson, 2003). Thus, systems analysts for GIS may be less well trained than for non-spatial IS. Regarding Step 4 (determine the material factors for costs and benefits), for a GIS system, the material benefits may be reduced or more difficult to measure, due to the greater presence of visualization. In Step 7 (comparative analysis), comparisons may be complicated by the problem discussed earlier of close linkages between GIS and other systems and technologies.

How does GIS differ from non-spatial IS in relative importance of different categories of costs and benefits, given in *Tables 1* and *2*? For GIS, these categories are influenced by the relatively greater costs for acquiring spatial data (Tomlinson, 2003). The capital costs of data acquisition vary by whether the data and coverages come from the public or private domains. Since historically GISs were mainly in the public domain, large banks of public data and boundary files have been available for free or very low cost. An example is the U.S. Census maps and associated data on population, housing, social characteristics, and the economy, distributed free over the Internet (U.S. Census, 2004). The challenges are larger if a GIS is to be designed and constructed utilizing data from two or more countries' censuses (Pick et al., 2000a, 2000b). In an era of globalization, there is greater demand for multi-census GISs. However, a problem in firms' utilizing these data is that they may not be well organized for business purposes (Ray, 2005). The time spent organizing the data, entering it into databases such as Oracle, and checking it for quality often makes up for the "free availability" from governments. Sometimes, third-party service firms will make the government data available in better-organized form, but at some cost. On the other hand, business proprietary data can be quite costly or unavailable for competitive reasons.

Comparison of Costs and Benefits Over Time

As was seen in *Figure 1*, the concluding step in a cost-benefit analysis is to compare the costs against the benefits in order to determine whether or not the IT investment has net benefit or not. Graphical and non-graphical methods are available to perform comparative analysis of cost-benefit results over time (King & Schrems, 1978; Kingma, 2001; Boardman, Greenberg, Vining, & Weimer, 2001). Among the more prominent ones are the following:

- *Break-even point.*
  - The total investment cost is divided by the annual benefit of the IT. This gives the number of years to break-even. However, this calculation may need to be adjusted for a multi-year investment due to changes in the time value of money.
- *Formulae.*
  - Costs and benefits are directly compared through a formula.
- *Baseline cost comparison chart.*
  - Use of a graph to compare the annual cost of running an enterprise/organization without IT and the cost of running the enterprise/organization with IT.
- *Discounted cash flow.*
  - Covered in the next section.

Usually the C-B analysis involves several alternatives having different values for costs and benefits. A criterion is used to compare them. Let's say that an analyst has prepared ten alternatives under different cost-benefit assumptions. A *criterion* can be used to select the most suitable one (King & Schrems, 1978). Among the criteria commonly used are:

- Maximize the present value of the benefits minus the present value of the costs.
- Maximize the ratio of benefits over costs.
- Assume a given level of costs for all alternatives and maximize benefits.
- Assume a given level of benefits for all alternatives and minimize costs.

Any of these criteria, or others, can be used. The choice depends on the particular problem and context of management decision-making.

A problem can occur if insufficient alternatives are examined for the problem at hand. For instance, for a small-scale GIS with a single data source, single boundary file, and clear uses, perhaps two or three alternatives would be appropriate. By contrast, for an enterprise-wide GIS involving millions of dollars in expenditure and hundreds of users, more alternatives are needed. For a complex GIS, it is not however possible to include all the alternatives. Experience has shown that results may be improved if the key interested parties are involved in determining which alternatives to include (King & Schrems, 1978).

These comparison methods do not differ for GIS versus conventional IS. Their purpose is to assist management and, hence, they are not influenced by the type of system.

Discounted Cash Flow

A standard aspect of cost-benefit analysis is to account for the discounted value of money over time. Both costs and benefits are discounted over time, assuming a regime of future discount rates for inflation. The present value (PV) for a cost or benefit for a particular year in the future may be estimated by:

$$PV = \sum_{i=0}^{n} \frac{x_t}{(1+d)^t}$$

where PV is the present value, x_t is the cost or benefit value during time period t, and d is the discount rate.

After choosing appropriate discount rate or rates, all the future costs and benefits can be estimated.

The net present value of the costs and benefits can be summarized together in the equation:

$$NPV = \sum_{i=0}^{n} \frac{B_t - C_t}{(1+d)^t}$$

where B_t represents the benefit at time t and C_t represents the cost at time t.

There are fine points in calculating the present value that are beyond the chapter's scope, but can be found in books on cost-benefit and financial accounting. They include at what point in a time period the inflation rate is calculated (beginning, middle, end); whether the inflation rate fluctuates (this equation assumes it remains steady); and whether costs are influenced differently by inflation than benefits (King & Schrems, 1978).

For GIS, the issues of discounting do not differ from those of standard IS applications. This is because the discount rate is extraneous to GIS and IS systems, but rather depends on interest rates and other factors in the outside economy (Obermeyer, 1999; Kingma, 2001; Boardman, Greenberg, Vining, & Weimer, 2001). However, since GIS projects tend to have long periods from start to break-even due to the higher investment in start-up and data acquisition (Tomlinson, 2003), the impact of present value calculations may be large.

Intangible Costs and Benefits

Intangible costs and benefits are prevalent in GIS applications experiences (Tomlinson, 2003). Some examples of intangible benefits are the following (Ahituv & Neumann, 1990; Tomlinson, 2003):

- Image improvement of the organization
- Better decision-making
- Enhanced employee morale
- Improved information to executives

After determining the intangible costs and benefits, they are presented in summary form to management, along with the tangible ones (Ahituv & Neumann, 1990). Management can then decide how much to weigh the intangibles.

What is distinctive for GIS vs. non-spatial IS concerning intangibles? First, the visualization aspects of GIS encourage a greater degree of intangibles than for a conventional IS application. Visual responses tend to be difficult to measure. For example, if there are two GIS systems, where the first one produces low resolution maps and the second one yields maps with five-fold better resolution, how can the advantage of this intangible benefit on business effectiveness and decision-making be measured? A second feature of GIS that has stimulated more intangibles is the tendency of GIS to move up in the hierarchy of business applications to become more strategic. At the strategic/competitive level, the benefits are less tangible versus at the operational or middle management levels (Ahituv, 1989).

The Feasibility Decision

Management must eventually weigh the results of a cost-benefit analysis and make a decision on one of the alternatives, including staying with the status quo (Ahituv & Neumann, 1990; Satzinger, Jackson, & Burd, 2002). Analysis of feasibility for IT systems is divided into three areas: (1) financial feasibility, (2) technical feasibility, and (3) institutional feasibility. Feasibility decisions are influenced by the total time period of commitment before the decision is revisited, for instance, one year, five years, or 10 years. If the feasibility decision needs to hold for ten years, then a much more in-depth study must be undertaken. No matter what depth, as the time frame extends out, it becomes increasingly difficult to accurately predict technological changes (Day & Schoemaker, 2000).

For financial feasibility, GIS is similar to a conventional IT application. It is based on the cost-benefit comparison described in the last section. Determination of technical feasibility for GIS follows standard methods detailed elsewhere (Satzinger et al., 2002), but may put more emphasis on spatial feasibilities or on the feasibility of linking with technologies such as GPS and mobile services. Institutional feasibility refers to the capability of the institution to support a GIS project. GIS requires specialized human workers, with sufficient knowledge to carry out a GIS project, either inside the institution or present in consultants or an outsourcer. Institutional feasibility is affected by GIS's distinctive features, especially visualization capability and presence of linked systems.

Stakeholders and Externalities

Cost-benefit results may vary from the vantage points of different stakeholders in an organization. For instance, GIS in an advertising firm will have different costs and benefits for the corporation itself, the firm's customers, its investors, and the general public. The effort of cost-benefit work is made more difficult and time-consuming by adding diverse stakeholder analyses. However, it may be worth the effort if there are large differences in the stakeholder outcomes. At the minimum, stakeholders should be mentioned in the cost-benefit analysis.

Positive and negative externalities of a GIS refer to the indirect impacts of system implementation. This can be seen by the analogy of environmental externalities of an industry processes. An industry process that has the purpose of manufacturing a product may cause indirect effects of pollution, noise, and human injuries. An example of an externality for GIS is the loss of the information security of a land property company, as its GIS land-property system becomes more widely deployed. By contrast, positive externalities may arise from GIS-driven websites of local governments that provide GIS analysis and mapping to the public. There may be unintended benefits, for instance, enhanced public image to local citizens or to more distant site website visitors.

A Case Study of GIS Costs and Benefits: Sears Roebuck Delivery

Sears Roebuck constitutes a case example of the benefits of GIS in a large-scale enterprise. Sears delivers a vast amount of merchandise nationally every day. Yearly its 1,000 delivery trucks perform more than four million deliveries to homes (Kelley, 1999). Its truck delivery system covers 70 percent of the U.S. territory (Kelley, 1999). However, prior to implementing GIS, the manual process at Sears was very time consuming and wasteful, with many hours each day spent by routing-center workers locating street addresses. There was an average time slippage rate of 20 percent of deliveries. In the early 1990s, an enterprise-wide GIS system was constructed by geocoding Sears' millions of customer addresses and setting up optimized delivery with the goal of 90 percent reliability in the promised delivery window (Kelley, 1999). The GIS calculates daily delivery routes, based on a model that includes "estimated travel times, in-home time, truck capacity, optimal stop sequence" (Kelley, 1999).

For tangible benefits, the system has increased efficiency by reducing the time for routing and addressing from an average of five hours daily to 20 minutes. The miles per delivery-truck stop were lowered by 0.6 mile, which allows four more stops per truck per day (Kelley, 1999). This has allowed reduction in the number of Sears national routing centers from 46 to 12. Delivery orders have expanded by 9 percent with the same-sized truck fleet. In all, equipment and facilities savings from the GIS-based networking enhancement are $30 million per year (Kelley, 1999).

The intangible benefits apply to Sears management and customers. Sears middle and top management are able to use the information in this system strategically to plan improved efficiency versus its competitors over long periods of time, a strategic efficiency approach resembling Wal-mart's well-known inventory and just-in-time delivery systems. At the customer level, reducing the 20 percent of missed deliveries down to less than five percent has had image-enhancement benefits that help in marketing other company products.

In this case, what was different vs. an enterprise-wide non-spatial system deployed in a large corporation? Several factors differ. First, the initial data acquisition was relatively expensive and time consuming, since over four million addresses had to be geocoded and then tediously corrected (Kelley, 2002). However, once this one-time massive data acquisition took place, the costs of subsequent data acquisition were within a normal range for IS applications. Another difference is the integration of GIS with GPS on the delivery trucks. Knowing the location of each truck in real-time allows more optimal management of the whole fleet. However, as we discussed, the necessity to couple GIS with other systems may increase the overall cost, as well as make cost-benefit analysis more difficult.

Value of IT Investment: Relevance to GIS

In the information systems field, a special evaluation methodology has evolved, referred to as the "value of IT investment" approach. These methods examine the level of IT investment and variety of returns on investment. The methodology started in the late 1970s and has produced hundreds of studies. Early studies tended to be pessimistic regarding the productivity resulting from investment. This led to the term, "productivity paradox," i.e., the paradox that companies or larger economic units invest in IT but fail to realize appropriate productivity gains (Brynjolffson, 1993).

This method is related to traditional cost-benefit analysis. The difference is that it is especially sensitive to IT measurement problems in the analysis. Hence, instead of the broad category of costs, it focuses on investment in IT, which implies monetary funds intentionally directed towards IT. Instead of encompassing the broad range of tangible and intangible benefits, it focuses on quantitative measures of value, productivity, and performance. Return on investment is regarded as best measured by multi-attributes, not by a single attribute (Ahituv, 1980, 1989). This approach has not solved the difficult problems of measuring benefits, since the difficulties remain in deciding which attributes to measure, how to measure them, and how to combine them in a multi-attribute function (Ahituv, 1989).

This section will briefly discuss some important ideas from the value of information approach, with particular focus on the value of GIS investments. The objective is to add these techniques to the group of methods that are available to people doing GIS cost and benefit evaluations.

Like cost-benefit analysis, the value of IT investment approach recognizes a variety of units of analysis, much the same list as that given earlier. Studies have been done on value of IT investment at the levels of the economy (Osterman, 1986; Baily & Chakrabarti, 1988; Roach, 1989), industry (Noyelle, 1990; Cron & Sobol, 1983, Barua et al., 1995; Strassmann, 1997, 1999; Navarrete & Pick, 2003), firm (Brynjolfsson & Hitt, 1993; Harris & Katz, 1989; Loveman, 1988; Barua, 1991; Ahituv et al., 1999), and units within firms (Ragowsky et al., 2000).

The value of information approach advocates that information value is multidimensional. An information item is valued depending, among other things, on its timeliness, relevance of contents, display format, and cost to produce it (Ahituv, 1980). The cost to produce it means that the value of the information item depends on the information system that produced it (Ahituv, 1980). These multiple attributes are combined in a multi-attribute function to estimate value.

The challenges with the attribute and multi-attribute function include the following (Ahituv, 1989): identifying which attributes to include, measuring the attributes, formulating the multi-attribute function, and tradeoffs between the attributes.

In creating a multi-attribute function to ascribe value to GIS information, the same challenges apply. It is important with GIS to stress the attributes of timeliness and the display format. Since a GIS often takes a long time to implement, its information may not be timely, and so would have reduced value. GISs need to be designed so they can be regularly refreshed with up-to-date information. The display format attribute needs also to be considered. First, measurement problems are difficult. How can one map display be valued in quantitative terms compared to another? Part of the problem here is that individuals differ in their perceptions and evaluations of visual displays. Second, the importance of display versus other attributes needs to be decided upon for such a function. Even though GIS produces impressive visual displays, other outputs, such as tables, graphs, and multimedia pictures, need to be weighed in importance as competing attributes.

There are three ways to value information (Ahitiv, 1989; Ahituv & Neumann, 1990).

- *Normative value of information.* This consists of a theoretical model that produces a quantitative value. The multi-attribute function is most commonly used in such a model.
- *Realistic value of information.* Here measurements are made in the actual organization before and after an information system has been put into use. The differences in value, productivity, or performance measures, before and after, is used to estimate the returns on investment.
- *Perceived value of information.* This valuation is entirely a subjective rating of the user regarding information value.

More detailed explanation for applying these methods appear elsewhere (Ahituv, 1989; Ahituv & Neumann, 1990; Brynjolfsson, 1993). However, experience has shown that the normative approach is mainly useful for prospective valuation of future systems. Realistic value of information may be useful for present or past systems, and is more appropriate for lower level operational systems, rather than management decision-making or strategic systems (Ahituv, 1989). Perceived value method is appropriate for higher-level systems.

The relevance for GIS is that perceived value method is the most applicable, since GISs are increasingly utilized for managing and decision-making, rather than for low-level operations. At the same time, the perceived value method has many more sources of error and should be utilized with caution. Permanent problems with perceived value are the following: (1) Individual respondents differ considerably in their subjective reactions. (2) It is hard to translate a subjective rating scale into tangible dollar amounts.

In spite of these problems, the perceived-value approach can be applied to analyzing the value of past, present, and future GIS systems. Care should be taken to achieve a large enough sample, and to have a panel of questions that may be partly convertible to tangible values.

A final issue in this section for the value of IT investment approach is that of appropriate lag times for the value or productivity to be realized. Consider that if you invest some of your own funds and purchase a desktop GIS system for your individual use. How many months would be needed before you are at the peak of realizing the benefits of the investment? The same type of lag time would occur for a large-scale enterprise GIS system. It may take many years after the firm's investment for such GIS payoffs to be realized (Deveraj & Kohli, 2002).

Because of the presence of lag times, the ideal systems for measuring IT investment payoff should gather longitudinal information, i.e., to keep track of investments and measure multiple attributes of value, productivity, and performance on a periodic basis, quarterly or annually, over many years. Another advantage to the longitudinal approach is that it can recognize the impacts of long-term intervening processes, such as economic ups and downs, and multi-year competitive impacts. If the value of IT investment analysis is applied for a single time point, it cannot recognize lag times and misses these concurrent patterns.

Since GIS investments have been observed to take longer times for the payoffs to be realized than non-spatial investments (Tomlinson, 2003), systems to track investments and payoffs longitudinally can account for the lags. Of course, the advantages need to be weighed against the added cost of setting up and maintaining such a tracking system.

Overall, the value of IT investment methods offer many potential opportunities for persons responsible for assessing the costs and benefits of GIS systems. Because of the expanding investments being made by organizations in GIS, these techniques should be seriously considered to complement or replace traditional cost-benefit methods that were presented earlier.

Conclusions

This chapter has examined the methods and procedures of cost-benefit analysis for information technology, and sought to identify the distinguishing aspects for GIS, compared to non-spatial IS. One major difference is that GIS has high data-collection costs. This problem is improving somewhat over time, as web-accessible digital libraries become available, often at lower cost and improved quality.

Another area of relatively higher GIS costs is for training and technical support, since training for new technologies may require extra expenditures — internally, by web-based training, or through outside training services. University training offerings, including in business schools, are currently limited.

The benefits for GIS may accrue somewhat later and be stretched out over a longer period than for the average IS. This is because (1) organizations often do not understand all the benefits, and (2) benefits often are enterprise-wide and often less visible by themselves. One benefit that is frequently available is to sell GIS project results as a product. This needs to be factored into the cost-benefit analysis at the beginning, even if it is somewhat speculative.

This study leads to the following practical suggestions to GIS managers and users on how to improve the quality of a GIS cost-benefit analysis:

- Follow a careful plan utilizing all the conventional methods and knowledge available for IS systems.
- Gather as much post-audit benchmarking information as possible from your organization.
- Give extra attention to estimating data acquisition costs. Examine what are the sources for spatial data and, if applicable, how they can be inexpensively converted to digital. How will the database be organized with its spatial and non-spatial components.
- Use a long enough cost-benefit timeframe that all the benefits can be realized.
- Consider the cost of training. Anticipate the delays in realizing benefits.
- Do a careful analysis of intangibles and include them in the report to management.

Returning to the chapter's research questions, they may be answered as follows:

1. *How can the costs and benefits of a GIS be estimated?*

   Costs and benefits can be estimated by supplementing the standard method of cost-benefit analysis for IT in business (King & Schrems, 1978; Kingma, 2001). This includes using material costs for tangible benefits and applying benchmarking, modeling and other methods for estimating intangibles, albeit less reliably. Present-value calculations need to be run, to take into account changing monetary values. The cost-benefit comparison can be done graphically or through break-even analysis or formula comparisons. The feasibility decision is performed, based on financial, technological, and institutional capabilities.

2. *What differentiates cost-benefit analysis of GIS from that of conventional information systems?*

   Cost-benefit analysis needs to be supplemented for GIS by examining more thoroughly the costs of data acquisition, costs and benefits of the relatively higher training, pro-rating of the GIS costs and benefits for tightly linked combinations of GIS and other systems and technologies, and developing better techniques to estimate the costs and benefits of visualization features. The feasibility decision for GIS requires increased attention to spatial technologies and to the institutional capacities that may extend well outside the conventional IS realm.

3. *What are the appropriate data and methods to measure costs and benefits of GIS?*

   Data may be drawn from dozens of standard categories of tangible costs and tangible and intangible benefits that apply for IS in general. Intangible benefits may be somewhat more important for GIS, given its higher-level decision support function and the presence of visual outputs. The methods include simple comparisons of tangible costs and benefits, present value analysis, break-even and other comparison techniques, and IT value of investment methods. A challenge in gaining use of the methods is to provide more training, especially that pertaining to GIS.

References

Aronoff, S. (1989). *Geographic information systems: A management perspective.* Ottawa: WDL Publications.

Ahituv, N. (1980, December). A systematic approach toward assessing the value of an information system. *MIS Quarterly*, 61-75.

Ahituv, N. (1989). Assessing the value of information: Problems and approaches. In *Proceedings of the International Conference on Information Systems* (pp. 315-325).

Ahituv, N., & Neumann, S. (1990). *Principles of information systems for management.* Dubuque, Iowa: W.C. Brown.

Ahituv, N., Lipovetsky, S., & Tishler, A. (1999). The relationship between firm's information systems policy and business performance: A multivariate analysis. In M.A. Mahmood & E.J. Szewczak (Eds.), *Measuring information technology investment payoff: Contemporary approaches* (pp. 62-82). Hershey, PA: Idea Group Publishing.

Allaway, A., Murphy, L., & Berkowitz, D. (2005). The geographical edge: Spatial analysis of retail program adoption. In J. B. Pick (Ed.), *Geographic information systems in business*. Hershey, PA: Idea Group Publishing.

Baily, M., & Chakrabarti, A. (1988). Electronics and white-collar productivity. In M. Baily & A. Chakrabarti (Eds.), *Innovation and the productivity crisis.* Washington, Brookings Institution.

Barua, A., Kriebel, C., & Mukhopadhyay, T. (1991, May). *Information technology and business value: An analytic and empirical investigation.* Working Paper. Austin, Texas: University of Texas.

Bertschek, I. (ed.). (2003). Information technology and productivity gains and cost savings in companies. In I. Bertschek (Ed.), *New Economy Handbook* (pp. 213-249). Amsterdam: Elsevier.

Boardman, A. E., Greenberg, D.H., Vining, A.R., & Weimer, D.L. (2001). *Cost-benefit analysis: Concepts and practice* (2nd ed.). Upper Saddle River, NJ: Prentice Hall.

Boyles, D. (2002). *GIS means business* (Vol. 2). Redlands, California: ESRI Press.

Breshahan, T.F., & Greenstein, S. (2001). The economic contribution of information technology: Towards comparative and user studies. *Journal of Evolutionary Economics*, 11, 95-118.

Brynjolfsson, E. (1993). The productivity paradox of information technology. *Communications of the ACM*, 36 (12), 67-77.

Brynjolfsson, E., & Hitt, L. (1993). Is information systems spending productive? In J. Degress, R. Bostrom, & D. Robey (Eds.), *Proceedings of the International Conference on Information Systems* (pp. 47-64). New York, Association for Computing Machinery.

Chau, P.Y.K. (1995). Factors used in the selection of packaged software in small businesses: Views of owners and managers. *Information and Management*, 29, 71-78.

Clarke, K. (2003). *Getting started with geographic information systems.* Upper Saddle River, NJ: Prentice Hall.

Cron, W.L., & Sobol, M.G. (1983). The relationship between computerization and performance: A strategy for maximizing the economic benefits of computerization. *Journal of Information Management*, 6, 171-181.

Davila de Icaza, A., & Pick, J.B. (1998). Quantitative model of size and complexity of prospective geographic information systems for regional governments in Mexico. In *Proceedings of Americas Conference on Information Systems* (pp. 369-398). Atlanta GA: Association for Information Systems.

Day, G.S., & Schoemaker, P.J.H. (2000). A different game. In G.S. Day, P.J.H. Schoemaker, & R.E. Gunther (Eds.), *Wharton on managing emerging technologies* (pp. 1-23). New York: John Wiley & Sons.

Devaraj, S., & Kohli, R. (2002). *The IT payoff: Measuring the business value of information technology investments.* New York: Pearson Education.

Dickinson, H.J., & Calkins, H.W. (1988). The economic evaluation of implementing a GIS. *International Journal of Geographical Information Systems*, 2, 307-237.

Dickinson, H.J., & Calkins, H.W. (1990). Concerning the economic evaluation of implementing a GIS. *International Journal of Geographical Information Systems*, *4*, 211-212.

Directions Magazine. (2003, August 9). Daratech reports GIS revenues forecast to grow 8% to $1.75 billion in 2003: Utilities and government increase spending. *Directions Magazine.* Retrieved from: *http://www.directionsmagazine.com.*

Doering, D.S., & Parayre, R. (2000). Identification and assessment of emerging technologies. In G. S. Day, P.J.H. Schoemaker, & R.E. Gunther (Eds.), *Wharton on managing emerging technologies* (pp. 75-98). New York: John Wiley & Sons.

ESRI Inc. (2003). *Hand-held product information.* Retrieved from: www.esri.com.

Exler, R.D. (1988). Integrated solutions for GIS/LIS data management. In *GIS/LIS '88 Proceedings* (Vol. 2, pp. 814-824). ACSM, ASPRS, AAG, URISA.

Gerlach, J., Neumann, B., Moldauer, E., Argo, M., & Firsby, D. (2002). Determining the cost of it services. *Communications of the ACM, 45* (9), 61-67.

Greene, R.P., & Stager, J.C. (2005). Techniques and methods of GIS for business. In J.B. Pick (Ed.), *GIS in business.* Hershey, PA: Idea Group.

Grimshaw, D.J. (2000). *Bringing geographical information systems into business* (2nd ed.). New York: John Wiley & Sons.

Harder, C. (1997). *ArcView GIS means business.* Redlands, CA: ESRI Press.

Harris, S.E., & J.L. Katz. (1989). Predicting organizational performance using information technology managerial control ratios. In *Proceedings of the Twenty-Second Hawaiian International Conference on System Science.* Honolulu, Hawaii.

Huxhold, W.E. (1991). *An introduction to urban geographic information systems.* New York: Oxford University Press.

Huxhold, W.E., & Levinsohn, A.G. (1995). *Managing geographic information system projects.* New York: Oxford University Press.

Jarupathirun, S., & Zahedi, F. (2001, August). A theoretical framework for GIS-based spatial decision support systems: Utilization and performance evaluation. In *Proceedings of the Seventh Americas Conference on Information Systems* (pp. 245-248). Boston, MA.

Jarupathirun, S., & Zahedi, F. (2005). GIS as spatial decision support systems. In J.B. Pick (Ed.), *Geographic information systems in business.* Hershey, PA: Idea Group Publishing.

Kauffman, R.J., & Weill, P. (1989). An evaluative framework for research on the performance effects of information technology investment. In *Proceedings of the Tenth International Conference on Information Systems.*

Kelley, T. (1999, April). Put your business on the map. *Transport Technology Today*, 20-23.

King, J.L., & Schrems, E.L. (1978). Cost-benefit analysis in information systems development and operation. *ACM Computing Surveys, 10* (1), 19-34.

Kingma, B.R. (2001). *The economics of information: A guide to economic and cost-benefit analysis for information professionals* (2nd ed.). Englewood, CO: Libraries Unlimited Inc.

Loveman, G.W. (1988, July). An assessment of the productivity impact on information technologies. *MIT Management in the 1990s Working Paper #88-054.*

Lucas, H.C. (1999). *Information technology and the productivity paradox: Assessing the value of investing in IT.* New York: Oxford University Press.

McCune, J.C. (1998, March). The productivity paradox. *American Management Association International*, 38-40.

Meeks, W.L., & Dasgupta, S. (2003). Geospatial information utility: An estimation of the relevance of geospatial information to users. *Journal of Decision Support Systems*, in press.

Meeks, W.L., & Dasgupta, S. (2005). The value of using GIS and geospatial data to support organizational decision making. In J.B. Pick (Ed.), *Geographic information systems in business*. Hershey, PA: Idea Group Publishing.

Mitchell, A. (1999). *The ESRI guide to GIS analysis* (Vol. 1). Geographic patterns and relationships. Redlands, CA: ESRI Press.

Mueller, B. (1997, November). Measuring ROI: can it be done? *AS/400 Systems Management*, 8-10.

Murphy, L.D. (1995). Geographic information systems: Are they decision support systems? *Proceedings of 28th Annual Hawaiian International Conference on Systems Sciences*, Hawaii (pp. 131-140).

Navarrete, C.J., & Pick, J.B. (2002). Information technology expenditure and industry performance: The case of the Mexican banking industry. *Journal of Global Information Technology Management*, *5* (2), 7-28.

Navarrete, C.J., & Pick, J.B. (2003). Information technology spending association with organizational productivity and performance: A study of the Mexican banking industry, 1982-1992. In N. Shin (Ed.), *Creating Business Value with Information Technology: Challenges and Solutions* (pp. 89-124).

Navarrete, C.J., & Pick, J.B. (2003). Information technology spending and the value of the firm: The case of Mexican banks. In N. Shin (Ed.), *Creating business value with information technology: Challenges and solutions* (pp. 146-165).

Niederman, F. (1999). Valuing the IT workforce as intellectual capital. In *Proceedings of the SIGCPR Conference, Association for Computing Machinery* (pp. 174-181).

Noyelle, T. (ed.). (1990). *Skills, wages, and productivity in the service sector*. Boulder, CO: Westview Press.

Obermeyer, N.J. (1999). Measuring the benefits and costs of GIS. In P.A. Longley, M.F. Goodchild, D.J. Maguire, & D.W. Rhind (Eds.), *Geographical information systems* (Vol. 2): *Management issues and applications* (pp. 601-610). New York: John Wiley & Son.

Obermeyer, N.J., & Pinto, J.K. (1994). *Managing geographic information systems*. New York: The Guilford Press.

Osterman, P. (1986). The impact of computers on the employment of clerks and managers. *Industrial and Labor Relations Review*, 39, 175-186.

Phoenix, M. (2003). Personal communications based on ESRI Inc. Marketing Department.

Pick, J.B. (1996). GIS-based economic and social cluster analysis applied to a giant city. In *Proceedings of Americas Conference on Information Systems* (pp. 515-517). Atlanta, GA: Association for Information Systems.

Pick, J.B., Hettrick, W.J, Viswanathan, N., & Ellsworth, E. (2000b). Intra-censal geographical information systems: Application to bi-national border cities. In *Proceedings of European Conference on Information Systems* (pp. 1175-1181). Vienna, Austria: ECIS.

Pick, J.B., Viswanathan, N., & Hettrick, W.J. (2000a). A dual census geographical information systems in the context of data warehousing. In *Proceedings of Americas Conference on Information Systems* (pp. 265-278). Atlanta, GA: Association for Information Systems.

Ragowsky, A., Ahituv, N., & Neumann, S. (2000). The benefits of using information systems. *Communications of the ACM*, 43 (11), 303-311.

Ray, J. (2005). Spatial data repositories: Design, implementation, and management issues. In J.B. Pick (Ed.), *Geographic information systems in business*. Hershey, PA: Idea Group Publishing.

Reeve, D.E., & Petch, J.R. (1999). *GIS, organisations, and people: A socio-technical approach*. London: Taylor and Francis.

Roach, S.S. (1989). America's white-collar productivity dilemma. *Manufacturing Engineering*, 104.

Sassone, P.G. (1988). Cost benefit analysis of information systems: A survey of methodologies. In *Proceedings of Conference on Supporting Group Work* (pp. 126-133). Association for Computing Machinery.

Satzinger, J.W., Jackson, R.B., & Burd, S. D. (2002). *Systems analysis and design in a changing world* (2nd Ed.). Course Technology.

Smith, D.A., & Tomlinson, R.F. (1992). Assessing the costs and benefits of geographical information systems: Methodological and implementation issues. *International Journal of Geographical Information Systems*, 6, 247-256.

Strassmann, P.A. (1997, September 15). Computer have yet to make companies more productive. *Computerworld*.

Strassmann, P.A. (1999). *Information productivity*. New Canaan, CT: Information Economics Press.

Strassmann, P.A. (1999). *Information productivity: Assessing the information management costs of U.S. industrial corporations*. New Canaan, CT: The Information Economics Press.

Tomlinson, R. (2003). *Thinking about GIS: Geographic information system planning for managers*. Redlands, CA: ESRI Press.

U.S. Census (2004). U.S. Census website, *http://www.census.gov*.

Willcocks, L.P., & Lester, S. (1997). In search of information technology productivity: assessment issues. *Journal of the Operational Research Society*, *48*, 1082-1094.

Section II

Conceptual Frameworks

Chapter V

Spatial Data Repositories: Design, Implementation and Management Issues

Julian Ray, University of Redlands, USA

Abstract

This chapter identifies and discusses issues associated with integrating technologies for storing spatial data into business information technology frameworks. A new taxonomy of spatial data storage systems is developed differentiating storage systems by the systems architectures used to enable interaction between client applications and physical spatial data stores, and by the methods used by client applications to query and return spatial data. Five distinct storage models are identified and discussed along with current examples of vendor implementations. Building on this initial discussion, the chapter identifies a variety of issues pertaining to spatial data storage systems affecting three distinct aspects of technology adoption: systems design, systems implementation and management of completed systems. Current issues associated with each of these three aspects are described and illustrated along with a discussion of emerging trends in spatial data storage technologies. As spatial data and the technologies designed to store and manipulate it become more prevalent, understanding potential impacts these technologies may have on other technology decisions within an organization becomes increasingly important. Furthermore, understanding how these technologies can introduce security risks and other vulnerabilities into a computing framework is critical to successful implementation.

Introduction

Various organizations and authors estimate that more than 80% of all data used by businesses has an inherent spatial component (Adler, 2001; Haley, 1999; ESRI, 1996). Street addresses, postal codes, city names, and telephone numbers are common components of business data which can be used by geographic information systems (GISs) to orient these data in space, revealing spatial patterns and relationships between records which might otherwise remain latent. Experience has shown that organizations that exploit these spatial patterns and relationships can reduce operating costs (Weigel & Cao, 1999; Ratliff, 2003), increase efficiency and manage risk (Murphy, 1996), and reduce the time required to make complex decisions (Mennecke et al., 1994).

Spatially Enabled Business Frameworks

In order to exploit spatial data, organizations need to integrate spatial data and spatial services with their traditional business applications. This integration can be achieved by developing a technology framework, which facilitates interaction between business applications, spatial services, and data management systems (Figure 1). Business applications such as Enterprise Resource Planning (ERP), Business Intelligence, Electronic Commerce, and Customer Relationship Management (CRM) systems interact with a layer of services designed to manage and exploit spatial dimensions of business data. Spatial services, in turn, interact with a layer of traditional and spatial data storage systems.

This three-tier architecture is typical for many leading spatially-enabled enterprise business applications, including Oracle's 11i Application Suite and systems available from SAP, Siebel and others. In Oracle's case, spatial services are delivered as part of the application suite and spatial data is stored along side traditional data in a relational database system (Oracle, 2001). In contrast, SAP and Siebel systems use third-party GIS software for managing and manipulating spatial data. These third-party components, often purchased separately, integrate with business applications through standardized application programming interfaces (APIs). Spatial and traditional business data in these applications are usually stored in different data management systems, often using very different storage technologies for managing spatial and traditional data elements.

Spatial Data Repositories

Organizations often purchase spatial services and GIS software from a variety of vendors resulting in heterogeneous collections of spatial data and spatial services within an organization. A typical business, for example, might use address geocoding services from one vendor, mapping solutions from another vendor, and use traditional workstation-based GISs to define, create and manage their intrinsic spatial. Different spatial services often have differing spatial data storage needs in terms of both data content and data organization, resulting in a variety of different spatial data storage formats on

Figure 1. Typical Spatially Enabled Business Application Framework

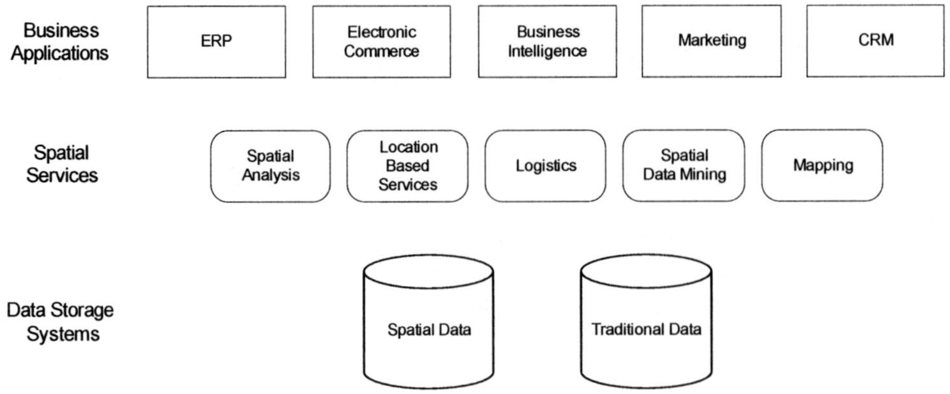

different data storage systems within the organization. In general, spatial data within an organization could be stored in commercial enterprise databases, in proprietary file structures on one or more physical storage devices, accessed from a remote server over the company's intranet, or downloaded on demand over the Internet.

Spatial data repositories (SDRs) are collections of possibly heterogeneous spatial data and spatial data-storage technologies, which provide spatial data management functions for spatially-enabled information systems. This chapter focuses on the issues that should be considered when organizations create SDRs by introducing spatial data into their enterprise information systems. The second section introduces spatial data storage technologies by developing a new taxonomy of spatial storage systems and identifying important issues pertaining to their adoption by organizations. The third, fourth and fifth sections examine some of the design, implementation, and management issues likely to be encountered as organizations introduce these spatial storage technologies into their information technology infrastructure. The sixth section provides insight into the future of spatial data storage by identifying trends occurring in spatial data storage systems, which are likely to affect how organizations deal with spatial data in the future. The last section provides a summary of these discussions.

Spatial Data Storage Technologies

Spatial data is often classified into two major forms: field-based and entity- or object-based models (Shekhar & Chawala, 2003; Rigaux et al., 2002). Field-based models impose a finite grid on the underlying space and use field-functions defined within the context of the application to determine attribute values at specific locations over the grid. Field data is most commonly associated with satellite imagery and raster data derived from grid-based collection methods. In contrast, object-based models identify discrete spatial objects by generalizing their shape using two or three dimensional coordinate systems. Spatial objects are a combination of non-spatial attributes describing each object's

Figure 2. Example Customer Data with Spatial Components

CustomerID	CustomerName	StreetAddress	City	State	Zip Code	Latitude	Longitude
123456	Alpha Supply Corp.	123 Main St	Boston	MA	02101	-84.1234	34.5678
234567	Beta Systems Inc.	321 Oak St	Quincy	MA	02169	-84.2345	35.6789

characteristics, and spatial or geometric components describing the relative location and geometric form of each object. Most data used by businesses today is stored as object data, as this form bears closest resemblance to traditional business data and can be stored in a variety of relational database management systems.

Figure 2 illustrates how business information representing customer addresses might be stored in a data table using an object-model approach. Each customer record contains a unique identifier, descriptive attributes, some of which contain a spatial component, and an explicit spatial location stored as a latitude and longitude.

Spatial references, such as the latitude and longitude data illustrated in *Figure 2* often are derived by geocoding business data containing spatial components using a GIS or spatial service. Depending on information system needs, derived geometric information might represent accurate or "real-world" spatial locations, for example, a customer's street address could be interpolated against a digital street database to provide an accurate latitude and longitude. Alternatively, a business record might be assigned a location representing a geometric center (centroid) of a larger area such as the city, state, zip code or sales area within which the business data record is logically located. Increasingly, global positioning systems (GPS) and other mobile technologies allow business data already containing accurate spatial references to be captured and used directly by spatial services, negating the need for deriving location.

Simple spatial data representing point locations, such as the customer address data illustrated in *Figure 2,* are easily managed in relational database management systems, as each geographical reference can be represented using a fixed number of data elements and stored using traditional numeric data types. More complex spatial data representing linear features such as streets and highways, and polygonal features such as geo-political divisions, sales areas and city blocks, however, are more difficult to represent in tabular form as each spatial object may contain many coordinate pairs. A city street or a sales-area, for example, might require hundreds or even thousands of coordinate pairs to accurately define its shape. Spatial objects requiring large numbers of coordinates to define their shape require innovative and efficient techniques to manage their storage.

A Taxonomy of Spatial Data Storage Models

Vendors of GIS software have developed a variety of methods to store spatial and non-spatial data. Adler (2001) identifies three generations of spatial data storage systems. First generation systems are primarily workstation-based and include some of the earliest

GIS and desktop mapping systems dating back to the 1970s. Second generation systems developed in the 1990s use spatial-middleware to process and manage spatial data stored in traditional RDBMSs. More recent third generation systems move all spatial processing and spatial data storage into a relational database. Today, GIS software and spatial services representing all three generations identified by Adler, as well as new Internet-based service-oriented models, are commercially available. Businesses that have already integrated spatial services into their information systems are likely to have examples of all three generations of spatial data storage supporting different spatial processes within their organizations.

An alternative taxonomy for spatial data storage systems is presented in this section. This new taxonomy differentiates spatial data storage systems by the technologies used to store spatial and non-spatial data as well as the methods used to access spatial objects by a spatial information system or GIS. Using these criteria, five distinct storage models are currently identifiable: the Hybrid Storage Model, the Unified Storage Model, Spatial Database Management Systems (SDBMSs), the Package-Specific Model and the Managed Service Model (*Figure 3*).

Hybrid Storage Model

The Hybrid Storage Model uses different storage systems for spatial and non-spatial data components. Spatial components, represented as variable length records, are stored in "geometry files" on a computer's file system, while non-spatial attributes are stored as fixed-length records in a relational database management system. Geometry files often use proprietary binary file structures accessible only by vendor-specific middleware. Non-spatial attributes are accessed from vendor-middleware using a database language such as SQL. Simple indexing mechanisms are used to logically link records in geometry files with records in RDBMSs (*Figure 4*).

Figure 3. Spatial Data Storage Models

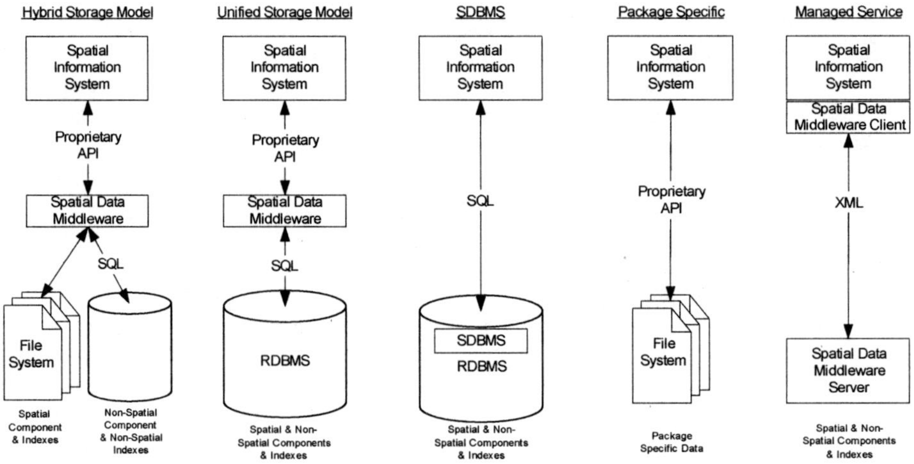

Figure 4. Hybrid Storage Model

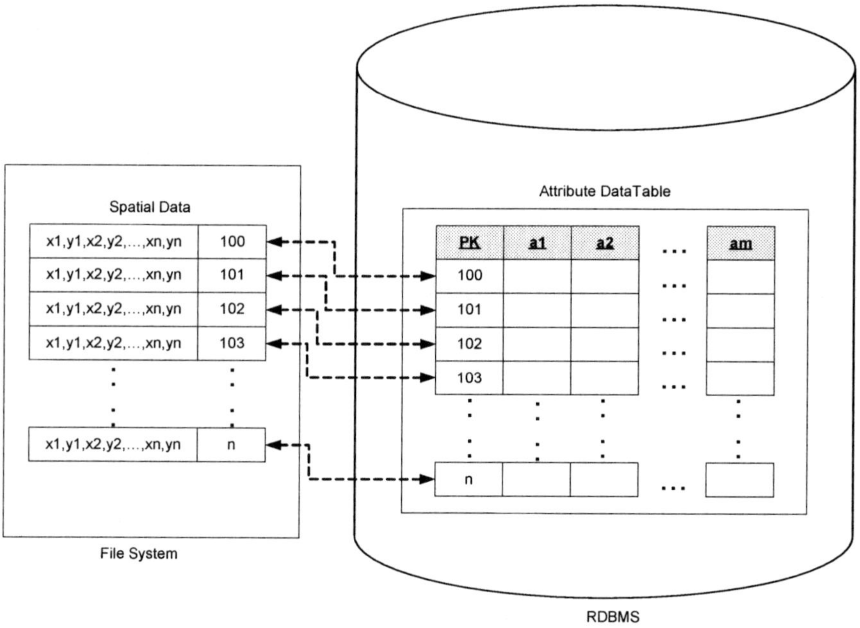

Vendor-supplied middleware extracts data from both RDBMS and geometry files and links logical records from both data stores together in memory in order to create complete spatial objects. This function is performed on behalf of client applications accessing the spatial middleware using a proprietary API. ESRI's Shapefile format is an example of a widely used spatial data storage system implementing the Hybrid Storage Model. ESRI provides various middleware components to enable access to spatial data from its desktop GISs. A complete discussion of the Shapefile format is provided in Rigaux et al. (2002, Chapter 8.3) and ESRI (1998).

Unified Storage Model

The Unified Storage Model, in contrast, uses a traditional RDBMS for both spatial and non-spatial data components (*Figure 5*). Spatial data is encoded into vendor-specific binary structures by spatial middleware and stored in columns of relational database tables as Binary Large Objects (BLOBS). BLOBs are stored, returned, and updated by RDBMSs at the request of client applications. Data within BLOBS, however, cannot be decoded and interpreted by the RDBMS itself. Spatial middleware is used to translate database BLOBS to and from geometric objects which can then be manipulated by GIS clients. All indexing and query operations on spatial data are performed by the spatial

Figure 5. Unified Storage Model

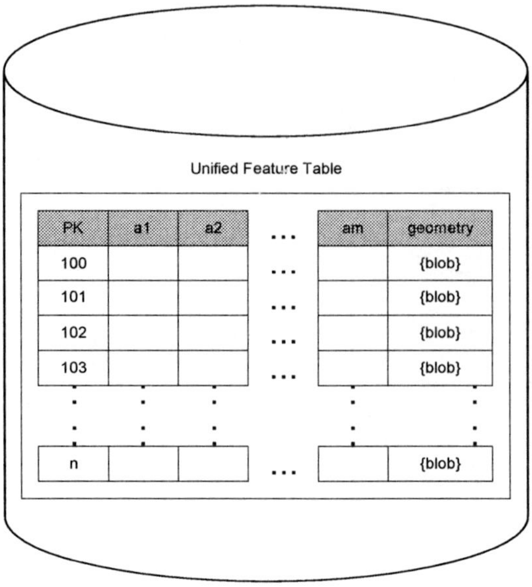

middleware rather than the RDBMS. Spatial indexes are often stored alongside spatial data in the RDBMS. Access to the RDBMS from the spatial middleware is usually via SQL while access to the spatial middleware by client application is via a proprietary API.

Spatial data storage systems implementing a Unified Storage Model inherit properties of the RDBMS, providing several advantages over file-based systems for managing spatial data. These advantages include:

- efficiently manage large volumes of data by allowing tables to span multiple logical files and devices,
- efficiently manage concurrent access by multiple clients,
- realize performance enhancements by caching tables, views, queries, and results-sets in memory,
- performing row and table locking during update processes,
- transaction management, and
- creating joins between spatial and non-spatial tables.

Additional security advantages for organizations can be realized, as facilities for managing, auditing, and restricting access to spatial data, as well as tools for exporting, archiving, and replicating spatial data, are normally provided by the RDBMS. More importantly, for an organization which has already standardized on a RDBMS such as Oracle or DB2, skills necessary to configure, deploy and protect these systems within the organization might already exist, thereby reducing implementation costs and minimizing risk caused by introducing new technologies into the enterprise.

Intergraph's GeoMedia suite of products uses a Unified Storage Model to manage spatial data in a variety of commercial RDBMSs including Microsoft SQL Server, Sybase, DB2, Informix, and Oracle. Intergraph provides a COM-based middleware technology called Geographic Data Objects (GDO) to enable client access to spatial data using a framework loosely based on Microsoft's Data Access Objects (DAO) API. GDO middleware is responsible for reading and writing geometry BLOBs and translating them into a form which can be used by GeoMedia client software. More information on GDO can be found at Intergraph (2003).

Spatial Database Management Systems

Similar to the Unified Storage Model, Spatial Database Management Systems (SDBMSs) combine functions of traditional RDBMSs with spatial data storage facilities. With SDBMSs, however, the database itself, rather than third-party middleware, provides the system for storing geometric data within the database using intrinsic, SQL compliant data types. Spatial features in a SDBMS are stored in tables with columns containing geometry information while non-spatial attributes are stored in columns containing standard SQL data types (*Figure 6*).

Spatial data in SDBMSs are stored as either database BLOBS or as structured User Defined Types (UDTs). Structured UDTs are defined in the SQL-3/SQL:1999 specification and provide a mechanism for defining and storing complex objects and their methods in a relational database (Melton, 2003). Along with spatial storage, SDBMSs provide services and functions enabling spatial data to be indexed, analyzed, and queried using SQL (Shekhar & Chawala, 2003). In order for this to work, generalized spatial objects have to be encoded in a form that is compatible with SQL. The OGIS specification provides two standardized formats for this process. Well Known Binary Format (WKBF) encodes spatial data into strings of binary digits and is designed primarily as an interface for applications. In contrast, Well Known Text Format (WKTF) provides a human-readable system for encoding spatial data in SQL statements. SDBMS often define an intrinsic geometry storage type conforming to OpenGIS's Simple Features Specification for SQL Revision 1.1 (OGIS, 1999). This specification defines a set of geometry types which can be stored in geometry valued columns and a set of spatial methods which operate on spatial objects and determine spatial relationships between them.

Modern enterprise databases such as Oracle's 9i Database and IBM's DB2 can be used to implement either a Unified Storage Model or a SDBMS. When used as an SDBMS, however, the basic database system is usually extended by installing a specialized software module or "database extender" which enhances the capabilities of the underlying RDBMS with spatial data management capabilities. These software extenders are usually licensed separately from the database software itself, as is the case for IBM's Spatial Data Blade and Oracle's Oracle Spatial.

Native spatial data storage capabilities of SDBMS provide all the benefits of Unified Storage Models as well as several additional advantages. The database engine can

Figure 6. Spatial Database Management System

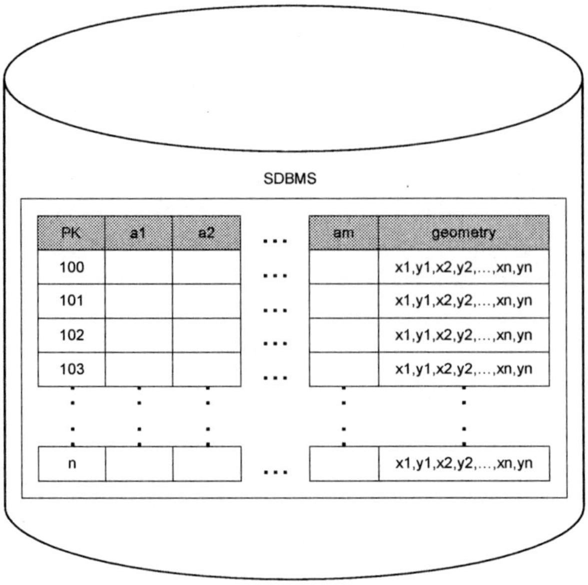

process spatial data within its kernel without having to export data to a GIS or spatial middleware platform to perform a spatial query. Spatial queries involving large datasets are therefore more efficient as less data is moved over the network and, more importantly, can be initiated by client applications that are not necessarily spatially aware. Layout of the data dictionary is usually unconstrained by requirements imposed by GIS-processing client and middleware systems, allowing SDRs to be designed according the needs of information systems rather than package-specific requirements of a GIS. Lastly, open standards including SQL, WKTF, and OGIS-compliant geometric structures remove constraints associated with GIS vendor-specific dependencies.

Package-Specific Storage Models

Package-Specific Storage Models are characterized by proprietary file structures and direct access of spatial and non-spatial data by client applications using proprietary APIs. Data files are usually stored on CDROM or local file systems, and in many cases might be distributed as part of the spatial software itself. There are three major classes of use for this type of storage:

- as a low-cost storage system for spatial and non-spatial data for GIS,
- as a distribution and protection system for proprietary data associated with a spatial service, and
- as a data cache to enhance performance of various spatial services.

Early GIS systems and some popular GISs in use today store spatial data in proprietary file formats and access it directly from the client, negating the need for spatial middleware. MapInfo's TAB format, for example, provides a general purpose read/write spatial data storage structure for their single-user desktop GISs as well as a spatial data cache for their web-based map generating software (MapInfo, 2002). Caliper Corporation's Compact Data Format (CDF) is an example of a read-only, package-specific format used to store large quantities of pre-processed spatial data on a single CDROM (Caliper, 1995, p. 326). Some spatial information services, particularly services for address geocoding and creating maps, often use specialized file-based data structures to optimize access to spatial and non-spatial data. Storage systems used by this group of technologies can be delivered either by the software vendor as part of the application itself or created by a run-time process of the software. There are several reasons for using this approach:

- create a dependency between software and data, thereby requiring users purchase data from specific data sources,
- increase performance of spatial services by extracting data from slower systems accessed indirectly using middleware to storage systems which can be accessed directly, and
- persist results of time consuming operations between invocations.

Map-generating services, for example, often use file-based data storage structures as spatial-data caches to increase run-time performance and speed at which a system can be restarted. Current versions of MapInfo's MapXtreme and Intergraph's GeoMedia WebMap products, for example, use this approach.

Managed Service Models

Managed Service Models extend capabilities of remote, proprietary spatial data repositories to business partners. Spatial data is transmitted between client and server middleware components over a computer network such as the Internet. Spatial and non-spatial data are encoded in standardized form, often as XML documents or as binary objects by spatial middleware for transmission. Client access to spatial services and spatial data on managed servers is via middleware APIs, thereby masking details associated with service implementations and data storage technologies used by the managed systems. Communication between client applications and managed systems can be implemented using a variety of technologies including web services leveraging XML and the Simple Object Access Protocol (SOAP), inter-application protocols such as RPCs and Java RMI, or distributed objects including CORBA and Microsoft's COM+.

Several managed service implementations are currently available including Microsoft's MapPoint .NET web service, ESRI's ArcWeb web services, as well as ESRI's Geography Network and ArcIMS server software. A variety of Internet mapping and analysis systems built using Internet architectures also fall in this category including Intergraph's GeoMedia Web Map and GeoMedia Web Enterprise products along with similar products by other GIS vendors.

Storage Technology Issues

Hybrid Storage Models and Package-Specific Storage Models provide efficient access to spatial data, as geometry components are typically stored as memory-mapped objects on a file system allowing spatial data to be quickly retrieved and organized internally by a GIS. These storage models, however, present several challenges to the design and performance of spatial information systems. Spatial data stored in proprietary formats is accessible only by vendor-provided software, ultimately limiting options for selecting hardware and operating system platforms, as only those platforms supported by the vendor's software are viable. The volume of spatial data that can be stored in a single file structure is often a function of the operating system. Windows-based systems, for example, usually restrict file sizes to less than 2 GB, often necessitating the decomposition of large logical data sets into multiple smaller files. Further, file-based storage often limits options for supporting concurrent reads and writes to data files based on the file management capabilities of the operating system. Lastly, Hybrid Storage Models require system administrators manage and protect two distinct groups of data: geometry files managed by an operating system and non-spatial data stored in an RDBMS. Controlling access and performing data archival and restore operations in these systems is, therefore, twice as involved relative to storage systems that use a single technology.

Unified Storage Models remove many limitations of Hybrid Storage Models. The amount of data that can be stored in a single table is constrained by the capability of the underlying RDBMS and physical storage capacity, rather than limitations imposed by the operating system hosting the RDBMS. Memory caches, edit buffers and locking mechanisms in the RDBMS enable simultaneous read and write operations and multi-user access against database tables. Spatial data, however, is still stored in proprietary structures requiring specialized spatial-data processing middleware, thereby resulting in vendor-specific dependencies, potentially limiting hardware and software platform options. Further, as spatial data is managed entirely by spatial middleware, design and organization of database schemas is constrained by design requirements imposed by spatial middleware, often limiting options for performing data integration across tables, schemas, and database instances. Materialized views in Oracle, for example, are unsupported by many GIS middleware systems. Lastly, spatial middleware provides a potential data-processing bottleneck as it is responsible for extracting data from the RDBMS and assembling spatial objects from storage BLOBS in the database. This architecture can negatively affect the performance of spatial queries against large datasets, as spatial queries are processed by the middleware often resulting in worst-case performance of the database, as full-table scans are usually required.

Spatial Database Management Systems combine benefits of RDBMS storage with design advantages afforded by open systems architectures. Spatial data in an SDBMS is accessed using SQL thereby eliminating problems associated with proprietary storage formats and middleware technologies. Moreover, native spatial indexing and spatial-query facilities of SDBMSs provide efficient methods for processing spatial queries as full-table scans can often be eliminated. Perhaps more important than either of these benefits, however, is the ability to design SDRs based solely on information system requirements without design-constraints imposed by spatial middleware.

There are, however, several factors affecting the efficacy of SDBMSs as enterprise spatial data storage platforms. Result-sets containing results of a spatial queries can be slow to create, as spatial data first has to be converted from internal storage into objects which can be operated on by the database kernel, and, second, into a form which can be transmitted over the network to a client. Executing spatial queries in a database increases database workload, since the database is now performing GIS functions as well as normal database operations. Spatial services requiring rapid and repetitive access to spatial data stores, such as those that create maps, can place additional processing burdens on database servers. Disk storage requirements for SDBMSs can be high, as databases are required to store spatial data as well as specialized spatial indexes. Spatial indexes for point-data, for example, can often require as much as four times the storage space as the spatial data itself. Thus, a database performing as a SDBMS would potentially require more disk storage than a similar database using a Unified Data Model. A SDBMS instance would also require more physical memory and faster processors to provide equivalent system throughput.

System administrators for SDBMS need to develop additional skills to effectively manage non-traditional data. Databases implementing SDBMSs often have to be configured differently to manage spatial data and achieve maximum performance. The amount of memory allocated for sorting result sets before they are returned to the client, for example, can be excessive depending on the size of geometries selected in a query. Lastly, cost of SDBMSs can be prohibitive as most SDBMS are licensed separately from the underlying RDBMS, increasing overall cost of implementation and ownership.

Managed Services provide basic spatial services and spatial data for client applications that are authorized to connect to and use these services. Organizations wishing to integrate basic spatial services with business applications, such as address geocoding and creating maps, can connect to and use these managed services directly, thereby negating many of the issues associated with managing spatial data. Costs for using these systems, however, can be prohibitive as managed services are typically billed by transaction. Simply generating a map display would normally require a number of separate transactions. Further, as managed services require access to remote services over a computer network, organizations implementing systems using these services create direct dependencies between their business systems and those of their business partners. Issues pertaining to managed services are explored more fully in the discussion of Geospatial web services at the end of this chapter.

Design Issues

There are several issues that must be taken into account when SDRs are being designed. These issues include ensuring minimum performance requirements for specific spatial services, the nature and volume of spatial data to be stored, methods for managing data updates, and vendor-specific storage technology requirements.

Performance

The organization of spatial data in a spatial database can affect the speed at which data can be retrieved by a GIS (Sloan et al., 1992). Laurini & Thompson (1992, p. 473) describe how normalization in relational databases tends to scatter data across tables, making spatial queries cumbersome and inefficient thereby negatively impacting performance of GISs. These performance issues are exacerbated as the amount of spatial data increases and the number and type of spatial services accessing a SDR also increases.

Understanding the spatial nature of client requests, particularly how spatial and non-spatial queries are formulated by different client applications, can often help in the design of more efficient spatial data storage systems. Best performing SDR designs are often most costly to maintain, as they often require introducing data redundancy leading to higher overall data maintenance and storage costs. To illustrate, consider the design of a spatial information system providing services to locate and generate maps of customers. The information system requires access to a digital database of U.S. streets. Streets in the U.S. often have alternate names; for example, a street named Worcester Road might have the alternate names RT. 9 and U.S. 123. If any alternate name is not accessible by the geocoding service problems could arise. 1001 Worcester Rd, for example, might be locatable, whereas 1001 RT. 9 or 1001 U.S. 123 might not, even though they are physically

Figure 7. Data Model Options for Street Databases

PK	Geometry	Street Name
103	x1,y1,x2,y2,...,xn,yn	Worcester Rd
104	x1,y1,x2,y2,...,xn,yn	Rt 9
105	x1,y1,x2,y2,...,xn,yn	U.S. 123

(a) Un-normalized

PK	Geometry
103	x1,y1,x2,y2,...,xn,yn

PK	FK	Street Name
1	103	Worcester Rd
2	103	Rt 9
3	103	U.S. 123

(b) Normalized Street Feature

PK	Geometry	Primary Street Name
103	x1,y1,x2,y2,...,xn,yn	Worcester Rd

PK	FK	Alternate Street Name
1	103	Rt 9
2	103	U.S. 123

(c) Partially Normalized Street Feature

PK	Geometry	Primary Street Name
103	x1,y1,x2,y2,...,xn,yn	Worcester Rd

PK	FK	Geometry	Street Name
1	103	x1,y1,x2,y2,...,xn,yn	Worcester Rd
2	103	x1,y1,x2,y2,...,xn,yn	Rt 9
3	103	x1,y1,x2,y2,...,xn,yn	U.S. 123

(d) Optimized Street Feature

the same location. Designing data structures to store street data can affect system performance in a variety of ways. *Figure 7* illustrates four alternative relational storage structures that can be used to store spatial and non-spatial data for the Worcester Rd. example.

Figure 7(a) illustrates a typical un-normalized representation of the data. Three separate records are created in the database, one for each alternate name. This simplistic organization is typical for address geocoding services, which predominantly search on non-spatial attributes. For services that generate predominantly spatial queries, such as mapping engines, this design can introduce performance issues. Spatial queries against (a) will return three times as much spatial data as is necessary to draw the street on a map. Further, labeling streets on a map display requires filtering alternate names to avoid labeling the same logical map feature multiple times. A normalized version of the street data is shown in (b). In this format, street names are stored in a separate table and a one-to-many relationship of street geometries to street names is maintained. This design is more efficient for data-storage and maintenance as data is not duplicated between tables. Services performing spatial queries are optimized; however, services performing non-spatial queries need to create database joins or perform multiple queries, ultimately resulting in performance degradation. Mapping services that display street names will be similarly impacted.

A compromise between (a) and (b) is shown in (c). This design provides a partially normalized organization defining a single geometry record for each street along with a primary street name. The design is efficient for map-generating services, as spatial queries return a single geometry for each street and filtering is not required for map labeling engines. Geocoding services, on the other hand, need to create a union between tables, which are typically slower to process and require additional system resources to maintain. Option (d) provides a design in which both mapping requirements and geocoding requirements are treated with equal weight. This option requires the most storage space and generates most data redundancy as the single physical road is now represented using four separate feature records in the data store. To aid maintenance and update operations, a common key is maintained in both tables to uniquely identify parentage of geometric features.

When spatial queries are formulated with respect to other geographies such as sales regions or political boundaries, storage of spatial data can be often be optimized to reflect the spatial organization of client requests. A company with four regional call centers, for example, can partition their spatial data into four separate regions using various techniques including creating separate data tables or even independent SDRs. If the underlying storage technology permits, regional views or partitioned indexes can be used to make geographically organized queries more efficient.

Physical Storage Requirements

The volume of spatial data to be stored in a SDR can influence storage technology selection, design of storage structures, and affect how data updates to the spatial data are managed over time. Modern spatial data storage systems can manage hundreds of thousands or even millions of records. There is, however, a point at which the number of records and amount of spatial data becomes an overarching design issue. Digital street databases used for a variety of spatial services including logistics, location based analyses, and address geocoding present such a problem. A typical street database for a large U.S. metropolitan area contains in the order of 500,000 to 1 million individual street records, well within the capabilities of most spatial storage technologies. A regional area, such as coverage for the Northeastern U.S., requires in the order of 4-5 million individual street records, while storage requirements for the entire U.S. increases to 20-40 million records depending on the street database vendor. Ancillary data, such as alternate street names, increase non-spatial data storage requirements. As data volumes increase, comparing capabilities of different spatial data storage technologies to manage both access to data and the spatial data itself becomes critical. Several important factors should be taken into account:

- performance characteristics as the number of records in a spatial table increases,
- performance characteristics as the average number of coordinates in a geometry increases,
- performance characteristics as the number of simultaneous accesses against a spatial table increases, and
- performance characteristics of read-only vs. read-write operations against a spatial table.

If spatial data is accessed via spatial middleware, as in the case of Hybrid and Unified Data Models, performance of the middleware becomes a critical factor, as the middleware is responsible for accessing the data stores, reassembling geometric objects in memory, and performing spatial query operations. In these cases, the spatial middleware should be tested rather than the underlying storage system.

Transactional Data

Transactional business data in a data warehouse is typically stored using traditional elements of relational database systems such as numbers (postal codes), and text (place names, addresses, etc.) (Mennecke & Higgins, 1999). In order to participate in spatial analyses, these implicit spatial components have to be transformed into explicit geographic references which can be directly used by a GIS or spatial service. Designers of SDRs have to develop strategies to perform these transformations. Two major strategies can be employed to achieve this goal: business data can be processed prior to participating in spatial analyses and an explicit geographic reference created and stored with each object; alternatively, data in a data warehouse can be joined with tables containing explicit geographic data by creating relationships between records in spatial and non-spatial tables.

Pre-processing business data and attaching explicit geographic references to each data object can be an effective strategy when the processes used to generate an explicit spatial reference from business data are computationally expensive and spatial references will be used repeatedly. Calculating the geographic location of a street address, for example,

Figure 8. Simple Database Join between Spatial and Non-Spatial Data

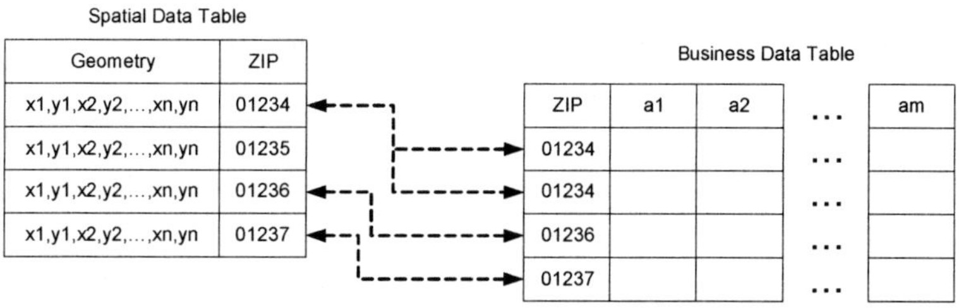

is a computationally expensive process requiring several steps: parsing and standardizing logical components of the street address, selecting candidate matching street records from a digital data base of streets using heuristic search methods, calculating a match value for the address against each candidate street record, and lastly, selecting a street record which maximizes the match criterion and calculating the parametric location of the address over the spatial representation of the street record. If the geocoding process fails, manual intervention is usually required to locate the address.

Pre-processing business data to calculate geographic references can occur at various times in the life-cycle of a data element depending on the needs of the information systems: geographic references can be generated and associated with a data element during data capture and before data is entered into a data warehouse, existing data within a warehouse can be batch-processed periodically to verify and update geographic references, and lastly, geographic references can be assigned as a transaction is entered into a data. The efficacy of these approaches in maintaining viable and accurate spatially-enabled data warehouses will vary with the spatial data storage technologies being used and level of interaction between information systems and spatial services.

Simple database joins between tables containing traditional business data with spatial components and tables of spatial data can be efficiently created using most relational database management systems and GIS software. *Figure 8* illustrates how a table containing business data with an implicit spatial component — zip code — could be related to a spatial table of zip codes using a simple foreign key.

Database joins between business data and spatial data are most efficient when a many-to-one relationship exists between business data elements and spatial elements and where join-keys are easily indexed and managed by the RDBMS. Depending on the storage technology used, joins such as these can be created and persisted as views in a RDBMS. Alternatively, using SQL statements, joins can be defined and created dynamically as a query is processed.

Real-Time Data

Data delivered in real-time requires specialized handling within a SDR. Many logistics companies, for example, use automatic vehicle location (AVL) services to track locations of vehicles as they move over transportation systems. Similarly, mobile phones and other "location-aware" devices can be used to locate users and assets in real time. The primary design questions associated with integrating this type of transient data into a SDR are:

- Should the SDR store these locations or should such data remain at the service layer?
- If data will be stored, how long should it be stored for? and
- What is the best form for storing these data?

In order to answer these questions, SDR designers first should determine the length of time for which each data record has tangible value. The location of a vehicle being tracked over a delivery route updated every five minutes, for example, can easily be stored in a spatial database as a geometric point and a time-stamp attribute denoting the time for which the vehicle's location was recorded. Once stored, this data can be used to fine tune route-finding algorithms by determining expected versus actual travel times for different portions of the route. Real-time data captured and used in this manner has significant potential value for refining optimization models, potentially leading to reduced logistics costs and efficiency gains. Maintaining this type of data for a large fleet of vehicles over an extended period of time, however, will likely pose more problems than it solves due to the sheer volume of data captured and the impact of the number of transactions on the database and service layers.

Other forms of real-time or near real-time spatial data can be captured and stored in SDRs presenting significant storage and performance issues. Weather data, such as areas of precipitation and lightning strikes, for example, can vary in quantity and density based on current weather conditions. Weather data is usually delivered in the form of polygons requiring more storage and processing than simple point data. This type of data, once captured, stored, and analyzed can enhance decision systems by generating models of how, for example, travel time over particular transportation links can be affected by specific types of weather patterns.

Data Updates

Spatial data can rapidly become outdated. The locations of mobile users and assets can change in real-time, houses and streets are built and added to the physical infrastructure, store locations and services they provide change over time, and customers relocate. Theses changes often occur faster than their digital spatial representations can be updated.

Analyzing the rate at which different data components change over time allows SDR designers to determine which storage strategies work best for different data components. Some data components of a SDR change slowly over time. Census and demographic data, for example, might be updated every one, five, or 10 years. Other types of data will change at a faster rate: updates to street databases are typically released quarterly; an organization's internal data such as sales areas, forecasts, and customer lists can be updated every few weeks or even days depending on need; service and delivery orders

Figure 9. Example Rates of Change for Spatial Data

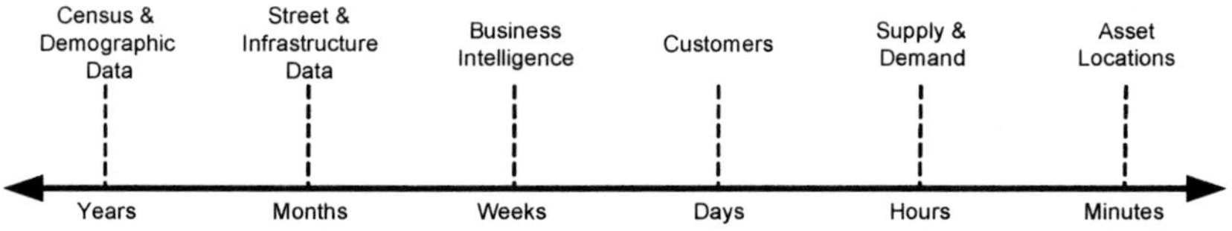

and locations of corporate assets such as delivery vehicles can change hourly or faster depending on the needs of the information system (*Figure 9*).

SDR designers have to develop a plan for updating various spatial data components such that inconsistencies across various data sources are minimized. When data from different sources is used to create a SDR, data inconsistencies are inevitable, however, planning and designing a data update methodology can minimize potential business impacts associated with them. Complications arise when two or more spatial services have data-dependencies across multiple data components within a SDR. This is particularly common for information systems, which use commercial address geocoding software relying on file-based data storage, and mapping generating software, which uses open data sources such as SDBMSs and other GIS formats. Inconsistencies at the spatial data level between geocoding and mapping services often result in addresses that cannot be located even though the street can be displayed by the mapping software. Conversely, customer locations that have been successfully geocoded can be located in what appears to be an undeveloped area by the map display.

Data Inter-Dependencies

When spatial locations for one set of data have been derived from other spatial data in an SDR, data dependencies between the two sets of spatial data have to be managed. Customer locations that have been geocoded against a digital street database, for example, need to be validated and possibly revised if the street data is updated, as changes in street names, address ranges, and geometries of the street data might affect the location of one or more of the customers. Additionally, newly added sub-divisions and streets might locate customers which were previously un-locatable.

Updating derived spatial data after a dependent data source has changed ensures that all spatial services using these data remain congruent. This process, however, can often involve cascading update operations across multiple tables and physical storage systems. Service vehicle routes optimized and stored with respect to previously defined customer locations and demands for service, for example, should be validated and possibly updated if any customer locations have changed, thereby ensuring that service routes remain optimal and reports such as service schedules derived from the data are correct (*Figure 10*). In this example, customer locations are affected when changes occur in the customer data itself or within the street database. Changes in either of these tables affect absolute locations of customers and distances over the street network between

Figure 10. Cascading Update Events for a Multi-Component SDR

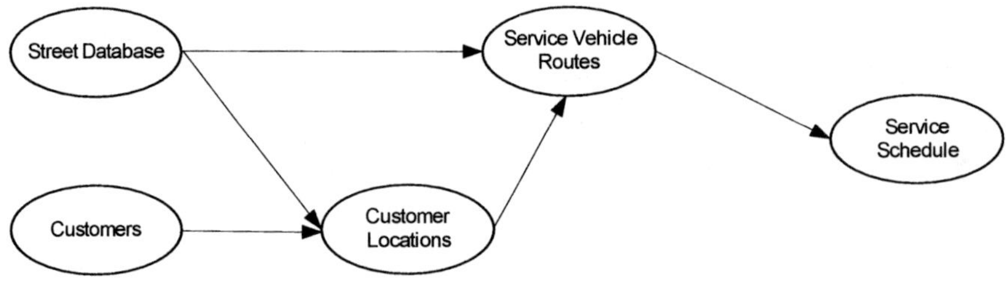

them, ultimately affecting vehicle routes and service schedules. When several dependencies of this type are evident, managing updates across an SDR has to be carefully controlled. Flow charts and PERT diagrams can be used to identify dependency chains and determine correct sequencing for updates.

Performing cascading updates across a SDR is time consuming, error prone, expensive, and requires validating the results of the update before the SDR can be brought back on line. In some cases update processes might take hours or even days, in which case updates are best performed at noncritical times or on redundant systems in order to minimize potential user impacts.

Vendor-Specific Data Requirements

Selecting a particular brand of GIS or spatial service can impose design constraints on a SDR and create issues associated with data concurrency and maintenance workflows, hardware selection, data compatibility, security, and disaster recovery. Spatial services using Hybrid and Application Specific data storage models require file-based storage for some or all of the spatial data. As file-storage is often dependent on the operating system, selecting a spatial service vendor might require also adopting particular hardware and operating system platforms. Many commercial geocoding solutions, for example, require a Microsoft Windows platform. Geographic Data Technology's Matchmaker address geocoding software, for instance, is only available on 32-bit Window's platforms and file systems. MapInfo's MapMarker V8 geocoding software has similar platform restrictions, as ODBC is required to provide access to remote databases (MapInfo, 2003).

GIS vendors often provide a range of spatial services that can be integrated across a common data storage platform. ESRI's Spatial Database Engine (SDE), for example, is spatial middleware designed to inter-operate between leading commercial DBMS packages such as Oracle and a variety of ESRI clients including their ArcIMS mapping server, Arc GIS desktop analysis and data maintenance software, and MapObjects Java and COM based programmable software libraries (ESRI, 2002). Similarly, Intergraph's Geographic Data Objects (GDO) is spatial middleware enabling the GeoMedia suite of products to access data stored in a variety of back-end data stores. Standardizing on spatial middleware such as these allows an organization to select a back-end database from a number of enterprise class database management systems which best meets their

IT needs. The flexibility afforded by this spatial middleware, however, can impose design constraints on the SDR. At the time of writing, SDE, for example, has several restrictions, which should be taken into account during design:

- constraints on the logical and physical design of the database,
- SDE middleware has to be co-located with the RDBMS on the same server,
- distributed and replicated data is not supported, and
- the use of default schema objects and naming conventions provides potential security risks.

Further, limited options for deploying the SDE middleware exist. SDE deployed on Solaris, for example, only supports Oracle as the back-end database. Intergraph's GDO middleware is even more restrictive as it only works on Windows systems. Although some of these restrictions will change as new versions of the software become available, organizations wishing to adopt such technology should be aware of the system design constraints imposed by their technology selections.

Implementation Issues

Once designed, a SDR has to be populated with spatial data and, once populated, the SDR has to be maintained and updated. There are a number of implementation issues arising from the form in which spatial data is delivered from vendors, how large volumes of data are physically loaded, and compatibility issues arising when spatial data is merged from different sources.

Spatial Data Distribution Formats

Organizations wishing to use spatial services have three basic options available for acquiring spatial data: purchase from commercial vendors, obtain from public domain sources, and create internally or have it created for them. A typical organization may use data from two or three of these sources. Spatial data such as digital street databases and business intelligence data are readily available from a variety of commercial data vendors and are generally more cost-efficient to purchase than create. Other spatial data such as geo-political data and census-related socio-economic data is readily available from a variety of public-domain resources such as the U.S. Census Bureau (http://www.census.gov) and can be downloaded free or for a nominal charge. Company-specific data such as sales areas, customer locations, and utility lines often need to be created by the organization itself or by a third-party specializing in GIS data capture. In each case, raw spatial data will be delivered to the SDR in a form that may not be directly compatible with its structure and definition due to differences in format, data dictionaries, or spatial representation of the raw data.

Third party data vendors as well as public-domain sources often provide several different distribution formats for their data. Geographic Data Technologies (http://www.gdt1.com), for example, provides street databases and boundary files in GIS format, supporting major GIS software vendors such as ESRI, Intergraph and Mapinfo. Although convenient for many GIS uses, there are implications for developers of large SDRs. In many cases data delivered in this manner requires further manipulation before it can be added to a SDR, this is particularly true if the data will be loaded into a SDBMS as few data vendors directly support these systems. Most GIS vendors provide one or more data-interchange formats, such as MapInfo's MIF/MID data structure and ESRI's E00 and Ungenerate formats. These data-interchange formats are non-proprietary and enable other GIS and data management software to read and write them, facilitating data interchange between disparate systems. Perhaps the most commonly used native vendor-specific GIS format is the Shapefile. Shapefiles were initially designed by ESRI, which later published the Shapefile format in a successful attempt to provide a de-facto GIS data standard (ESRI, 1998). Today, most GIS data can be purchased in Shapefile form.

For large data-sets, however, manually exporting thousands of GIS-tables distributed on a large number of CDROMs to intermediate formats using a GIS interface and subsequently importing the intermediate files using another vendor's import tools is a laborious and error prone task. This process is best accomplished using specialized loader programs which access vendor specific GIS data directly, reformat raw data and subsequently upload the reformatted data directly to the SDR. These programs are often built to specification by in-house development teams or consultants. Alternatively, specialized spatial data translation software such as Safe Software's Feature Manipulation Engine (FME) Suite (http://www.safe.com) can be purchased and used. FME, for example, accommodates a wide variety of input and output spatial data distribution formats including European formats, as well as SDBMS systems. FME can translate between data dictionaries as well as perform coordinate transformation functions if necessary.

Loading Spatial Data

Loading spatial data into an SDR often requires multiple steps and creation of intermediate data. A typical sequence of steps required to load a large data set is illustrated in *Figure 11*. Raw data is initially loaded from its vendor-distribution format to a location and form where it can be further manipulated. This initial load is followed by a series of transformations, which standardize and rationalize staged data resulting in a prototypical SDR. The final stage consists of a series of post-load procedures resulting in a completed SDR.

Large spatial datasets such as street databases and socio-economic datasets are usually delivered based on common geographic areas: states, counties, or regions. The U.S. street database from GDT, for example, consists of 3,111 individual county-based data sets. These individual data sets need to be aggregated into composite tables within the SDR. Depending on storage systems selected for the SDR, loading processes often require developing specialized software which can extract spatial data from vendor-

Figure 11. Typical Loading Process for Spatial Data

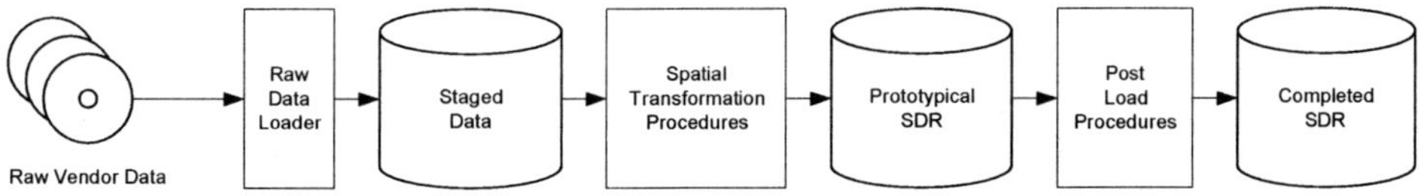

specific distribution formats and create new tables in an intermediate storage system or staging area, possibly within a database or GIS. Once raw spatial data has been loaded to a staging area, GIS tools and data-manipulation procedures can be used to further transform the spatial data to meet SDR design requirements. This second transformation process can involve a series of steps including:

- rationalizing data to remove issues associated with the discrete geographies of the raw data that was loaded,
- deriving spatial references,
- coordinate system transformations,
- data dictionary standardization, and
- validation and remediation.

Rationalization tasks often have to be performed when data is imported from geography-based data sources such as those distributed by U.S. CENSUS, GDT, TeleAtlas and others. Rationalizing these types of data typically involves removing duplicate boundaries on adjacent counties and re-constructing polygon and network topologies for data spanning county borders. During the loading process, data that does not have a direct spatial reference will need to have spatial references derived from other data sources in the SDR. Customer lists, for example, need to be geocoded and spatial joins created between other data sets to allow each data element to have a spatial reference. When data from different sources is imported into the SDR, some or all of the data might need to be transformed to fit the coordinate system of the SDR. Most commercial data available in the U.S., for example, is distributed using a geographic coordinate system, but variation in geographic datum is common among vendors necessitating coordinate projection for non-conforming data sets. Standardization of data elements within the data dictionary is best performed during initial data load or using batch-processing techniques if raw data has been staged to a RDBMS. Common standardizations include creating geographical abbreviations for administrative and geo-political areas and reformatting postal addresses and telephone numbers. Once data has been processed the final data set should be validated and remediation processes used to fix data that fails validation tests. For example, customer addresses that could not be geocoded should be examined and re-geocoded using manual techniques.

After all data has been transformed, a prototypical SDR is generated. The final stage of the loading process involves creating spatial and non-spatial indexes, creating spatial metadata, initiating data access controls, and enabling data protection systems.

Data Compatibility

When purchasing spatial data, it is important to ensure that the format and content of the spatial data is consistent with the requirements of the SDR. Spatial data is usually collected and distributed to support a specific set of spatial services. Street databases, for example, are often sold in two forms: a form suitable for address geocoding and a more expensive form suitable for street-by-street vehicle navigation systems. Attempting to perform street-based navigation on a format that has been designed for purely geocoding purposes might prove impossible, as topology of the underlying street network is often inconsistent and data elements required to perform successful navigation, such as the identification of one-way streets and turn restrictions, might be missing.

Laurini & Thompson (1992, pp. 102-103) identify a number of basic issues with spatial data, which affect spatial information systems. Among the issues identified are:

- quality of data, including identification and measurement of error,
- forms and sources of data,
- treatment of the time dimension, and
- impact of scale.

Understanding how these factors influence spatial information systems is important for both system designers and users, as inference based on incorrect or "fuzzy" data can lead to erroneous conclusions. Spatial data is well-known for its compatibility issues due to differences in representation of geographies, scale, temporality, collection methods, and planned use. For many organizations, populating a SDR involves merging spatial data from multiple, often disparate, data sources and manipulating it to meet SDR needs. These processes can exacerbate differences between data sets and introduce incompatibilities and error, which has to be resolved. Dutton (1989) maintains differences in accuracy, precision and uncertainty in the location of spatial objects are potential sources of error between data sets. Brusegard & Menger (1989), discussing development of large spatial databases for market research analyses, identify several additional issues that can affect outcomes of spatial analyses when spatial data has been merged from different sources. Issues identified include:

- age of different data sets, especially in areas of rapid growth,
- spatial representation of data caused by aggregation and disaggregation procedures; for example, assigning centroids to polygons or assigning point-values to areas,
- responsibilities of data providers to alert users to potential sources of error especially in the representation and encoding of spatial phenomena such as boundaries of "fuzzy" areas, and

- definition of error itself which is dependent on the context for which the data was created in the first place.

The time at which spatial data is collected is important as spatial data can change quickly. Other issues raised are related to a generalized notion of spatial error, which has been variously studied by a number of authors. A comprehensive treatment of error in spatial data is provided in Goodchild & Gopal (1989).

Management Issues

In a discussion of management issues associated with implementing GIS software and data in an organization, Schott (1998) identifies financial and vendor support considerations as paramount. Green & Bossomaier (2002, p. 173) similarly identify cost as the biggest issue in enabling enterprise spatial information systems. Financial considerations include not only initial costs of systems but also future costs necessary to support ongoing use of those systems. Maintenance and training costs usually exceed initial investment costs and are, accordingly, important management issues. Other important management issues include licensing of spatial data, and systems security including protecting access to data storage systems.

Financial Issues

Goldstein (2003) estimates that the cost of creating a two-person GIS department for an organization between $154,375 and $227,215, not including staff salaries estimated to be between $50,000 and $100,000 per year. The major cost burden identified by Goldstein is cost of the business data, with estimates ranging between $124,875 and $176,042. For large SDRs, however, back-end storage costs can quickly outweigh costs of purchasing data. Oracle Spatial, for example, costs $10,000 per processor in addition to a $40,000 per-processor license for the basic RDBMS (http://www.oracle.com). Software storage licenses for a mid-range database server with eight or 16 processors would then be expected to cost between $400,000 and $800,000.

Replicated and parallel database architectures can increase overall database performance by balancing workload across multiple database servers and removing single points of failure within a system. These architectures, however, can significantly increase storage cost. Replication of spatial data from a SDBMS requires that each replicated database has the SDBMS installed, requiring additional software licenses for each replicated instance. Parallel storage options similarly require each database instance to have the SDBMS loaded. Parallel systems also incur additional costs as software and hardware for parallelizing the database has to be purchased.

Staffing and Management

Schott (1998) identifies several staffing issues associated with successful implementation of a GIS. Staffing options available for companies intending to implement GIS include internal training programs, new hires of qualified personnel, use of consultants, and lastly, use of value-added resellers (VARs). In a discussion on the skill-sets required by successful GIS employees in companies, Goldstein (2003) argues such staff needs to be experts in programming, web development, database management, and GIS techniques, as well as understand the industry space they are working within.

Designing and implementing SDRs presents even more of an issue for companies. Often SDRs use multiple data storage technologies and multiple hardware and software environments. Employees then need systems integration skills and experience with a variety of heterogeneous hardware and software environments in addition to the general GIS skills identified above. Companies electing to use a SDBMS, for example, have options to hire staff with SDBMS skills on a particular platform, train existing database staff in GIS, or out-source to third parties. Unless these skills are already available within the organization, a significant investment in both cost and time is required to make these skills available.

Licensing of Data

One backlash of the rise of Internet mapping has been stricter licensing restrictions for commercial spatial data. Companies such as GDT, NavTech, and TeleAtlas, which produce high-quality spatial data, now have restrictive licensing agreements for their data products if they are to be used for applications that disseminate derived information over the World Wide Web. Derived information, such as vector-based maps and driving directions, commonly used in location-based services and in vehicle navigation systems, can potentially be used to serve millions of customers from a single information system, effectively circumventing standard multi-user licensing. Many commercial data vendors have negotiable licensing fees, which vary based on amount of data being licensed, length of license period, and type of applications being implemented. Licenses for internal use where derivable content is not disseminated to non-licensed entities are usually more reasonable.

Security

Protecting and restricting access to spatial data is often more difficult than protecting and restricting access to traditional data. Many enterprise RDBMS systems use additional schema accounts to store and manage spatial metadata. These accounts, unless controlled, can provide serious potential security issues. Oracle Spatial when installed, for example, creates an account called MDSYS in the database. The MDSYS account is used to store metadata about spatial data stored in each schema as well as provide a central location for stored procedures and data. Access to MDSYS schema objects has to be provided by for any schema that requires Oracle Spatial. Earlier versions of Oracle Spatial installed this schema with a default password with the same name as the schema, providing an obvious security hole. Other companies that use Oracle for spatial data storage use similar schemes wherein a default-named schema is created to store system-dependent information. Examples include SDE (ESRI), GDOSYS (Intergraph), and MAPCATALOG (MapInfo).

Although these system-dependent accounts can be secured by changing default passwords created during installation and restricting access to these schemas, current versions of the software require these accounts be present with default account names. A hacker attempting to gain access to a database containing spatial data need only identify the GIS software vendor and the hacker is guaranteed that a particular named account, often with administrator level access, is present in the database system. Once an account name is known it provides an obvious point of attack for password-cracking software and, thus, represents a potential high-risk vulnerability for a company.

A critical and often overlooked concern associated with implementing a SDR is providing a robust system, which protects against loss associated with critical systems failure. The difficulty associated with implementing such a system depends largely on the storage technologies used. File-based storage is perhaps most straightforward to manage. File archive and restore systems such as tape libraries and redundant disk-storage can be used to archive and restore individual or groups of files as required. Current backup and restore software systems are able to archive files that are open and in use, negating need to quiesce SDRs during backup operations.

Hybrid Data Models using a combination of system files and RDBMS storage require specialized handling. Most RDBMSs provide facilities to backup and archive data stored within them. Microsoft's SQL Server 2000 Enterprise Manager is an example of a GUI tool, which can export data from databases to disk files for subsequent archiving to offline devices. Creating an effective, automated backup strategy for a SDR which uses hybrid data storage often involves creating a specialized backup agent, which sequentially runs a database export process and then archives files and exported data together to off-line devices. When SDRs contain volatile data components, archiving operations are best performed while the system is quiescent to avoid potential data concurrency issues.

SDRs using Unified and SDBMS storage technologies are perhaps easier to protect than other storage technologies, as effective data archiving and restoration facilities are built-in to the storage systems. Replication provides an alternative to export/import operations for protecting spatial data. Database replication is a technique that allows one database, the master database, to transfer changes to tables made within it to other databases over a network, thereby keeping two or more database instances concurrent. Replication can be scheduled by database administrators, allowing changes in master databases to be replicated to client databases on a schedule. The main advantage of replication over sequential import/export processes is that one or more redundant database instances can be kept online and, in the event of a failure of the master database, can be switched to with minimal loss of data or systems down-time.

Replication as a means of maintaining concurrency across two or more database instances does present challenges. Configuring a distributed database, which shares data across one or more database instances, can greatly increase costs of SDRs as redundant hardware and software licenses are required. In some circumstances, replication may not be an available option. Earlier versions of Oracle's Internet Database, for example, could not replicate Oracle Spatial objects, which are stored in the database as UDTs. The same version of Oracle could replicate data stored as BLOBs, therefore, an Oracle database used to support a Unified Data Model would replicate, whereas the same database using Oracle Spatial as an SDBMS would not. Oracle has since resolved replication issues associated with UDTs in current versions of its database. When replication is not available for a RDBMS, customized replication agents can be created and used to maintain concurrency between database instances. These replication agents are designed to monitor changes in an SDR and perform batch or transaction-based updates against one or more remote database instances as needed.

Future of Spatial Data Management Systems

Several trends are occurring in spatial data management systems which will affect how spatial data is stored and managed in future systems. Spatial data storage technologies are evolving, providing more storage options. Data collected in near real-time from mobile systems and other sources will become more important in decision-making, affecting the design of future spatial storage systems. Lastly, the Internet and World Wide Web will play an increasingly important role by providing access to spatial data through subscription updates and Geospatial web services.

Spatial Data Storage

Issues with storage and management of spatial data are becoming more prevalent as new and innovative applications that utilize spatial data are adopted by organizations. This move towards spatially-enabled information systems is providing a driving force which is rapidly re-shaping the database industry as vendors attempt to accommodate spatial data and data-access capabilities. Sonnen (2003) identifies four factors that are expected to re-shape the spatial database industry in the next three to five years:

- availability of basic spatial functionality in data access and database management,
- substantially lower cost of entry for spatially-enabled database vendors that want to enter new vertical markets,
- lower costs for IT vendors that wish to include location-specific functionality into their applications, and
- delivery of spatial functionality both as software and services.

Today, spatial data can be stored and manipulated in a number of enterprise-class database systems including Oracle, IBM, and Informix, as well as some of the popular open source database systems including MySQL and PostgresSQL. Other database systems will eventually follow suit. This proliferation and competition will reduce costs and lead to cheaper, more functional, and more importantly, more inter-operable storage solutions in the future. Oracle, for example, now includes basic spatial data storage and manipulation services into its latest enterprise database free of charge. In the business application space, SAP, Siebel and other vendors of CRM and ERP systems are working to add additional spatial capability to their products.

Managing Real-Time Data

As the ability to process and analyze spatial data in real-time increases, the numbers and types of business applications that can make use of these capabilities can also be expected to increase. Today, business assets can be tracked and customers located in real-time using combinations of increasing more ubiquitous, accurate and cost-effective technologies such as GPS and mobile communications. Other types of data, such as traffic conditions and weather forecasts, are available in near real-time in the form of streams of data pushed over the Internet. Given a suitable decision analysis framework, real-time and near real-time data can greatly enhance decision-making capabilities. There are, however, several factors that have to be resolved before it can be used in real-time:

- geocoding,
- format,
- persistence,
- reliability and availability, and
- decision modeling.

The ability to accurately geocode data in real-time is perhaps the biggest barrier to using this type of data today. Locations of traffic incidents and traffic congestion are often referenced indirectly, relative to a particular road or landmark for example, rather than using explicit spatial references such as latitude and longitude. To use this data efficiently, indirect location information has to be converted to a form which can be manipulated by a GIS or spatial service. Recently, vendors of traffic data and street databases have been collaborating to deliver travel time and traffic information pre-registered to one or more commercial street databases based on unique feature IDs (GDT, 2003).

The format in which data is delivered from real-time systems can affect the speed and ability for data to be processed. Input from GPS systems, data streams over the Internet and other sources of real-time data have to be incorporated into the information system framework in a way which facilities processing by a spatial service. Today, much real-time data is delivered using vendor-specific formats. Weather data, for example, is typically distributed as ESRI Shapefiles, often requiring transformation through multiple stages before it can be used in a modeling context. Vendors delivering real-time data are moving towards Extensible Markup Language (XML) as a distribution format. Open standards for geographical and mobile data, such as Geography Markup Language (GML) and other specialized XML formats, are rapidly being developed and adopted, thereby facilitating the process of transferring spatial data between inter-operable systems.

Use of real-time data by organizations requires three additional factors: predictive models which can use real-time data to facilitate decision making, a decision analysis framework which can operate in real time, and reliable data sources with sufficient spatial coverage. The last point is perhaps most important in the near future as many real-time data streams such as traffic and weather are available for specific collection areas such as major population centers. Without access to ubiquitously available data in a reliable, cost effective, and efficient manner, models and decision frameworks designed to use them have limited general utility.

Subscription-Based Update Services

As spatial data becomes prevalent, a more appropriate means of updating large spatial databases has to be developed. Change-logs and "delta" files are one means of identifying changes to large data sets. Change-logs provide lists of features, usually identified by unique primary keys, which have been altered, added, or deleted since the last version of the database. Change-logs are distributed along with an update of the spatial data source. To be effective, change-logs should be readable by computer programs in such a way that they can be incorporated into an automatic update system. Delta files are similar to change-logs, except the files, rather than containing a list of features that have been updated, contain actual updated features themselves. Delta files can be processed more efficiently as the updated data is encapsulated and does not need to be extracted from large, often compressed files.

One approach to managing data updates for large databases being considered by some data vendors is to allow subscribers to automatically download change logs and extract changed data from online databases over the Internet and World Wide Web. Street database vendors, for example, are considering providing subscription-based updates to corporate users over the Internet. Users of these services will be able to download change logs or delta files directly from the vendor's database server, extract specific records from a master database, and process each change as a transaction against an active SDR. Users will be able to access changes in batch mode or in real-time as changes are committed to the master database.

Geospatial Web Services

Online access to spatial data and spatial services, such as Microsoft's MapPoint Web Service (http://www.microsoft.com/mappoint/net) and ESRI's ArcWeb Services (http://www.esri.com/software/arcwebservices/index.html), is a recent trend. These online

Figure 12. Spatial Data Updates over the Internet

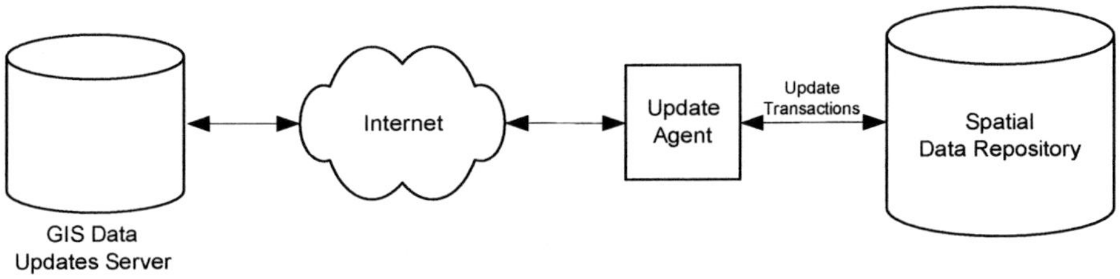

geospatial services provide an alternative for companies who wish to subscribe to third-party spatial data sources and services rather than creating and managing their own. Service oriented architectures based on the Web Service Application Service Provider (ASP) model provide XML-based Simple Object Access Protocol (SOAP) APIs and allow client applications programmatic access to spatial data and spatial services over the Internet. Organizations are charged a per-use fee to access services, normally based on the number of transactions made against the services for a specific account. Processes such as generating a map or locating a customer, for example, each generate a transaction. Organizations subscribing to geospatial web services gain access to data and services provided by service hosts. At the time of writing, Microsoft's MapPoint Web Service, for example, provides data sets for Western Europe, North America and Canada. Data sets include street data, topographic data, and other contextual data necessary to generate high quality, visually pleasing digital maps. Microsoft allows companies to upload and store their own spatial data on the Map Point servers and merge it with data stored in the remote databases.

Geospatial web services provide several advantages for organizations. First, the hosting company creates and manages the SDR and is responsible for maintaining data quality and quality of service. Second, access to spatial data is made indirectly through a middleware API rather than direct access. This approach masks the implementation details of the spatial data as it is stored in the SDR, thereby reducing both cost and time required to create client applications. Third, storage of client-specific data is performed using an API, again negating issues associated with implementation details. Lastly, the XML/SOAP specification is platform and vendor independent allowing traditional (thick) or lightweight (thin) client applications written in a variety of industry-standard programming languages such as Visual Basic, C++, Java, and C# to quickly connect to and use these services.

Organizations requiring limited geo-spatial functionality will find geospatial web services a cost-effective alternative. There are, however, several issues that can limit their general utility and cost-effectiveness. First, APIs provided by web services are usually highly abstracted and limited. For example, methods might be available to create maps at different scales, but the ability to alter map symbology will generally be limited. Similarly, methods for locating business objects such as customers and sales areas might be restricted. Microsoft's MapPoint Web Service, for example, limits customer-specific data to point geometries only. Therefore, business objects represented as linear or aerial features cannot be added. Further, business objects often cannot be used in geospatial analyses. Calculating the number of customer objects in a sales area, for example, might not be feasible if both sales area and customer objects are uploaded by the client. Second, APIs provide limited geo-spatial methods, usually restricted to basic location-based services functionality such as geocoding, spatial searches, route generation, and mapping tools. Third, accounting methods used to charge for use of services can affect the overall cost effectiveness. Each time a map is generated, for example, a transaction is recorded and additional transactions are generated each time the map is re-scaled or panned. Lastly, control is relinquished by subscribing to remote third-party services rather than having data and services available locally. Mission critical spatially-enabled business processes such as call centers or logistics planning software require maintaining access to spatial services and spatial data, which in this case, must rely on consistent and uninterrupted services provided by third-party companies over the Internet.

Summary

When properly managed and manipulated, spatial data can be an extremely powerful business tool enabling organizations to effect greater efficiencies, service customers more effectively, and reduce risk and uncertainty. Spatial data, however, differs from traditional business data requiring specialized handling and storage systems to manage it effectively. Over the years, GIS and enterprise-database vendors have developed a variety of methods for overcoming these storage challenges, resulting in an eclectic mix of technologies and methods available for today's businesses. In an effort to aid decision makers and researchers to understand potential effects of these storage systems from a systems integration perspective, a new taxonomy for spatial data storage systems is described. Focusing on system architectures, this new taxonomy differentiates spatial data storage systems based on inter-relationships between client, middleware, and data storage components, providing a basis for comparing disparate spatial data storage systems and understanding how internal organization of these systems and can affect other technology decisions in an organization.

Existing spatial data storage systems have potential to create vulnerabilities in an organization's information technology platform, ultimately making spatially-enabled information systems vulnerable to attack or failure. Understanding issues pertaining to three distinct aspects of technology adoption: systems design, systems implementation and management of completed systems, can help create more secure, efficient and effective spatial information systems; an important first-step as spatially-enabled business systems become commonplace and organizations increasingly rely on the advantages well-designed and implemented spatial information systems can afford them.

Acknowledgments

The author would like to thank the three anonymous referees for their comments and suggestions during the final preparation of this document.

References

Adler, D.W. (2001). IBM DB2 Spatial Extender – Spatial Data within the RDBMS. *Proceedings of the 27th VLDB Conference.* Roma, Italy.

Bruegard, D., & Menger, G. (1989). Real data and real problems: Dealing with large spatial databases. In M. Goodchild & S. Gopal (Eds.), *Accuracy of Spatial Databases.* London: Taylor and Francis.

Caliper. (1995). *Maptitude users guide v. 3.0 for Windows.* Newton, MA: Caliper Corp.

Dutton, G. (1989). Modeling locational uncertainty via hierarchical tessellation. In M. Goodchild & S. Gopal (Eds.), *Accuracy of Spatial Databases.* London: Taylor and Francis.

ESRI. (1996). *Mapping for data warehousing.* Retrieved August 8, 2003 from: *http://saaslin.petech.ac.za/CONTENT/PDFS/SDE.PDF.*

ESRI. (1998). *ESRI shapefile technical description.* Retrieved March 13, 2003 from: *http://www.esri.com/library/whitepapers/other_lit.html#shapefiles.*

ESRI. (2002). *ArcSDE: The GIS application server.* Retrieved August 8, 2003 from: *http://www.esri.com/library/brochures/pdfs/arcsde-server.pdf.*

GDT. (2003). *GDT and TrafficCast to deliver real-time traffic information.* Retrieved April 19, 2003 from: *http://www.directionsmag.com/pressreleases.php?press_id=6715.*

Green, D., & Bossomaier, T. (2002). *Online GIS and spatial metadata.* London: Taylor and Francis.

Goldstein, D. (2003). The five major roadblocks to GIS gaining corporate marketshare. *Directions Magazine.* Retrieved April 27, 2003 from: *http://www.directionsmag.com/article.php?article_id=332.*

Goodchild, M., & Gopal, S. (1989). *Accuracy of spatial databases.* Taylor & Francis: London.

Haley, M. (1999). *Data visualization: Adding spatial components to data.* Proceedings GITA. Retrieved August 8, 2003 from: *http://www.gisdevelopment.net/proceedings/gita/1999/dda/dda002.shtml.*

Intergraph. (2003). *GeoMedia: Architectural Overview.* Retrieved May 10, 2003 from: *http://www.intergraph.com/gis/geomedia/geo_objects.asp.*

Laurini, R., & Thompson, D. (1992). *Fundamentals of spatial information systems.* London: Academic Press.

MapInfo. (2002). *MapInfo professional user's guide v 7.0.* Troy, NY: MapInfo Corp.

MapInfo. (2003). *MapMarker U.S. – Overview.* Retrieved Sept 16, 2003 from: *http://www.mapinfo.com/products/Overview.cfm?productid=139.*

Melton, J. (2003). *Advanced SQL: 1999 understanding object relational and other advanced features.* San Francisco: Morgan Kauffman.

Mennecke, B. E., & Higgins, G. (1999). Spatial data in the data warehouse: A nomenclature for design and use. *Proceedings of the Americas Conference on Information Systems.*

Mennecke, B. E., Crossland, M. D., & Killingsworth, B. (1997). An experimental examination of spatial decision support system effectiveness: The roles of task complexity and technology. *Proceedings of the Americas Conference on Information Systems.*

Murphy, L. D. (1996). Competing in space: The strategic roles of geographic information systems. *Proceedings of the Americas Conference on Information Systems.*

Open GIS Consortium, Inc. (1999). *OpenGIS simple features specification for SQL: Revision 1.1.* Wayland, MA: Open GIS Consortium.

Oracle. (2001). *Leveraging location based services for mobile applications: An Oracle white paper.* Retrieved August 15, 2003 from: *http://otn.oracle.com/products/spatial/pdf/location_based_svcs_twp.pdf.*

Ratliff, D. (2003). *10 rules for logistics optimization.* Atlanta, GA: Velant Corp.

Rigaux, P., Scholl M., & Voisard, A. (2002). *Spatial databases with application to* GIS. San Francisco: Elsevier Science.

Schott, L. R. (1998). Software, data, and management considerations in establishing a GIS/Mapping system. In G. H. Castle (Ed.), *GIS in Real Estate: Integrating, Analyzing, and Presenting Locational Information.* IL: Appraisal Institute.

Shekhar, S., & Chawala, S. (2003). *Spatial databases: A tour.* NJ: Pearson Education Inc.

Sloan, T. M., Dowers, S., Gittings, B. M., Healy, R. G., & Waugh, T.C. (1992). Exploring GIS performance issues. *Proceedings 5th International Symposium on Spatial Data Handling,* (Vol. 1, pp. 154-165).

Sonnen, D. (2003). *Oracle's geospatial capabilities for enterprise information systems.* Retrieved April 19, 2003 from: *http://www.directionsmag.com/article.php?article_id=350.*

Weigel, D., & Cao, B. (1999). Applying GIS and OR techniques to solve Sears technician-dispatching and home-delivery problems. *Interfaces, 29* (1), 112-130.

Chapter VI

Mining Geo-Referenced Databases:
A Way to Improve Decision-Making

Maribel Yasmina Santos, University of Minho, Portugal

Luís Alfredo Amaral, University of Minho, Portugal

Abstract

Knowledge discovery in databases is a process that aims at the discovery of associations within data sets. The analysis of geo-referenced data demands a particular approach in this process. This chapter presents a new approach to the process of knowledge discovery, in which qualitative geographic identifiers give the positional aspects of geographic data. Those identifiers are manipulated using qualitative reasoning principles, which allows for the inference of new spatial relations required for the data mining step of the knowledge discovery process. The efficacy and usefulness of the implemented system — Padrão — has been tested with a bank dataset. The results support that traditional knowledge discovery systems, developed for relational databases and not having semantic knowledge linked to spatial data, can be used in the process of knowledge discovery in geo-referenced databases, since some of this semantic knowledge and the principles of qualitative spatial reasoning are available as spatial domain knowledge.

Introduction

Knowledge discovery in databases is a process that aims at the discovery of associations within data sets. Data mining is the central step of this process. It corresponds to the application of algorithms for identifying patterns within data. Other steps are related to incorporating prior domain knowledge and interpretation of results.

The analysis of geo-referenced databases constitutes a special case that demands a particular approach within the knowledge discovery process. Geo-referenced data sets include allusion to geographical objects, locations or administrative sub-divisions of a region. The geographical location and extension of these objects define implicit relationships of spatial neighborhood. The data mining algorithms have to take this spatial neighborhood into account when looking for associations among data. They must evaluate if the geographic component has any influence in the patterns that can be identified.

Data mining algorithms available in traditional knowledge discovery tools, which have been developed for the analysis of relational databases, are not prepared for the analysis of this spatial component. This situation led to: (i) the development of new algorithms capable of dealing with spatial relationships; (ii) the adaptation of existing algorithms in order to enable them to deal with those spatial relationships; (iii) the integration of the capabilities for spatial analysis of spatial database management systems or geographical information systems with the tools normally used in the knowledge discovery process.

Most of the geographical attributes normally found in organizational databases (e.g., addresses) correspond to a type of spatial information, namely qualitative, which can be described using indirect positioning systems. In systems of spatial referencing using geographic identifiers, a position is referenced with respect to a real world location defined by a real world object. This object is termed a *location*, and its identifier is termed a *geographic identifier*. These geographic identifiers are very common in organizational databases, and they allow the integration of the spatial component associated with them in the process of knowledge discovery.

This chapter presents a new approach to the analysis of geo-referenced data. It is based on qualitative spatial reasoning strategies, which enable the integration of the spatial component in the knowledge discovery process. This approach, implemented in the PADRÃO system, allowed the analysis of geo-referenced databases and the identification of implicit relationships existing between the geo-spatial and non-spatial data.

The following sections, in outline, include: (i) an overview of the process of knowledge discovery and its several phases. The approaches usually followed in the analysis of geo-referenced databases are also presented; (ii) a description of qualitative spatial reasoning presenting its principles and the several spatial relations — direction, distance and topology. For the relations, an integrated spatial reasoning system was constructed and made available in the Spatial Knowledge Base of the PADRÃO system. The rules stored enable the inference of new spatial relations needed in the data mining step of the knowledge discovery process; (iii) a presentation of the PADRÃO system describing its architecture and its implementation achieved through the adoption of several technologies. This section continues with the analysis of a geo-referenced database, based on the several steps of the knowledge discovery process considered by the PADRÃO system; and (iv) a conclusion with some comments about the proposed research and its main advantages.

Knowledge Discovery in Databases

Large amounts of operational data concerning several years of operation are available, mainly from middle-large sized organizations. Knowledge discovery in databases is the key to gaining access to the strategic value of the organizational knowledge stored in databases for use in daily operations, general management and strategic planning.

The Knowledge Discovery Process

Knowledge Discovery in Databases (KDD) is a complex process concerning the discovery of relationships and other descriptions from data. Data mining refers to the application algorithms used to extract patterns from data without the additional steps of the KDD process, e.g., the incorporation of appropriate prior knowledge and the interpretation of results (Fayyad & Uthurusamy, 1996).

Different tasks can be performed in the knowledge discovery process and several techniques can be applied for the execution of a specific task. Among the available tasks are *classification*, *clustering*, *association*, *estimation* and *summarization*. KDD applications integrate a variety of data mining algorithms. The performance of each technique (algorithm) depends upon the task to be carried out, the quality of the available data and the objective of the discovery. The most popular Data Mining algorithms include *neural networks*, *decision trees*, *association rules* and *genetic algorithms* (Han & Kamber, 2001).

The steps of the KDD process (*Figure 1*) include data selection, data treatment, data pre-processing, data mining and interpretation of results. This process is interactive, because it requires user participation, and iterative, because it allows for going back to a previous phase and then proceeding forward with the knowledge discovery process. The steps of the KDD process are briefly described:

- *Data Selection.* This step allows for the selection of relevant data needed for the execution of a defined data mining task. In this phase the minimum sub-set of data to be selected, the size of the sample needed and the period of time to be considered must be evaluated.
- *Data Treatment.* This phase concerns with the cleaning up of selected data, which allows for the treatment of corrupted data and the definition of strategies for dealing with missing data fields.
- *Data Pre-Processing.* This step makes possible the reduction of the sample destined for analysis. Two tasks can be carried out here: (i) the reduction of the

Figure 1. Knowledge Discovery Process

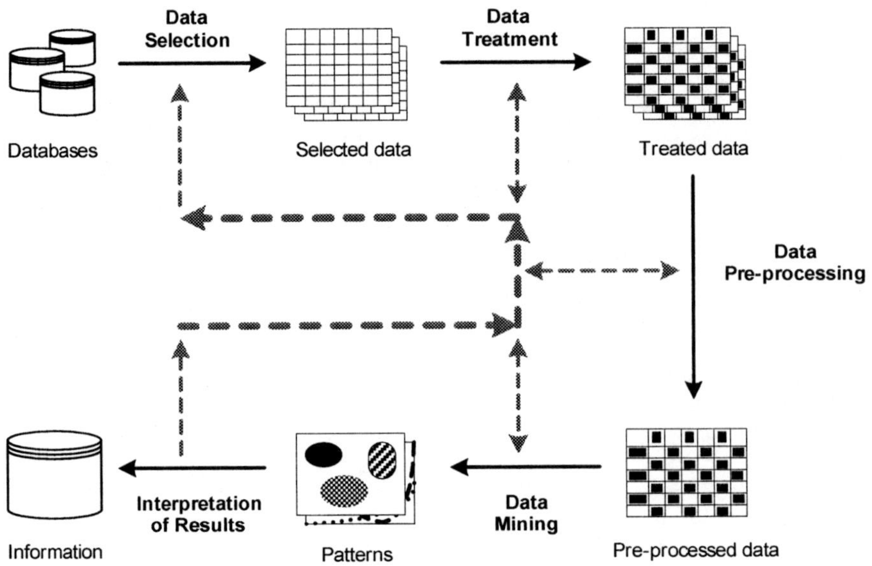

number of rows or, (ii) the reduction of the number of columns. In the reduction of the number of rows, data can be generalized according to the defined hierarchies or attributes with continuous values can be transformed into discreet values according to the defined classes. The reduction of the number of columns attempts to verify if any of the selected attributes can now be omitted.

- *Data Mining.* Several algorithms can be used for the execution of a given data mining task. In this step, various available algorithms are evaluated in order to identify the most appropriate for the execution of the defined task. The selected one is applied to the relevant data in order to find implicit relationships or other interesting patterns that exist in the data.
- *Interpretation of Results.* The interpretation of the discovered patterns aims at evaluating their utility and importance with respect to the application domain. It may be determined that relevant attributes were ignored in the analysis, thus suggesting that the process should be repeated.

Knowledge Discovery in Spatial Databases

The main recognized advances in the area of KDD (Fayyad, Piatetsky-Shapiro, Smyth & Uthurusamy, 1996) are related with the exploration of relational databases. However, in most organizational databases there exists one dimension of data, the *geographic* (associated with addresses or post-codes), the semantic of which is not used by traditional KDD systems.

Knowledge Discovery in Spatial Databases (KDSD) is related with "*the extraction of interesting spatial patterns and features, general relationships that exist between spatial and non-spatial data, and other data characteristics not explicitly stored in spatial databases*" (Koperski & Han, 1995).

Spatial database systems are relational databases with a concept of spatial location and spatial extension (Ester, Kriegel & Sander, 1997). The explicit location and extension of objects define implicit relationships of spatial neighborhood. The major difference between knowledge discovery in relational databases and KDSD is that the neighbor attributes of an object may influence the object itself and, therefore, must be considered in the knowledge discovery process. For example, a new industrial plant may pollute its neighborhood entities depending on the distance between the objects (regions) and the major direction of the wind. Traditionally, knowledge discovery in relational databases does not take into account this spatial reasoning, which motivates the development of new algorithms adapted to the spatial component of spatial data.

The main approaches in KDSD are characterized by the development of new algorithms that treat the position and extension of objects mainly through the manipulation of their coordinates. These algorithms are then implemented, thus extending traditional KDD systems in order to accommodate them. In all, a quantitative approach is used in the spatial reasoning process although the results are presented using qualitative identifiers.

Lu, Han & Ooi (1993) proposed an attribute-oriented induction approach that is applied to spatial and non-spatial attributes using conceptual hierarchies. This allows the discovery of relationships that exist between spatial and non-spatial data. A spatial concept hierarchy represents a successive merge of neighborhood regions into large regions. Two learning algorithms were introduced: (i) non-spatial attribute-oriented induction, which performs generalization on non-spatial data first, and (ii) spatial hierarchy induction, which performs generalization on spatial data first. In both approaches, the classification of the corresponding spatial and non-spatial data is performed based on the classes obtained by the generalization. Another peculiarity of this approach is that the user must provide the system with the relevant data set, the concept hierarchies, the desired rule form and the learning request (specified in a syntax similar to SQL – Structured Query Language).

Koperski & Han (1995) investigated the utilization of interactive data mining for the extraction of spatial association rules. In their approach the spatial and non-spatial attributes are held in different databases, but once the user identifies the attributes or relationships of interest, a selection process takes place and a unified database is created. An algorithm, implemented for the discovery of spatial association rules, analyzes the stored data. The rules obtained represent relationships between objects, described using spatial predicates like *adjacent to* or *close to.*

These approaches are two examples of the efforts made in the area of KDSD. One approach uses two different databases, storing spatial and non-spatial data separately. Once the user identifies the attributes of interest, an interface between the two databases ensures the selection and treatment of data without the creation of a new integrated repository. The other approach also requires two different databases, but the selection phase leads to the creation of a unified database where the analysis of data takes place. In both approaches new algorithms were implemented and the user is asked for the specification of the relevant attributes and the type of results expected.

Two approaches for the analysis of spatial data with the aim of knowledge discovery have been presented. Independently of the adopted approach, several tasks can be performed in this process, among them: *spatial characterization*, *spatial classification*, *spatial association* and *spatial trends analysis* (Koperski & Han, 1995; Ester, Frommelt, Kriegel & Sander, 1998; Han & Kamber, 2001).

A *spatial characterization* corresponds to a description of the spatial and non-spatial properties of a selected set of objects. This task is achieved analyzing not only the properties of the target objects, but also the properties of their neighbors. In a characterization, the relative frequency of incidence of a property in the selected objects, and their neighbors, is different from the relative frequency of the same property verified in the remaining of the database (Ester, Frommelt, Kriegel & Sander, 1998). For example, the incidence of a particular disease can be higher in a set of regions closest or holding a specific industrial complex, showing that a possible *cause-effect* relationship exists between the disease and the industry pollution.

Spatial classification aims to classify spatial objects based on the spatial and non-spatial features of these objects in a database. The result of the classification, a set of rules that divides the data into several classes, can be used to get a better understanding of the relationships among the objects in the database and to predict characteristics of new objects (Han, Tung & He, 2001; Han & Kamber, 2001). For example, regions can be classified into *rich* or *poor* according to the average family income or any other relevant attribute present in the database.

Spatial association permits the identification of spatial-related association rules from a set of data. An association rule shows the frequently occurring patterns of a set of data items in a database. A spatial association rule is a rule of the form "X → Y (s%, c%)," where X and Y are sets of spatial and non-spatial predicates (Koperski & Han, 1995). In an association rule, s represents the support of the rule, the probability that X and Y exist together in the data items analyzed, while c indicates the confidence of the rule, i.e., the probability that Y is true under the condition of X. For example, the spatial association rule "is_a (x, House) ∧ close_to (x, Beach) → is_expensive (x)" states that houses which are close to the beach are expensive.

A *spatial trend* (Ester, Frommelt, Kriegel & Sander, 1998) describes a regular change of one or more non-spatial attributes when moving away from a particular spatial object. Spatial trend analysis allows for the detection of changes and trends along a spatial dimension. Examples of spatial trends are the changes in the economic situation of a population when moving away from the center of a city or the trend of change of the climate with the increasing distance from the ocean (Han & Kamber, 2001).

After the presentation of two approaches and some of the most popular tasks associated with the analysis of spatial data with the aim of knowledge discovery, this chapter posits a new approach to the process of KDSD (more specifically in geo-referenced datasets). This approach integrates qualitative principles in the spatial reasoning system used in the knowledge discovery process. Since the use of coordinates for the identification of a spatial object is not always needed, this work investigates how traditional KDD systems (and their generic data mining algorithms) can be used in KDSD.

Qualitative Spatial Reasoning

Human beings use qualitative identifiers extensively to simplify reality and to perform spatial reasoning more efficiently. *Spatial reasoning* is the process by which information about objects in space and their relationships are gathered through measurement, observation or inference and used to arrive at valid conclusions regarding the relationships of the objects (Sharma, 1996). *Qualitative spatial reasoning* (Abdelmoty & El-Geresy, 1995) is based on the manipulation of qualitative spatial relations, for which composition[1] tables facilitate reasoning, thereby allowing the inference of new spatial knowledge.

Spatial relations have been classified into several types (Frank, 1996; Papadias & Sellis, 1994), including *direction relations* (Freksa, 1992) (that describe order in space), *distance relations* (Hernández, Clementini & Felice, 1995) (that describe proximity in space) and *topological relations* (Egenhofer, 1994) (that describe neighborhood and incidence). Qualitative spatial relations are specified by using a small set of symbols, like *North, close, etc.*, and are manipulated through a set of inference rules.

The inference of new spatial relations can be achieved using the defined qualitative rules, which are compiled into a composition table. These rules allow for the manipulation of the qualitative identifiers adopted. For example, knowing the facts, A North, very far from B and B Northeast, very close to C, it is possible, by consulting the composition table for integrated direction and distance spatial reasoning (presented later), to infer the relationship that exists between A and C, that is A North, very far from C.

The inference rules can be constructed using quantitative methods (Hong, 1994) or by manipulating qualitatively the set of identifiers adopted (Frank, 1992; Frank, 1996), an approach that requires the definition of axioms and properties for the spatial domain.

Later in this section the construction of the qualitative spatial reasoning system used by PADRÃO is presented. The qualitative system integrates *direction*, *distance* and *topological* spatial relations. Its conception was achieved based on the work developed by Hong (1994) and Sharma (1996). The application domain in which this qualitative reasoning system will be used is characterized by objects that represent administrative subdivisions.

Direction Spatial Relations

Direction relations describe where objects are placed relative to each other. Three elements are needed to establish an orientation: two objects and a fixed point of reference (usually the North Pole) (Frank, 1996; Freksa, 1992). Cardinal directions can be expressed using numerical values specifying degrees (0°, 45°...) or using qualitative values or symbols, such as North or South, which have an associated acceptance region. The regions of acceptance for qualitative directions can be obtained by projections (also known as half-planes) or by cone-shaped regions (*Figure 2*).

A characteristic of the cone-shaped system is that the region of acceptance increases with distance, which makes it suitable for the definition of direction relations between

Figure 2. Direction Relations Definition by Projection and Cone-Shaped Systems

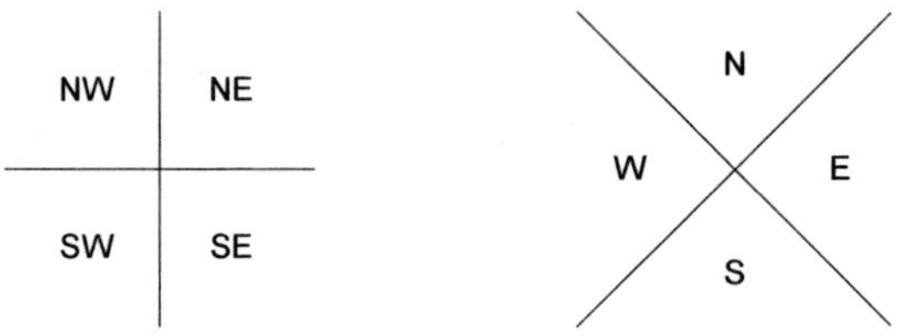

Figure 3. Cone-Shaped System with Eight Regions of Acceptance

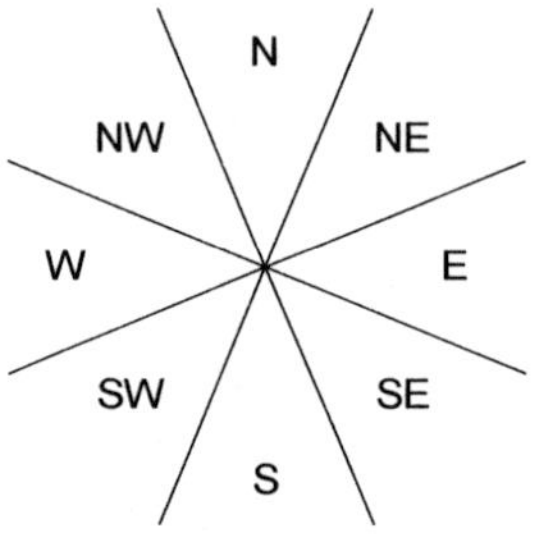

extended objects[2] (Sharma, 1996). It also allows for the definition of finer resolutions, thus permitting the use of eight (*Figure 3*) or 16 different qualitative directions. This model uses triangular acceptance areas that are drawn from the *centroid* of the reference object towards the primary object (in the spatial relation A North B, B represents the reference object, while A constitutes the primary object).

Distance Spatial Relations

Distances are quantitative values determined through measurements or calculated from known coordinates of two objects in some reference system. The frequently used definition of distance can be achieved using the Euclidean geometry and Cartesian coordinates. In a two-dimensional Cartesian system, it corresponds to the length of the shortest possible path (a straight line) between two objects, which is also known as the Euclidean distance (Hong, 1994). Usually a metric quantity is mapped onto some qualitative indicator such as very close or far for human common-sense reasoning (Hernández et al., 1995).

Qualitative distances must correspond to a range of quantitative values specified by an interval and they should be ordered so that comparisons are possible. The adoption of

Figure 4. Qualitative Distances Intervals

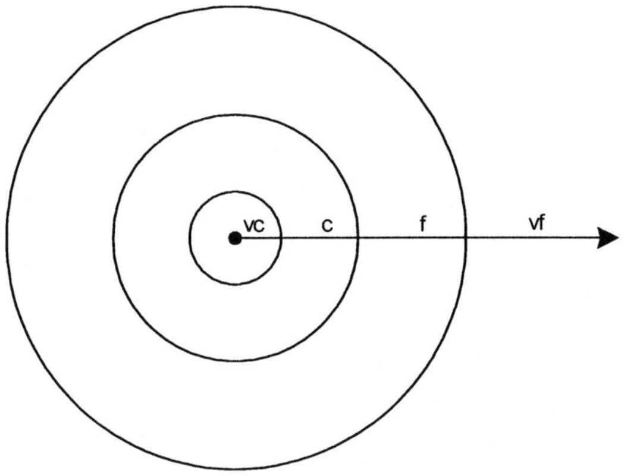

the qualitative distances very close – vc, close – c, far – f and very far – vf, intuitively describe distances from the nearest to the furthest. An order relationship exists among these relations, where a lower order (vc) relates to shorter quantitative distances and a higher order (vf) relates to longer quantitative distances (Hong, 1994). The length of each successive qualitative distance, in terms of quantitative values, should be greater or equal to the length of the previous one (*Figure 4*).

Topological Spatial Relations

Topological relations are those relationships that are invariant under continuous transformations of space such as rotation or scaling. There are eight topological relations that can exist between two planar regions without holes[3]: disjoint, contains, inside, equal, meet, covers, covered by and overlap (*Figure 5*). These relations can be defined considering intersections between the two regions, their boundaries and their complements (Egenhofer, 1994). These eight relations, which can exist between two spatial regions without holes, will be the exclusive focus of topological relations in this chapter.

Figure 5. Topological Spatial Relations

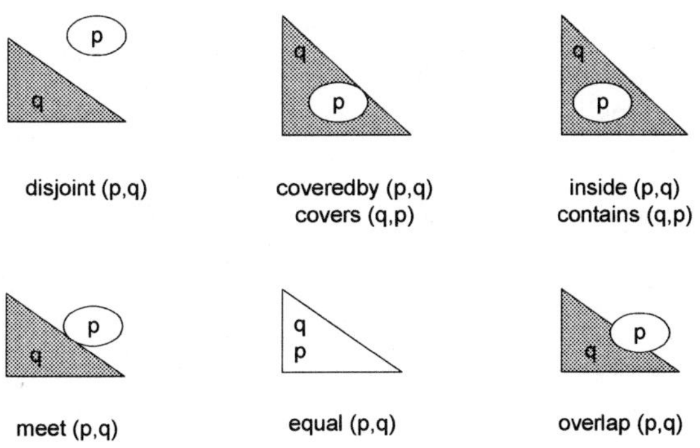

In some exceptional cases, the geographic space cannot be characterized, in topological terms, with reference to the eight topological primitives presented above. One of these cases is related with application domains in which the geographic regions addressed are administrative subdivisions. Administrative subdivisions, represented in this work by full planar graphs[4], can only be related through the topological primitives disjoint, meet and contains (and the corresponding inverse inside), since they cannot have any kind of overlapping. The topological primitives used in this chapter are disjoint and meet, since the implemented qualitative inference process only considers regions at the same geographic hierarchical level.

Integrated Spatial Reasoning

Integrated reasoning about qualitative directions necessarily involves qualitative distances and directions. Particularly in objects with extension, the size and shape of objects and the distance between them influence the directions. One of the ways to determine the direction and distance[5] between regions is to calculate them from the *centroids* of the regions. The extension of the geographic entities is somehow implicit in the topological primitive used to characterize their relationships.

Integration of Direction and Distance

An example of *integrated spatial reasoning* about qualitative distances and directions is as follows. The facts A is very far from B and B is very far from C do not facilitate the inference of the relationship that exists between A and C. A can be very close or close to C, or A may be far or very far from C, depending on the orientation between B and C.

For the integration of qualitative distances and directions the adoption of a set of identifiers is required, which allows for the identification of the considered directions and distances and their respective intervals of validity. Hong (1994) analyzed some possible combinations for the number of identifiers and the geometric patterns that should characterize the distance intervals. The *localization system* (*Figure 6*) suggested by Hong is based on eight symbols for direction relations (North, Northeast, East, Southeast, South, Southwest, West, Northwest) and four symbols for the identification of the distance relations (very close, close, far and very far).

In the case of direction relations, for the cone-shaped system with eight acceptance regions, the quantitative intervals adopted were: [337.5, 22.5), [22.5, 67.5), [67.5, 112.5), [112.5, 157.5), [157.5, 202.5), [202.5, 247.5), [247.5, 292.5), [292.5, 337.5) from North to Northwest respectively.

The definition of the validity interval for each distance identifier must obey some rules (Hong, 1994). In these systems, as can be seen in *Table 1*, there should exist a constant ratio (ratio = length ($dist_i$)/length ($dist_{i-1}$)) relationship between the lengths of two neighboring intervals. The presented simulated intervals allow for the definition of new distance intervals by magnification of the original intervals. For example, the set of values for ratio 4[6] can be increased by a factor of 10 supplying the values $dist_0$ (0, 10], $dist_1$

Figure 6. Integration of Direction and Distance Spatial Relations

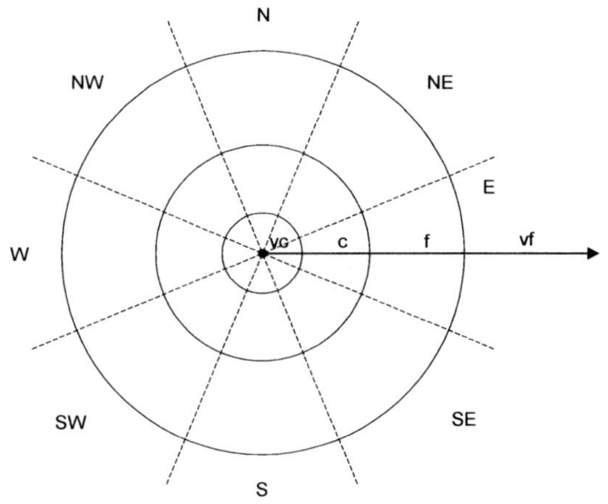

Table 1. Simulated Intervals for Four Symbolic Distance Values

Ratio	$dist_0$	$dist_1$	$dist_2$	$dist_3$
1	(0, 1]	(1, 2]	(2, 3]	(3, 4]
2	(0, 1]	(1, 3]	(3, 7]	(7, 15]
3	(0, 1]	(1, 4]	(4, 13]	(13, 40]
4	(0, 1]	(1, 5]	(5, 21]	(21, 85]
5	(0, 1]	(1, 6]	(6, 31]	(31, 156]
...	...	...	...	...

(10, 50], $dist_2$ (50, 210] and $dist_3$ (210, 850]. Since the same scale magnifies all intervals and quantitative distance relations, the qualitative compositions will remain the same, regardless of the scaled value.

It is important to know that the number of distance symbols used and the ratio between the quantitative values addressed by each interval play an important role in the robustness of the final system, i.e., in the validity of the composition table for the inference of new spatial relations (Hong, 1994).

The final composition table, a 32x32 matrix for the localization system adopted, was constructed following the suggestions made by Hong (1994) and it is presented in this work through an iconic representation (*Figure 7*). This matrix represents part of the knowledge needed for the inference of new spatial information in the localization system used. Due to its great size, *Figure 8* exhibits an extract of the final matrix. An example of the composition operation: suppose that A North, close B and that B Southeast, very close C. Consulting the composition table (this example is marked in *Figure 8* with two traced arrows) it is possible to identify the relation that exists between A and C: A North, close C. For the particular case of the composition of opposite directions with equal qualitative distances, the system is unable to identify the direction between the objects. For this

Figure 7. Graphical Representation of Direction and Distance Integration

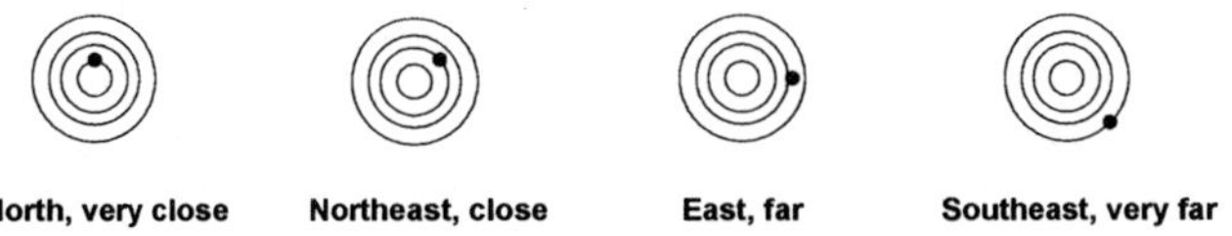

Figure 8. Extract of the Final Composition Table — Integration of Direction and Distance

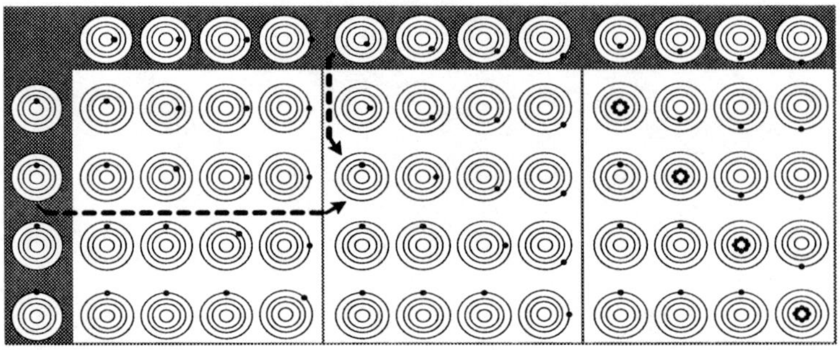

reason, the composition of these particular cases presents all the qualitative directions as possible results of the inference (*Figure 8*).

Integration of Direction and Topology

The relative position of two objects in the bi-dimensional space can be achieved through the dimension and orientation of the objects. Looking at each of these characteristics separately implies two classes of spatial relations: *topological*, which ignores orientations in space; and *direction* that ignores the extension of the objects.

The integration of these two kinds of spatial relations enables the definition of a system for qualitative spatial reasoning that describes the relative position existing between the objects and how the limits (frontiers) of them are related.

Sharma (1996) integrated direction and topological spatial relations using the principles of qualitative temporal reasoning defined by Allen (1983). The approach undertaken by Sharma (1996) was possible through the adaptation of the temporal principles to the spatial domain. The 13 temporal primitives (Allen, 1983) are: before, after, during, contain, overlap, overlapped by, meet, met by, start, started by, finish, finished by and equal (*Figure 9*).

The temporal primitives (that are one-dimensional) were analyzed by Sharma (1996) along two dimensions (axes *xx* and *yy*) allowing their use in the spatial domain (restricted in this case to a two-dimensional space).

The construction of the composition tables was facilitated by the knowledge representation framework adopted for the integration of direction and topology. Topological

Figure 9. Temporal Primitives

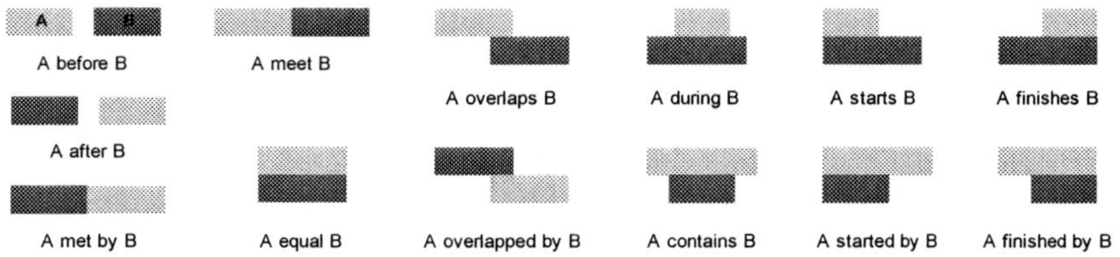

Adapted from Allen (1983, p. 835)

relations are independent of the order existing between the objects when analyzed along a given axis. Direction relations depend on the order and are defined by verifying the objects position along a specific axis.

The representation of each pair (direction, topology) is accomplished through temporal primitives. The transformation of the one-dimensional characteristics to the two-dimensional space is achieved analyzing the pair of temporal primitives that represent the behavior of the pair (direction, topology) along x and y (*Figure 10* supplies three examples of selection of the appropriate pair of temporal primitives, verifying the position of A and B along x and along y, for the characterization of the pair (direction, topology)).

Restricting the integration domain to objects that represent administrative subdivisions without overlap between them, the two topological relations considered were disjoint and meet. These two topological relationships can be represented by the temporal primitives before and meet, and by the corresponding inverses (after and met by). Attending to the direction relations, all the temporal primitives defined by Allen (1983) can be used in their characterization. *Figure 11* shows how the temporal primitives are used in the definition of a particular direction relation.

For the identification of the inference rules it is necessary to identify the temporal primitives that characterize each pair (direction, topology) and then do their composition

Figure 10. Integration of Direction and Topological Relations

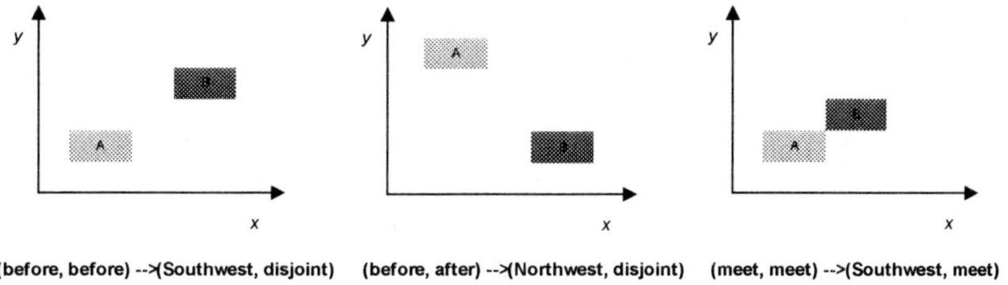

Figure 11. Interval Relations for Direction Relations Representation

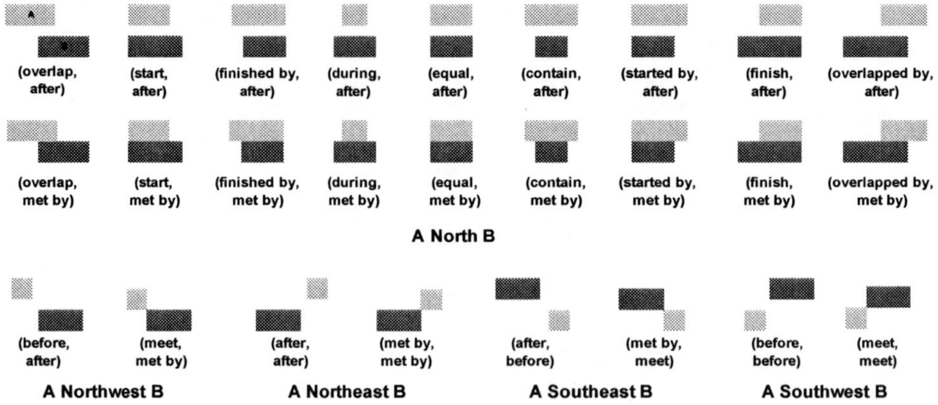

Adapted from Sharma (1996, p. 83)

Table 2. An Extract of the Composition Table for Temporal Intervals

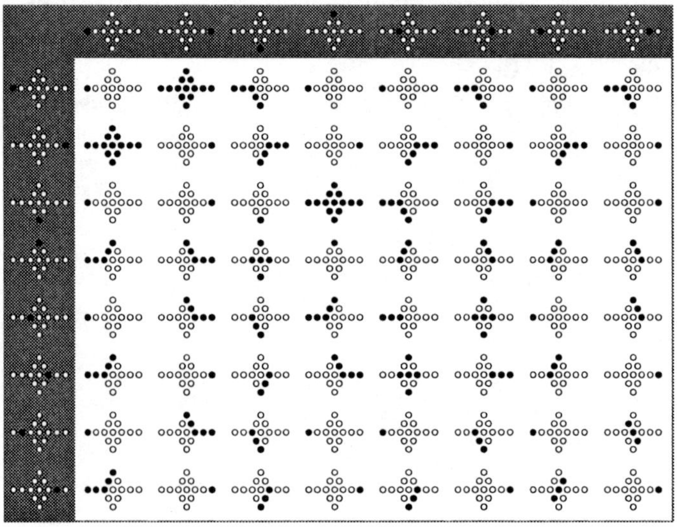

Adapted from Allen (1983, p. 836)

Figure 12. Temporal Relations — Graphical Representation

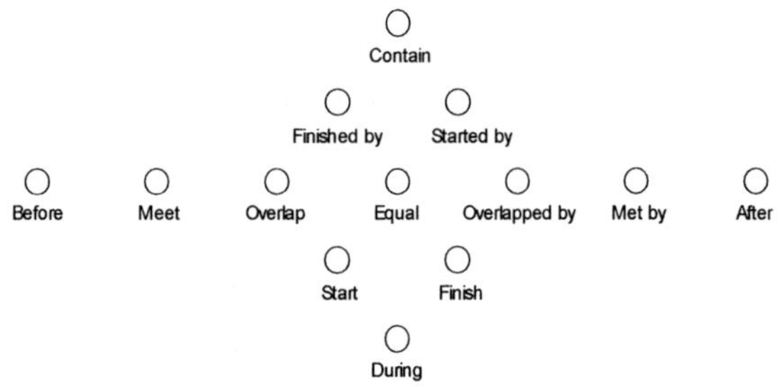

Notation suggested by Sharma (1996)

to achieve the result. *Table 2* presents an extract of the composition table for the temporal domain. This table, graphically presented using the notation showed in *Figure 12*, will be afterwards used for the spatial domain.

The composition of pairs of relations (direction, topology) is performed consulting *Table 2*. An example of the composition[7] operation for the spatial domain is the composition of the pair (Northeast, disjoint) with the pair (Northeast, disjoint). The result of the composition is achieved by the steps:

(Northeast, disjoint) ; (Northeast, disjoint)	=	*(after, after) ; (after, after)*
	=	*(after; after) x (after; after)*
	=	*(after) x (after)*
	=	*(after, after)*
	=	*(Northeast, disjoint)*

Following this composition process, Sharma obtained the several composition tables that integrate direction with the several topological pairs disjoint;disjoint, disjoint;meet, meet;disjoint and meet;meet. *Figure 13* presents the graphical symbols used in this chapter to represent the integration of direction and topology. *Table 3* shows one of the composition tables of Sharma, integrating direction with the topological pair disjoint;disjoint.

Figure 13. Graphical Representation of Direction and Topological Spatial Relations

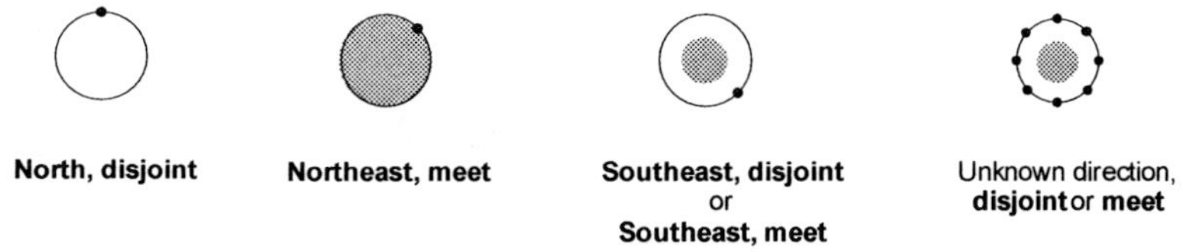

Table 3. Composition Table for the Integration of Direction with the Topological Pair disjoint;disjoint

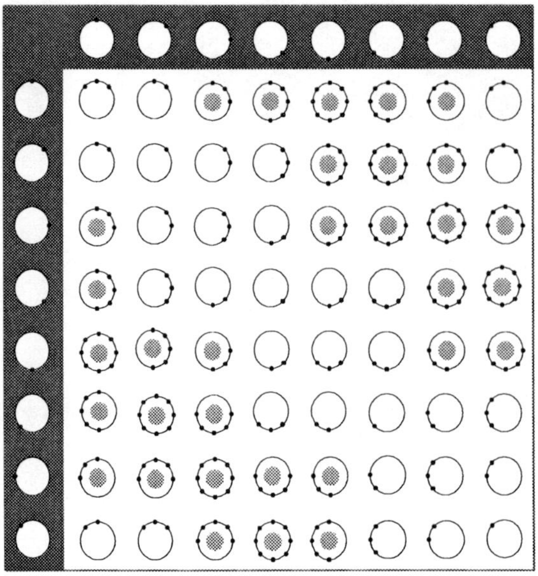

Adapted from Sharma (1996, p. 117)

Integration of Direction, Distance and Topological Spatial Relations

With the integration of direction and distance spatial relations a set of inference rules were obtained. These rules present a unique pair (direction, distance) as outcome, with the exception of the result of the composition of pairs with opposite directions and equal qualitative distances. In the integration of direction and topological spatial relations some improvements can be achieved, since several inference rules present as the result a set of outcomes.

Looking at the work developed by Hong and Sharma it was evident that the integration of the three types of spatial relations, direction, distance and topology, would lead to more accurate composition tables.

Since Hong adopted a cone-shaped system in the definition of the direction relations, and Sharma used a projection-based system for the same task, the integration of the three types of spatial relations was preceded by the adaptation[8] of the principles used by Sharma and the construction of new composition tables for the integration of direction and topology.

In the characterization of the integration of direction and topological relations, for the particular case of administrative subdivisions, new temporal pairs were defined, which allowed for the identification of new inference rules. *Figure 14* shows the several pairs of temporal primitives adopted according to the direction relations and the topological primitives disjoint and meet.

Figure 14. Temporal Intervals for the Characterization of Direction and Topology for Administrative Subdivisions

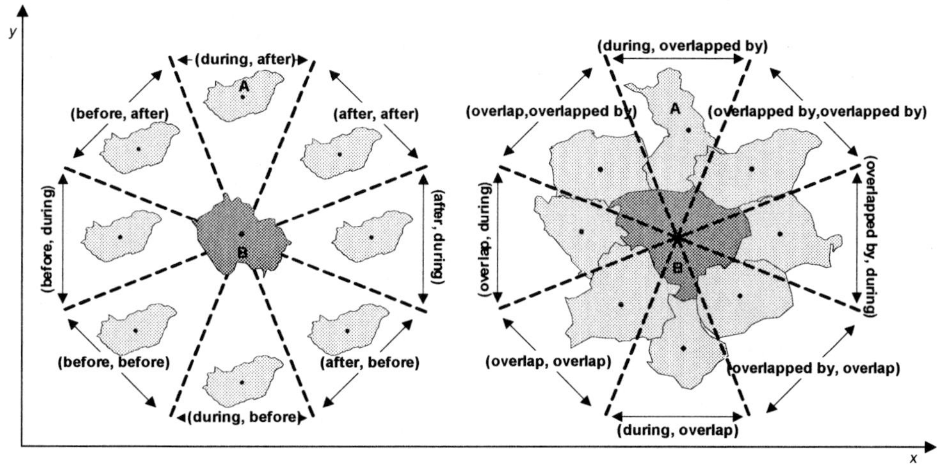

The adoption of the temporal intervals shown in *Figure 14* was motivated by the fact that administrative subdivisions have irregular limits, which impose several difficulties in the identification of the *correct* direction between two regions. Sometimes the *centroid* is positioned in a place that suggests one direction, although the administrative region may have parts of its territory at other acceptance areas in the cone-shaped system. The adoption of the during temporal primitive for the characterization of North, East, South and West directions was motivated by the assumption that the *centroid* of the primary object is located in the zone of acceptance for those directions, as defined by the reference object.

In the case of adjacency it is clear by an analysis of *Figure 14* that some overlapping between the regions can exist, when analyzed in a temporal perspective. This fact influenced the adoption of the overlap and overlapped by primitives instead of the meet and met by primitives adopted by Sharma.

Following the assumptions described above new composition tables were constructed. *Table 4* shows the particular case of integration of direction with the topological pair disjoint;disjoint. The other composition tables, for the topological pairs disjoint;meet, meet;disjoint and meet;meet, are available in Santos (2001).

After the identification of the composition tables that integrate direction and topology under the principles of the cone-shaped system, it was possible to integrate these tables with the composition table proposed by Hong (1994), with respect to direction and distance. This step was preceded by a detailed analysis of the application domain in which the system will be used, in particular the composition of regions that represent administrative subdivisions that cover all the territory considered, without any gap or overlap (Santos, 2001). Concerning the distance spatial relation, it was defined that the qualitative distance very close is restricted to adjacent regions. When the qualitative distance is close the regions may be, or may not be, adjacent. The far and very far qualitative distances can only exist between regions that are disjoint from each other.

Table 4. Composition Table for the Integration of Direction with the Topological Pair disjoint;disjoint *(particular case of administrative subdivisions)*

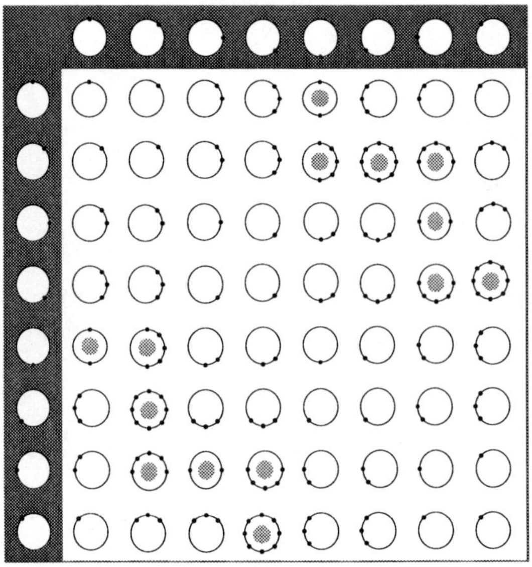

The basic assumption for the integration process was that the outcome direction in the integration of direction and distance is the same outcome direction in the integration of direction and topology, or it belongs to the set of possible directions inferred by the last one. The direction that guides the integration process is the direction suggested by the composition table of direction and distance (it is more accurate since it considers the distance existing between the objects).

The final composition table, which is shown with the graphical symbols expressed in *Figure 15*, was obtained through an integration process that is diagrammatically demonstrated in *Figure 16*. For example, the composition of (North, very close) with (North, very close) has as result (North, very close). The composition of (North, meet) with (North, meet) has as the result (North, disjoint or meet). The integration of the three spatial relations leads to (North, very close, disjoint or meet). As the qualitative distance relation very close was restricted to adjacent regions, the result of the integration is (North, very close, meet). Another example explicit in *Figure 16* is the integration of the result of (North, close);(North, close) with (North, disjoint);(North, disjoint). The result of the first composition is (North, far)

Figure 15. Graphical Representation of Direction, Distance and Topological Spatial Relations

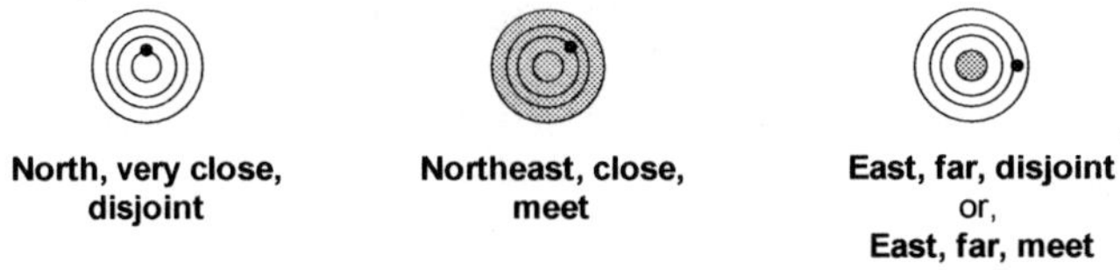

Figure 16. Integration of Direction, Distance and Topological Spatial Relations

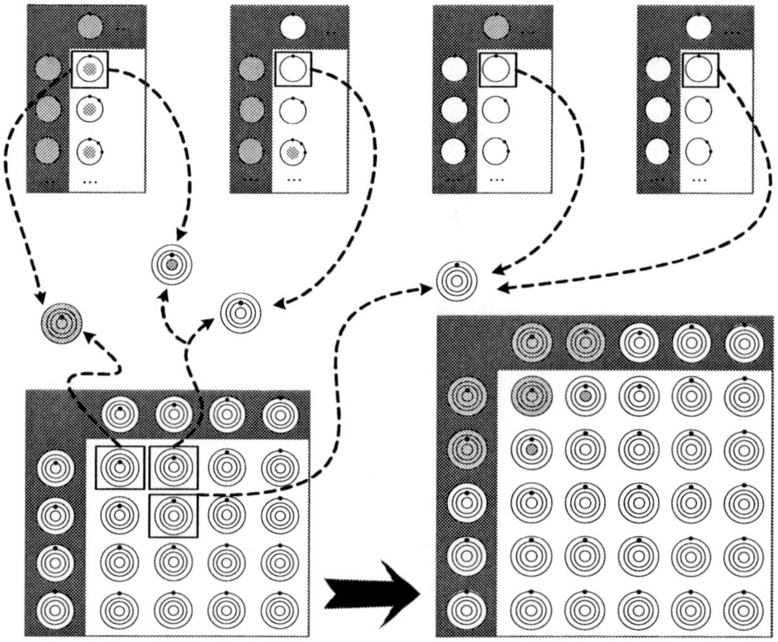

Figure 17. Influence of the Regions Dimension in the Inference Result

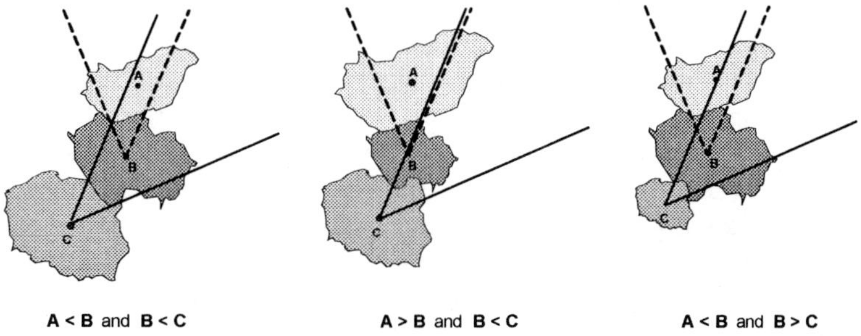

while the result of the second is (North, disjoint). The integration generates the value (North, far, disjoint), which matches the principles adopted in this work for the distance relation: if the regions are far from each other, then topologically they are disjoint.

In the evaluation of the composition table constructed it was realized that the dimensions of the regions influenced (sometimes negatively) the results achieved. Qualitative reasoning with administrative subdivisions is a difficult task, which is influenced not only by the irregular limits of the regions but also by their size. As can be noted in *Figure 17*[9], if the dimension of A is less than the dimension of B, and the dimension of B is less than the dimension of C, then the inference result must be A Northeast C. But if the dimension of A is greater than the dimension of B and the dimension of B is less than the dimension of C, then the inference result must be A North C. A detailed analysis of these

Figure 18. Municipalities of the Braga District

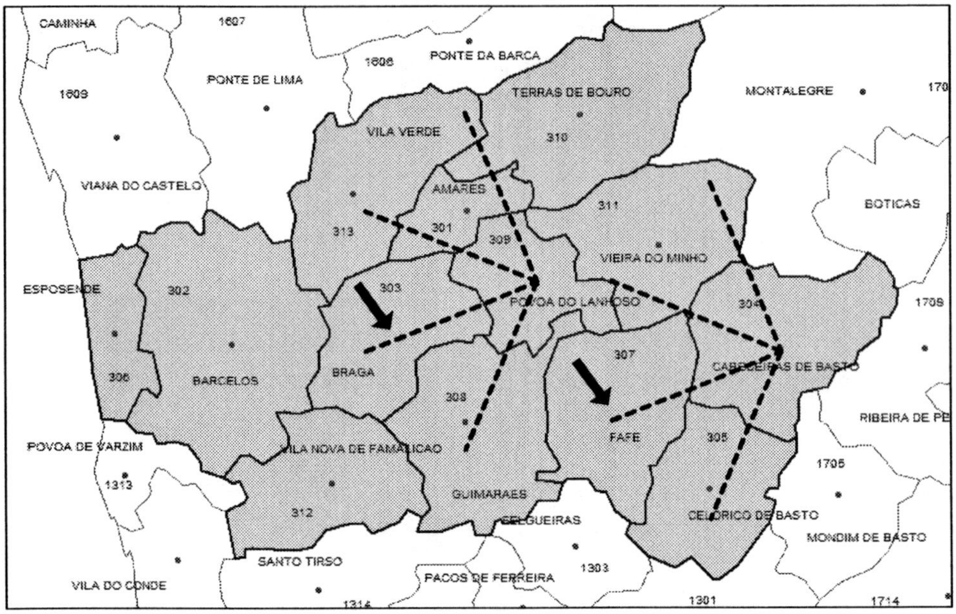

situations was undertaken, allowing the identification of several rules that integrate the dimensions of the regions in the qualitative reasoning process of the PADRÃO system. Through this process, the reasoning process was improved, and more accurate inferences were obtained.

The performance of the qualitative reasoning system was evaluated (Santos, 2001). The approach followed in this performance test was to compare the spatial relations obtained through the qualitative inference process with the spatial relations obtained by quantitative methods. A Visual Basic module was implemented for the execution of this task. This module calculated quantitatively all the spatial relations existing between the Municipalities of three districts of Portugal, looking at the position of the respective *centroids*. This information was stored in a table and compared with the spatial relations inferred qualitatively. The results achieved were, in the worst scenario, exact[10] for 75% of the inferences obtained in Districts with higher differences between the dimensions of their regions (two of the analyzed Districts). For the Braga District, a District that integrates regions with homogeneous dimensions, the inferences obtained were 88% exact for direction and 81% exact for distance. For topology, the inferences were in all cases 100% exact. The approximate inferences obtained were verified in regions that have parts of their territory in more than one acceptance area for the direction relation. For these cases, the *centroid* of the region is sometimes positioned in one acceptance area, although the region has parts of its territory in other acceptance areas. Another situation, as shown in *Figure 18* for two Municipalities, is verified when the *centroid* is positioned in the line that divides the acceptance areas, which makes even more difficult the identification of the direction between the regions and, as a consequence, the qualitative reasoning process.

After the evaluation of the qualitative reasoning system implemented and the analysis of the inferences obtained, which provided a good approximation to the reality, the system will later be used in the knowledge discovery process.

The PADRÃO System

PADRÃO is a system for knowledge discovery in geo-referenced databases based on qualitative spatial reasoning. This section presents its architecture, gives some technical details about its implementation and tests the system in a geo-referenced data set.

Architecture of PADRÃO

The architecture of PADRÃO (*Figure 19*) aggregates three main components: Knowledge and Data Repository, Data Analysis and Results Visualization. The Knowledge and Data Repository component stores the data and knowledge needed in the knowledge discovery process. This process is implemented in the Data Analysis component, which allows for the discovery of patterns or other relationships implicit in the analyzed geo-spatial and non-spatial data. The discovered patterns can be visualized in a map using the Results Visualization component. These components are described below.

Figure 19. Architecture of PADRÃO

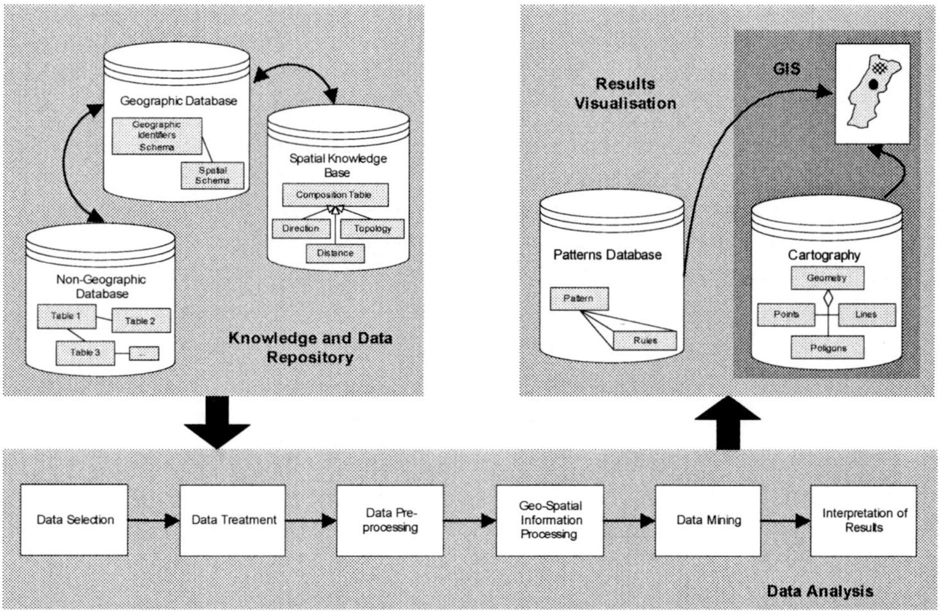

The Knowledge and Data Repository component integrates three central databases:

1. A *Geographic Database* (GDB) constructed under the principles established by the European Committee for Normalization in the CEN TC 287 pre-standard for Geographic Information. Following the pre-standard recommendations it was possible to implement a GDB in which the positional aspects of geographic data are provided by a *geographic identifiers system* (CEN/TC-287, 1998). This system characterizes the administrative subdivisions of Portugal at the municipality and district level. Also it includes a geographic gazetteer containing the several geographic identifiers used and the concept hierarchies existing between them. The geographic identifiers system was integrated with a *spatial schema* (CEN/TC-287, 1996) allowing for the definition of the *direction*, *distance* and *topological* spatial relations that exist between adjacent regions at the Municipality level.
2. A *Spatial Knowledge Base* (SKB) that stores the qualitative rules needed in the inference of new spatial relations. The knowledge available in this database aggregates the constructed composition table (integrating direction, distance and topological spatial relations), the set of identifiers used, and the several rules that incorporate the dimension of the regions in the reasoning process. This knowledge base is used in conjunction with the GDB in the inference of unknown spatial relations.
3. A *non-Geographic Database* (nGDB) that is integrated with the GDB and analyzed in the Data Analysis component. This procedure enables the discovery of implicit relationships that exist between the geo-spatial and non-spatial data analyzed.

The Data Analysis component is characterized by six main steps. The five steps presented above for the knowledge discovery process plus the Geo-Spatial Information Processing step. This step verifies if the geo-spatial information needed is available in the GDB. In many situations the spatial relations are implicit due to the properties of the spatial schema implemented. In those cases, and to ensure that all geo-spatial knowledge is available for the data mining algorithms, the implicit relations are transformed into explicit relations through the inference rules stored in the SKB.

The Results Visualization component is responsible for the management of the discovered patterns and their visualization in a map (if required by the user and when the geometry[11] of the analyzed region is available). For that PADRÃO uses a Geographic Information System (GIS), which integrates the discovered patterns with the geometry of the region. This component aggregates two main databases:

1. The *Patterns Database* (PDB) that stores all relevant discoveries. In this database each discovery is catalogued and associated with the set of rules that represents the discoveries made in a given data mining task.
2. A *Cartographic Database* (CDB) containing the cartography of the region. It aggregates a set of points, lines and polygons with the geometry of the geographical objects.

Implementation of PADRÃO

PADRÃO was implemented using the relational database system Microsoft Access, the knowledge discovery tool Clementine (SPSS, 1999), and Geomedia Professional (Intergraph, 1999), the GIS used for the graphical representation of results.

The databases that integrate the Knowledge and Data Repository and the Results Visualization components were implemented in Access. The data stored in them are available to the Data Analysis component or from it, through ODBC (Open Database Connectivity) connections.

Clementine is a data mining toolkit based on visual programming[12], which includes machine learning technologies like rule induction, neural networks, association rules discovery and clustering. The knowledge discovery process is defined in Clementine through the construction of a *stream* in which each operation on data is represented by a *node*.

The workspace of Clementine comprises three main areas. The main work area, the Stream Pane, constitutes the area for the streams construction. The palettes area in which the several available icons are grouped according to their functions: links to sources of information, operations on data (rows or columns), visual facilities and modeling techniques (data mining algorithms). The models area stores the several models generated in a specific stream. These models can be directly re-used in other streams or they can

Figure 20. Workspace of Clementine

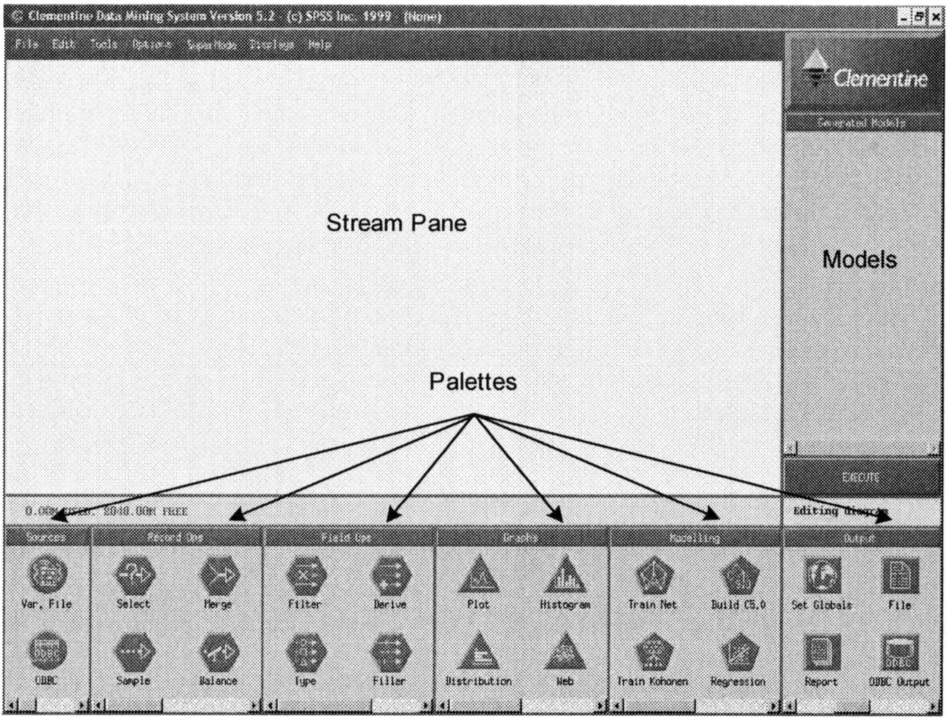

be saved providing for their later use. *Figure 20* shows the work environment of Clementine and presents some of the several nodes available according to their functionality. Circular nodes represent links to data sources and constitute the first node of any stream. Nodes with a hexagonal shape are for data manipulation, including operations on records (lines of a table) or operations on fields (columns of a table). Triangular nodes allow for data exploration and visualization, providing a set of graphs that can be used to get a better understanding of data. Nodes with a pentagonal shape are modeling nodes, i.e., data mining algorithms that can be used to identify patterns in data. The last group of square-shaped nodes is related to the output functions, which make available a set of nodes for reporting, storing or exporting data.

The Data Analysis component of PADRÃO is based on the construction of several streams that implement the knowledge discovery process. The several models obtained in the data mining phase represent knowledge about the analyzed data and can be saved or reused in other streams. In PADRÃO, these models can be exported through an ODBC connection to the PDB. The integration of the PDB with the CDB allows the visualization of the rules explicit in the models in a map. The visualization is achieved through the VisualPadrão application, a module implemented in Visual Basic. VisualPadrão manipulates the library of objects available in Geomedia. This application was integrated in the Clementine workspace using a *specification file*, i.e., a mechanism provided by the Clementine system that allows for the integration of new capabilities in its environment. This approach provides an integrated workspace in which all tasks associated with the knowledge discovery process can be executed.

Analysis of a Geo-Referenced Database

Several datasets have been analyzed by the PADRÃO system. Among them are demographic databases storing the Parish records of several Municipalities of Portugal (Santos & Amaral, 2000a; Santos & Amaral, 2000b; Santos & Amaral, 2000c). Another dataset analyzed was a component of the Portuguese Army Database (Santos & Ramos, 2003). The several data mining objectives defined allowed for the identification of the implicit relationships existing between the geo-spatial data and non-spatial data.

The dataset selected for description in this chapter integrates data from a financial institution, which supplies credit for the acquisition of several types of goods. To overcome confidentiality issues with the data and the several identifiable patterns, the data was manipulated in order to create a random data set. Through this process the confidentially is ensured and the knowledge discovery process in the PADRÃO system can be described.

The bank database aggregates a set of 3,031 records that characterize the behavior of the bank clients. The following data mining objective was defined: "*identify the profile of the clients in order to minimize the institutional risk of investment.*" This profile will be identified for the Braga District of Portugal.

The knowledge discovery process is preceded by the business understanding phase in which the meaning and importance of each attribute for this process is evaluated. The attributes integrated in the database are: identification number (ID), VAT number (VAT_number), client title (Title), name (Name), good purchased (Acquisition), contract

Figure 21. Data Exploration

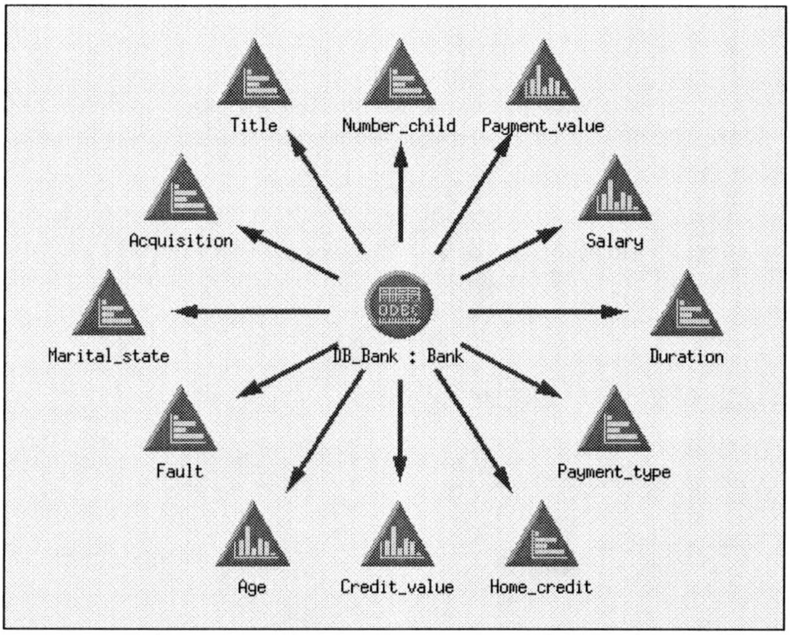

Figure 22. Distribution of Categorical Data

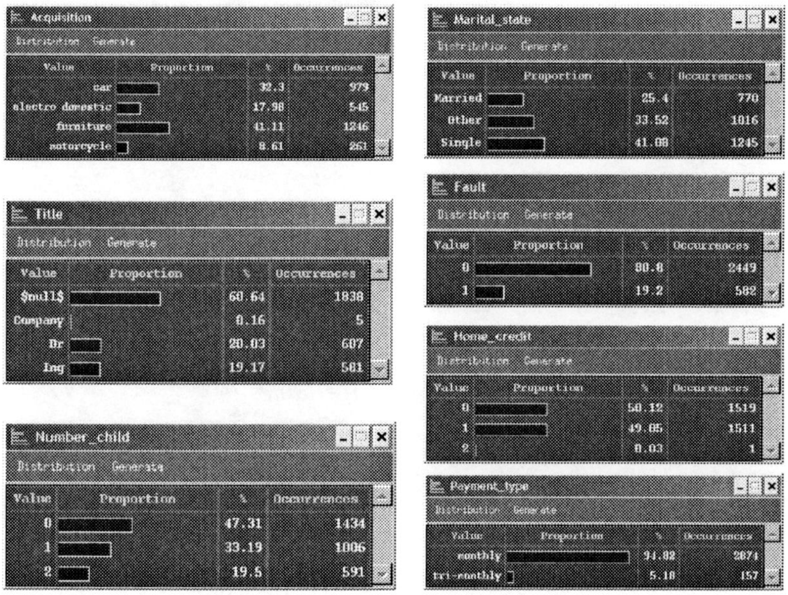

duration (Duration), income (Salary), overall value of credit (Credit_value), payment type (Payment_type), credit for home acquisition (Home_credit), lending value (Payment_value), marital state (Marital_state), number of children (Number_child), age (Age) and the accomplishment or not of the credit (Fault).

At this phase Distribution and Histogram nodes of Clementine were used to explore the several attributes, identify their values, and distribution, and determine if any of them present anomalies. *Figure 21* shows the stream constructed for this exploration phase.

The results obtained by each Distribution[13] graph are showed in *Figure 22*. It can be seen that the majority of attributes present a distribution of values that are the normal operation of the organization. However, exceptions were verified for the Home_credit and Title attributes. Namely:

- The attribute Home_credit, which shows if the client has or does not have a credit for home acquisition (values 1 and 0 respectively), also includes a record with the 2 value. As this value constitute an error, the respective record must be removed from the dataset;

Figure 23. Distribution of Continuous Values

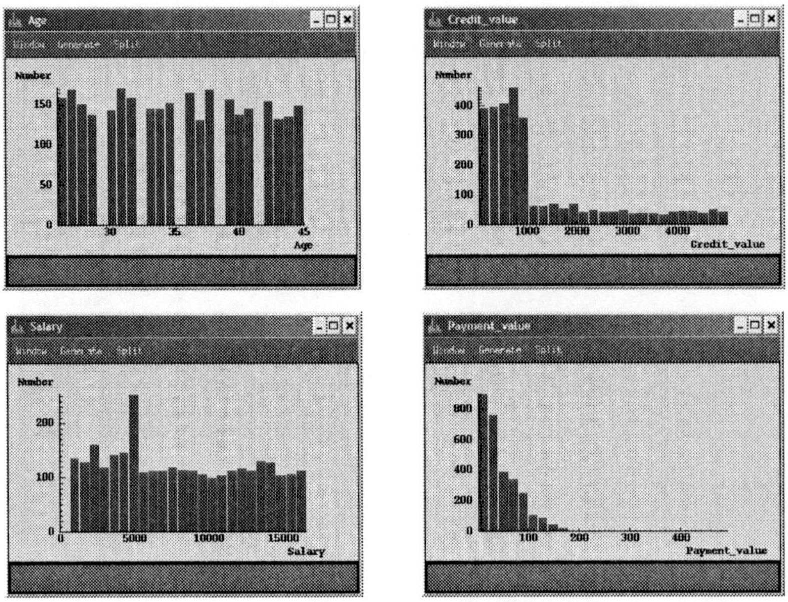

Table 5. Classes for Attributes with Continuous Values

Attributes	Classes
Age	(25..31] → '26-31', (31..38] → '32-38', (38..45] → '39-45'
Credit_value	(0..350] → '0-350', (350..650] → '351-650', (650..900] → '651-900', (900..2500] → '901-2500', (2500..5000] → '2501-5000'
Salary	(0..4500] → '0-4500', (4500..8000] → '4501-8000', (8000..12500] → '8001-12500', (12500..17000] → '12501-17000'
Payment_value	(0..17] → '0-17', (17..30] → '18-30', (30..50] → '31-50', (50..80] → '51-80', (80..500] → '81-500'

- The Title attribute integrates five cases of credit for organizations (value Company). As a result, these records must be removed from the database since they represent a minority class[14] in the overall set.

Figure 23 shows the Histograms with the distribution of attributes with continuous values. The analysis of the distributions allows for the verification of the several classes that will be created in order to transform continuous values into discreet values. The defined classes are presented in *Table 5*. Their definition is based on the assumption that the data available for analysis must be distributed homogeneously across the several classes.

This exercise of exploration and comprehension of the available data allowed the identification of the attributes for analysis and the definition of the several classes that will be used in the pre-processing step, i.e., to transform continuous values into discreet values. Next, the six steps considered in the PADRÃO system for the knowledge discovery process (Data Selection, Data Treatment, Data Pre-processing, Geo-spatial Information Processing, Data Mining and Interpretation of Results) are described.

Data Selection and Data Treatment

The data selection step allows for the exclusion of attributes that have no influence in the knowledge discovery process. Among them are ID, VAT_number, Title and Name, since they only have an informative role. The other attributes will be considered in order to evaluate the contribution of each one to the definition of the profile of the clients.

Figure 24 shows the stream constructed for the data selection and data treatment steps. The stream integrates a source node (DB_Bank:Bank) that makes the data available to the knowledge discovery process through an ODBC connection. The select node discards

Figure 24. Data Selection and Data Treatment Steps

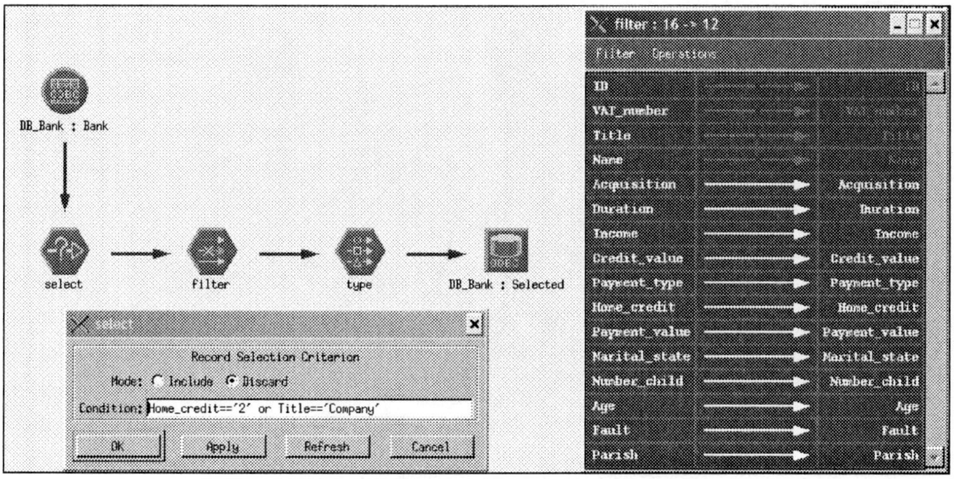

records with anomalies. As previously mentioned, the record with the value 2 in the Home_credit attribute must be deleted. All records associated with the value Company in the Title attribute also need to be removed. The filter node is used to select the attributes that will be excluded from the process. The type node allows for the specification of the data type (numeric, character ...) of the attributes that will be exported to the database. As result of the several tasks undertaken, a new table (DB_Bank:SelectedData) is created in the bank database.

Data Pre-Processing

The data pre-processing step (*Figure 25*) allows for the transformation of the attributes with continuous values into attributes with discreet values (nodes SalaryClass, CreditClass, PaymentClass and AgeClass), according to the classes presented in *Table 5*. In this step, web nodes, exploration graphs available in Clementine, are also used for the identification of associations[15] among the analyzed attributes (nodes Acquisition x Fault, SalaryClass x AgeClass x Fault, Marital_state x Number_child x Fault and PaymentClass x Fault). The last task undertaken is associated with the creation of the two datasets (nodes DB_Bank:Training and DB_Bank:Test) that will be used from now on. They are the Training and the Test datasets, and in which the original data is randomly distributed. The Training file is used in the model construction (data mining step) while the Test dataset evaluates the model confidence when applied to unknown data.

The web nodes constructed are shown in *Figure 26*. They combine several attributes and through the analysis of them it is possible to identify associations between attributes. Strong associations between attributes are represented by bold lines, while weak associations are symbolized by dotted lines. For the several acquisitions that can be effected, *Figure 26a* points out that no association exists between the good furniture and the value 1 of the Fault attribute, indicating that faults were not usual with the credit

Figure 25. Data Pre-Processing Step

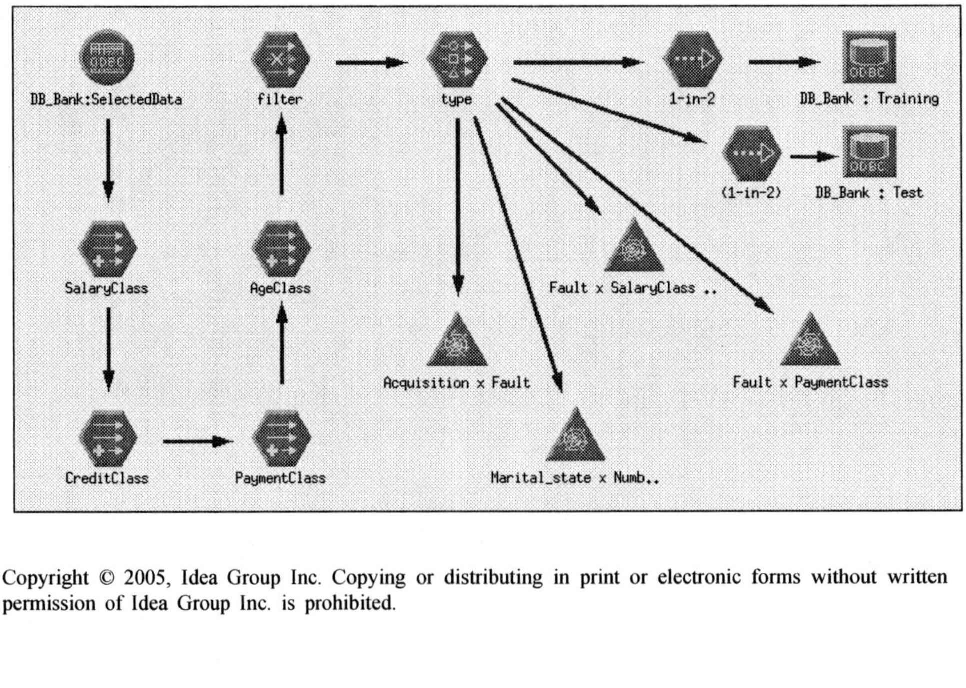

Figure 26. Data Exploration with Web Nodes

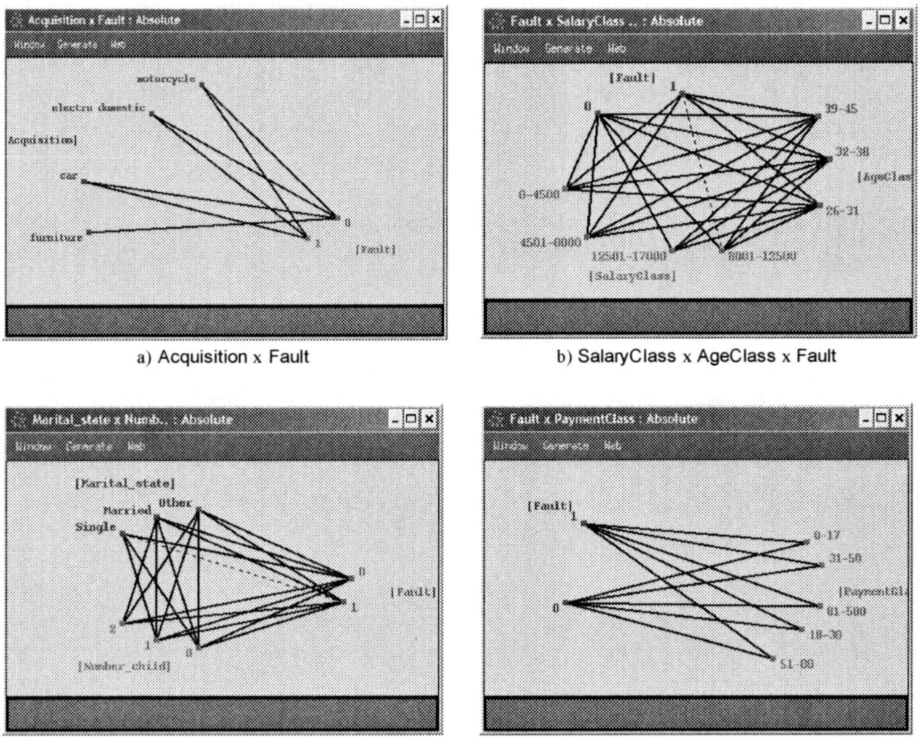

a) Acquisition x Fault

b) SalaryClass x AgeClass x Fault

c) Marital_state x Number_child x Fault

d) PaymentClass x Fault

supplied for this specific acquisition. Analyzing the income and age attributes with Fault in *Figure 26b*, it is evident that individuals with a higher income honor their payments, since the value 12501-17000 of the SalaryClass attribute presents no association with value 1 of Fault. Between value 8001-12500 and value 1 of Fault there exists a weak association, which indicates that this specific group may or may not be able to honor its credit payments. Similarly, a weak association is verified between the marital state Single and value 1 of Fault, *Figure 26c*. PaymentClass and Fault present strong connections between all attribute values as seen in *Figure 26d*, thus indicating that all type of payment values are associated with *good* and *bad* clients.

Geo-Spatial Information Processing

As the GDB only stores spatial relations for adjacent regions and, as it is necessary to analyze if the geographical component has any influence in the identification of the profile of the clients, all the other relationships that exist between non-adjacent regions and needed in the data mining step will be inferred. In Clementine, a rule induction[16] algorithm is able to learn the inference rules available in the composition table stored in the SKB. That enables the inference of new spatial relations.

Figure 27. Geo-Spatial Information Processing Step

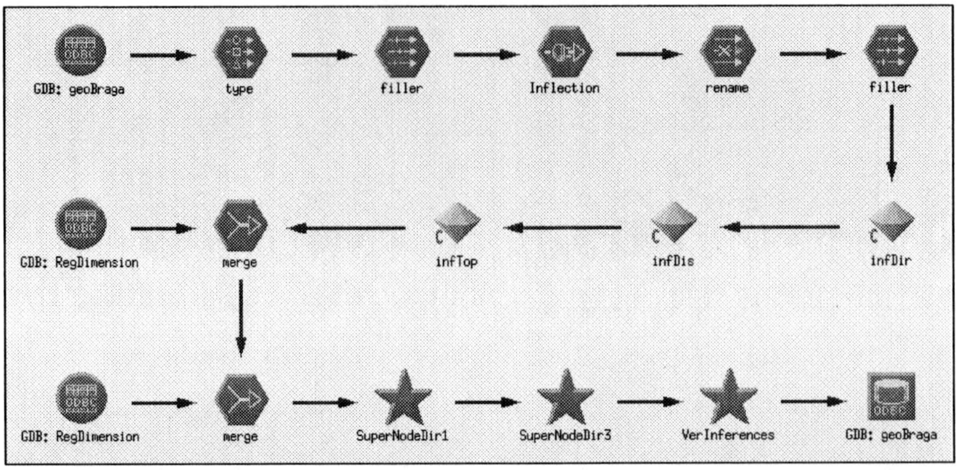

The models created, nodes infDir, infDis and infTop, can now be used in the inference process. With these models and as shown in *Figure 27* it is possible to infer the unknown spatial relationships existing in the Municipalities of the Braga District. The spatial relations for adjacent regions stored in the GBD are gathered through the source node (GDB:geoBraga) of the stream and combined (node Inflection) in order to obtain new associations between regions. The spatial relations existing among these new associations are identified by the models infDir, infDis and infTop. After the inferential process, the knowledge obtained is recorded in the GDB (output node GDB:geoBraga). In the stream of *Figure 27*, the super nodes SuperNodeDir1 and SuperNodeDir3 are responsible for the integration of the dimension of the regions in the reasoning process. In this process, there is validation if the several inferences obtained for a particular region agree independently of the composed regions. Several paths can be followed in order to infer a specific spatial relation. For example, knowing the facts A North B, A East D, B East C and D North C, the direction relation existing between A and C may be obtained composing A North B with B East C or combining A East D with D North C. If several compositions can be effected and if the results obtained from each one do not match, then the super node VerInferences excludes those results from the set of accepted ones.

Data Mining

In the data mining step (*Figure 28*) an appropriate algorithm is selected to carry out a specific data mining task. Three different tasks were undertaken (see *Figure 28*). First, a decision tree (node Fault_NG) that characterizes the profile of the clients without considering the location of the clients was generated. Second, the training set (DB_Bank:Training) was integrated with the spatial relations for the District in analysis (GDB:GeoBraga) in order to include the geographical component in the analysis of the profile of the clients (node Fault_G). Third, the geographical model of the District was created. This latter model (Direction) indicates the direction of each Municipality in the

Figure 28. Data Mining Step

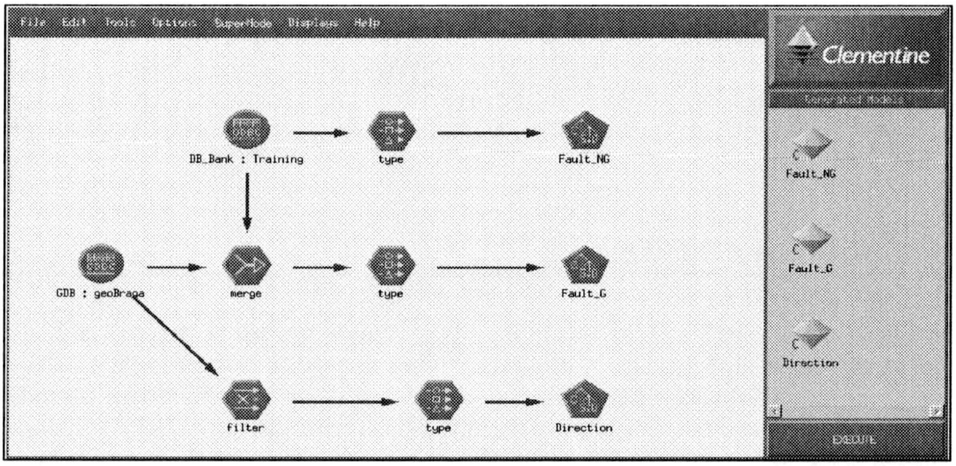

Figure 29. Generated Models for the Profile of the Clients

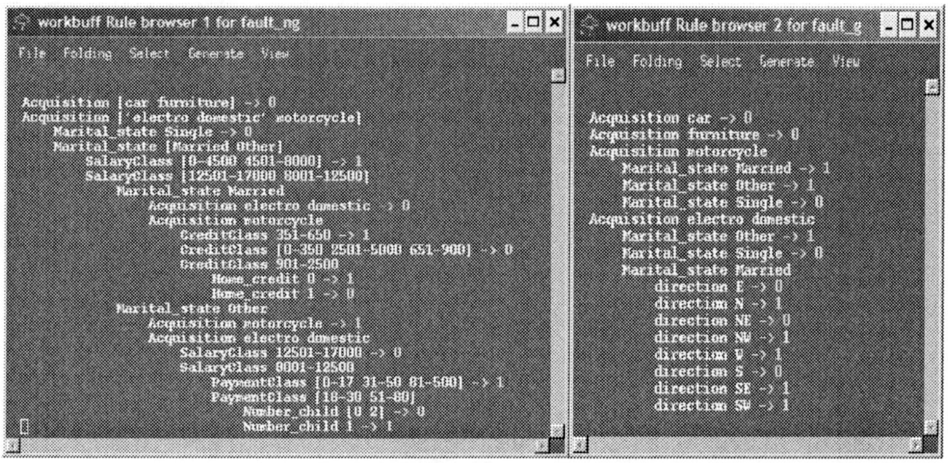

District and was obtained by analyzing the spatial relations inferred in the geo-spatial information processing step. All models were obtained with the C5.0 algorithm that allows for the induction of decision trees. *Figure 28* highlights the stream constructed for the generation of the three models. These models are available in the Generated Models palette and have the shape of a diamond (right hand side of *Figure 28*).

The Fault_NG model (*Figure 29*, left side) integrates a set of rules that are represented in a decision tree, which characterizes the profile of the clients. Through the analysis of the model it is possible to verify that the acquisition of car and furniture is traditionally associated with clients that honor their payments, while the acquisition of electro domestic and motorcycle have other attributes (Marital_state, Salary_class ...) that influence the profile of the clients. One explicit rule in the model for clients that the institution has no interest in supplying with credit, is: IF SalaryClass = '12501-17000' and Marital_state= 'Married' and Acquisition = 'motorcycle' and CreditClass = '351-600' THEN 1. The Fault_G model (*Figure 29*, right hand side) allows for the verification of geographic zones that have associated clients with a higher incidence of faults in credit payments. These zones are represented in directions, which partition the District into eight areas. The analysis of the model points out that Northeast (NE), East (E) and South (S) are associated with clients that pay the credit assumed.

Interpretation of Results

The Test set (DB_Bank:Test) is used in the interpretation of results step to verify the confidence of the models built in the Data Mining step. With respect to the Fault_NG model, *Figure 30* shows a percentage of confidence of 94.18%. The Fault_G model presents a percentage of confidence of 93.26%. This decrease in the model confidence, when considering the geographical component, may be caused by the aggregation of Municipalities into eight regions (the Cardinal directions), which represents a loss of specificity in favor of generality. Although the Direction model was obtained through the analysis of spatial relations inferred by qualitative rules, the results obtained in the Fault_G model maintain a high level of confidence.

The PADRÃO system permits the visualization of the results of the knowledge discovery process on a map. In this system the several rules that integrate a model are recorded in the PDB (Santos & Amaral, 2000b). At the same time the user has the option to run the VisualPadrão tool and visualize the desired model (*Figure 31*). As can be noted in the figure, Municipalities located at Northeast, East and South of the District contain clients mainly associated with no faults in their credit payments (information explicit in the

Figure 30. Percentage of Confidence of the Generated Models

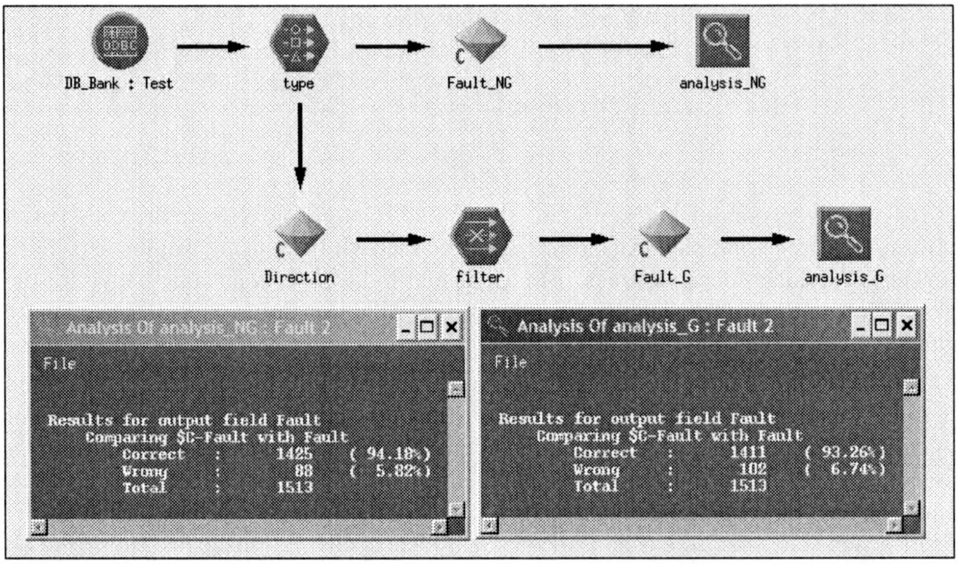

Figure 31. Visualization of Results

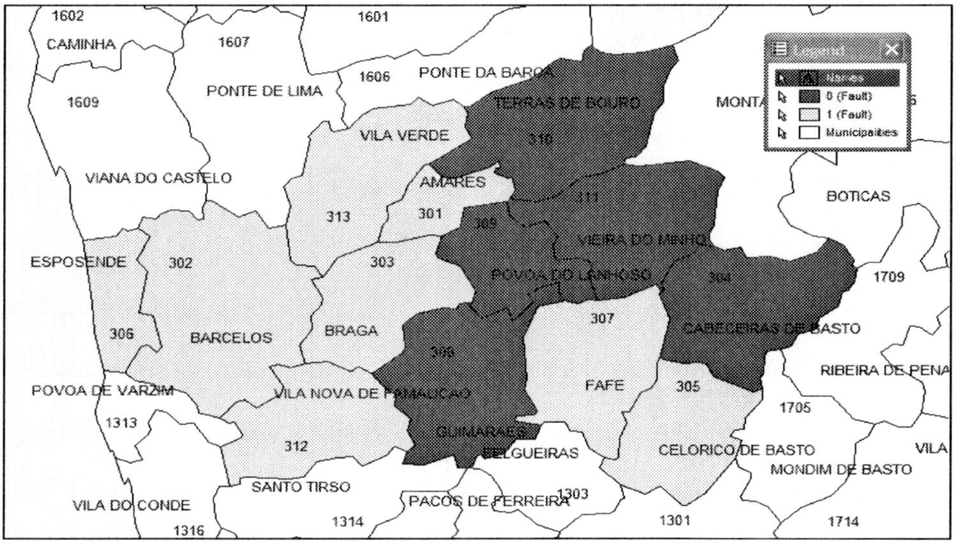

Fault_G model obtained in the data mining step). This geographic characterization enabled the identification of regions where the relative incidence of clients with faults is higher than elsewhere in the District.

Risk zones were identified, aggregating together regions that have clients with similar behavior. The geographic segments can be cataloged by the bank, looking at similarities like proximity with other regions, population density, population qualification and other relevant issues.

The models obtained in the data mining step define the profile of the bank clients. They integrate the attributes and the corresponding values related to the classification of the clients bearing in mind the risk of investment in specific classes of clients. For the available segments, the several rules identified can support managers in the decision-making process. In the granting of new credit, the organization is now supported by models that track the previous behavior of its clients, indicating groups of clients in which the organization has to pay more attention in the granting of credit and those groups without difficulties in the assignment of credit.

Suppose that 10 new potential clients request credit to the organization. *Figure 32* shows the relevant data on each client and the classification (column $C-Fault) of the model according to the rules explicit in it. The column $CC-Fault indicates the confidence of the classification, which is equal or superior to 94%. Looking at the classification achieved, for seven clients the decision of the model is 0 in the $C-Fault attribute, which means that, based on the past experience of the organization, these are *good* clients. For 3 clients the result was 1 in the $C-Fault attribute, labeling these clients as *risk* clients and suggesting that a more detailed analysis must be undertaken in order to identify the appropriate decision (grant credit or not).

The use of predictive models assumes that the past is a good predictor of the future. However, there are situations where the past may not be a good predictor, if the facts

Figure 32. Classification of New Clients by the Model

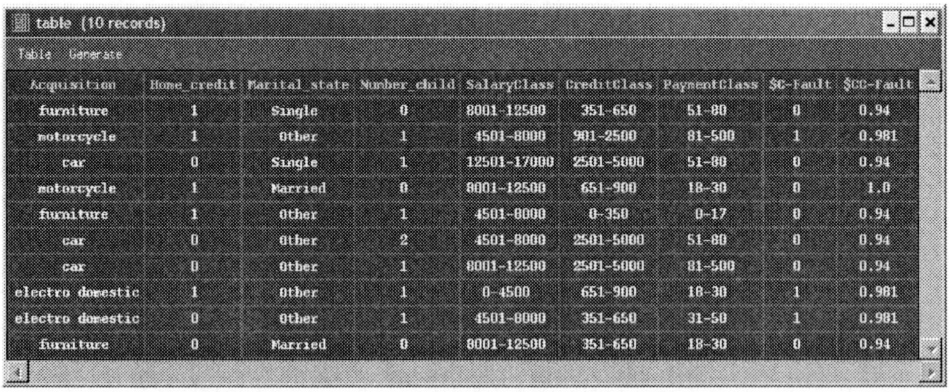

table (10 records)

Table Generate

Acquisition	Home_credit	Marital_state	Number_child	SalaryClass	CreditClass	PaymentClass	$C-Fault	$CC-Fault
furniture	1	Single	0	8001-12500	351-650	51-80	0	0.94
motorcycle	1	Other	1	4501-8000	901-2500	81-500	1	0.981
car	0	Single	1	12501-17000	2501-5000	51-80	0	0.94
motorcycle	1	Married	0	8001-12500	651-900	18-30	0	1.0
furniture	1	Other	1	4501-8000	0-350	0-17	0	0.94
car	0	Other	2	4501-8000	2501-5000	51-80	0	0.94
car	0	Other	1	8001-12500	2501-5000	81-500	0	0.94
electro domestic	1	Other	1	0-4500	651-900	18-30	1	0.981
electro domestic	0	Other	1	4501-8000	351-650	31-50	1	0.981
furniture	0	Married	0	8001-12500	351-650	18-30	0	0.94

occurred were influenced by external events not present in the analyzed data (Berry & Linoff, 2000). For this reason, in making predictions, the organization cannot only be supported by the models obtained from the knowledge discovery process, in order to avoid penalizing new potential *good* clients as a result of the behavior of past clients. The models obtained should be seen as tools that support the decision-making process, not as the decision-maker.

The knowledge discovery process should support the creation of organizational knowledge through the incorporation of the information expressed in the several models in its daily activities. This procedure will contribute to fulfill the information requirements of the bank and help in the accomplishment and improvement of its mission.

Conclusion

This chapter presented an approach for knowledge discovery in geo-referenced databases based on qualitative spatial reasoning, where the position of geographical data was provided by qualitative identifiers.

Direction, distance and topological spatial relations were defined for a set of Municipalities of Portugal. This knowledge and the composition table constructed for integrated spatial reasoning enabled the inference of new spatial relations analyzed in the data mining step of the knowledge discovery process.

The integration of a bank database with the GDB (storing the administrative subdivisions of Portugal) made possible the discovery of general descriptions that exploit the relationships that exist between the geo-spatial and non-spatial data analyzed. The models obtained in the data mining step define the profile of the clients, bearing in mind the risk of investment of the organization for specific segments of clients. For the available classes, the several rules identified support the managers of the organization in the decision-making process. The latter represents one of the organizational processes that can benefit from data mining technology through the incorporation of its results in the evaluation of critical and uncertain situations.

The results obtained with the PADRÃO system support that traditional KDD systems, which were developed for the analysis of relational databases and that do not have semantic knowledge linked to spatial data, can be used in the process of knowledge discovery in geo-referenced databases, since some of this semantic knowledge and the principles of qualitative spatial reasoning are available as domain knowledge. Clementine, a KDD system, was used in the assimilation of the geographic domain knowledge such as composition tables, in the inference of new spatial relations, and in the discovery of spatial patterns.

The main advantages of the proposed approach, for mining geo-referenced databases, include the use of already existing data mining algorithms developed for the analysis of non-spatial data; an avoidance of the geometric characterization of spatial objects for the knowledge discovery process; and the ability of data mining algorithms to deal with geo-spatial and non-spatial data simultaneously, thus imposing no limits and constraints on the results achieved.

Acknowledgments

We thank NTech – Sistemas de Informação, Lda. for making the database available for analysis. We thank Tony Lavender for his help in improving the English writing of this chapter.

References

Abdelmoty, A. I., & El-Geresy, B. A. (1995). A general method for spatial reasoning in spatial databases. *Proceedings of the Fourth International Conference on Information and Knowledge Management* (pp. 312-317). Baltimore, Maryland.

Allen, J. F. (1983). Maintaining knowledge about temporal intervals. *Communications of the ACM, 26*(11), 832-843.

Berry, M., & Linoff, G. (2000). *Mastering data mining: The art and science of customer relationship management*. New York: John Wiley & Sons.

CEN/TC-287. (1996). *Geographic information: Data description, spatial schema* (prENV 12160). European Committee for Standardization.

CEN/TC-287. (1998). *Geographic information: Referencing, geographic identifiers* (prENV 12661). European Committee for Standardization.

Egenhofer, M. J. (1994). Deriving the composition of binary topological relations. *Journal of Visual Languages and Computing, 5* (2), 133-149.

Ester, M., Frommelt, A., Kriegel, H.-P., & Sander, J. (1998). Algorithms for characterization and trend detection in spatial databases. *Proceedings of the Fourth International Conference on Knowledge Discovery and Data Mining*. AAAI Press.

Ester, M., Kriegel, H.-P., & Sander, J. (1997). Spatial data mining: A database approach. *Proceedings of the Fifth International Symposium on Large Spatial Databases* (pp. 47-68). Germany.

Fayyad, U., & Uthurusamy, R. (1996). Data mining and knowledge discovery in databases. *Communications of the ACM, 39* (11), 24-26.

Fayyad, U. M., Piatetsky-Shapiro, G., Smyth, P., & Uthurusamy, R. (eds.). (1996). *Advances in knowledge discovery and data mining.* MA: The MIT Press.

Frank, A. U. (1992). Qualitative spatial reasoning about distances and directions in geographic space. *Journal of Visual Languages and Computing, 3*, 343-371.

Frank, A. U. (1996). Qualitative spatial reasoning: Cardinal directions as an example. *International Journal of Geographical Information Systems, 10* (3), 269-290.

Freksa, C. (1992). Using orientation information for qualitative spatial reasoning. In A. U. Frank, I. Campari, & U. Formentini (Eds.), *Theories and methods of spatio-temporal reasoning in geographic space* (Lectures Notes in Computer Science 639). Berlin: Springer-Verlag.

Han, J., & Kamber, M. (2001). *Data mining: Concepts and techniques.* CA: Morgan Kaufmann Publishers.

Han, J., Tung, A., & He, J. (2001). SPARC: Spatial association rule-based classification. In R. Grossman, C. Kamath, P. Kegelmeyer, V. Kumar & R. Namburu (Eds.), *Data mining for scientific and engineering applications* (pp. 461-485). Kluwer Academic Publishers.

Hernández, D., Clementini, E., & Felice, P. D. (1995). Qualitative distances. *Proceedings of the International Conference COSIT'95* (pp. 45-57). Austria.

Hong, J.-H. (1994). *Qualitative distance and direction reasoning in geographic space.* Unpublished doctoral dissertation, University of Maine, Maine.

Intergraph. (1999). *Geomedia professional v3* (Reference Manual). Intergraph Corporation.

Koperski, K., & Han, J. (1995). Discovery of spatial association rules in geographic information systems. *Proceedings of the 4th International Symposium on Large Spatial Databases* (pp. 47-66). Maine.

Lu, W., Han, J., & Ooi, B. (1993). Discovery of general knowledge in large spatial databases. *Proceedings of the 1993 Far East Workshop on Geographic Information Systems* (pp. 275-289). Singapore.

Papadias, D., & Sellis, T. (1994). On the qualitative representation of spatial knowledge in 2D space. *Very Large Databases Journal, Special Issue on Spatial Databases, 3* (4), 479-516.

Santos, M., & Amaral, L. (2000a). Knowledge discovery in spatial databases through qualitative spatial reasoning. *Proceedings of the 4th International Conference and Exhibition on Practical Applications of Knowledge Discovery and Data Mining* (pp. 73-88). Manchester.

Santos, M., & Amaral, L. (2000b, November). Knowledge discovery in spatial databases: The PADRÃO's qualitative approach. *Cities and Regions, GIS special issue,* 33-49.

Santos, M., & Amaral, L. (2000c). A qualitative spatial reasoning approach in knowledge discovery in spatial databases. *Proceedings of Data Mining 2000: Data Mining Methods and Databases for Engineering, Finance and Others Fields* (pp. 249-258). Cambridge.

Santos, M. Y. (2001). *PADRÃO: Um sistema de descoberta de conhecimento em bases de dados geo-referenciadas (in Portuguese)*. Unpublished doctoral dissertation, Universidade do Minho, Portugal.

Santos, M. Y., & Ramos, I. (2003). Knowledge construction: The role of data mining tools. *Proceedings of the UKAIS 2003 Conference "Co-ordination and Co-operation: the IS role"*. Warwick.

Sharma, J. (1996). *Integrated spatial reasoning in geographic information systems: Combining topology and direction*. Unpublished doctoral dissertation, University of Maine, USA.

SPSS. (1999). *Clementine* (user guide, version 5.2). SPSS Inc.

Endnotes

1 GISs allow for the storage of geographic information and enable users to request information about geographic phenomena. If the requested spatial relation is not explicitly stored in databases, it must be inferred from the information available. The inference process requires searching relations that can form an *inference path* between the two objects where the relation is requested (Hong, 1994). The composition operation combines two contiguous paths in order to infer a third spatial relation. A composition table integrates a set of inference rules used to identify the result of a specific composition operation.

2 Extended objects are not point-like, so represent objects for which their dimension is relevant (Frank, 1996). In this work, extended objects are geometrically represented by a polygon, indicating that their position and extension in space are relevant.

3 In IR^2, there are eight topological relations between two planar regions without holes (two-dimensional, connected objects with connected boundaries); 18 topological relations between spatial regions with holes; 33 between two simple lines and 19 between a spatial region without holes and a simple line (Egenhofer, 1994).

4 The topology of a full planar graph refers to a planar graph that integrates regions completely covering the plane without any gap or overlap. Regions are topologically represented by faces, which are defined without holes (CEN/TC-287, 1996).

5 Defining distances between regions is a complex task, since the size of each object plays an important role in determining the possible distances. Sharma (1996) gives the following ways to define distances between regions: (i) taking the distance between the *centroids* of the two regions; (ii) determining the shortest distance between the two regions; or (iii) determining the furthest distance between the two regions.

6 Other validity intervals, for different ratios, can by found in Hong (1994).

7 The symbol used to represent the composition operation is “;”.

8 Since the system will be used with administrative subdivisions, the orientation between the several regions is calculated according to the position of the respective *centroids*.

9 The dotted lines define the acceptance area defined for the North direction (designed from the *centroid* of B), while the whole lines represent the acceptance area defined for the Northeast direction (designed from the *centroid* of C).

10 In this work, an inference is considered exact if the result achieved with the correspondent qualitative rule is the same as if the data was translated to quantitative information and manipulated through analytical functions. Otherwise, it is considered approximate.

11 The geometry is not required in the knowledge discovery process, since the manipulation of the geographic information is undertaken by a qualitative approach.

12 Visual programming involves placing and manipulating icons representing processing nodes.

13 Distribution nodes are used for the analysis of categorical data.

14 The data mining algorithms may be negatively influenced by classes with a great number of values.

15 The several associations identified anticipate the importance of each attribute in the definition of the profile of the clients.

16 A rule induction algorithm creates a decision tree aggregating a set of rules for classifying the data into different outcomes. This technique only includes in the rules the attributes that really matter in the decision-making process.

Chapter VII

GIS as Spatial Decision Support Systems

Suprasith Jarupathirun, University of Wisconsin, Milwaukee, USA

Fatemeh "Mariam" Zahedi, University of Wisconsin, Milwaukee, USA

Abstract

This chapter discusses the use of geographic information systems (GIS) for spatial decision support systems (SDSS). It argues that the increased availability in spatial business data has created new opportunities for the use of GIS in creating decision tools for use in a variety of decisions that involve spatial dimensions. This chapter identifies visualization and analytical capabilities of GIS that make such systems uniquely appropriate as decision aids, and presents a conceptual model for measuring the efficacy of GIS-based SDSS. The discussions on the applications of SDSS and future enhancements using intelligent agents are intended to inform practitioners and researchers of the opportunities for the enhancement and use of such systems.

Introduction

Geographic information systems (GIS) have been used by government agencies, researchers, and business as a tool to support a wide range of decisions that have location dimensions (Groupe, 1990; Wilson, 1994; Dawes & Oskam, 1999). Over the last 10 years, the popularity of using GIS among business organizations has increased due to a number of factors: (1) the belief that the use of GIS would improve decision making (Attenucci et al., 1991; Bracken & Webster, 1989; Dennis & Carte, 1998; Murphy, 1995; Robey & Sahay, 1996), (2) the fact that about 80% of data used in making business decisions has geographical dimensions (Worrall, 1991), (3) the increased availability of spatial data (Gagne, 1999; Heikkila, 1998), and (4) the availability and declining cost of the required hardware and software. Furthermore, with the ever-increasing popularity of the web and improvement in its technologies, web-based GIS are widely available to web-users in helping them in making decisions involving geographic information or *spatial decisions*. Like traditional DSS, the bottom line of using GIS is to improve the quality of decision-making. The issue explored in this chapter is the role of GIS in decision-making and its impact on improving decisions with spatial dimensions.

In examining this issue, we will briefly discuss the nature and functionalities of GIS, and contrast the parallel development of GIS and IT technologies and their main foci in order to bring out the aspects of GIS helpful in various decision-making tasks, hence making a case for spatial decision support systems (SDSS). Next, we discuss the critical role of visualization in decision-making as an important cognitive aid. In contrasting the nature of visualization in traditional DSS and SDSS, we highlight the potential contributions of SDSS in decision-making. We then report on the existing research in the use of GIS in business and note the absence of a theoretical framework for evaluating the efficacy of SDSS. This gap motivates the conceptualization of a theoretical-based framework for evaluating the efficacy of SDSS, which is presented next. The chapter ends with a discussion of the existing limitations and future directions.

Nature of GIS

Before we get into the discussion about the role and impact of the GIS in the spatial decision-making process and its impact on improving such decisions, we first need to discuss the nature of GIS and the unique features that make it different from traditional IT used in business. Although there have been a number of attempts to define GIS, there is no consensus about a single general definition of GIS. Most definitions are focused on either the technology or on the problem solving aspects of GIS (Malczewski, 1999).

The confusion about the definition of GIS may be due to the evolution and the diffusion of the technology. During the 1960s, the early GIS were initially developed to better manage geographic information by providing tools for the storage, retrieval and display of both spatial and attribute information in the form of maps, tables and graphs. The development of GIS applications can be traced back to Canada Geographic Information System and software from the laboratory at Harvard University. The early GIS such as SYMAP and ODYSSEY developed in the Harvard lab were applications used to produce geographic representation or maps with simple functionalities such as the overlay function. The outputs of the system were in the form of simple maps that were produced off-line. As the GIS technology progressed, various disciplines adopted the technology for their own specific purposes in order to take advantage of the flexible capability to visualize geographic information. Some GIS definitions reflect these limited capabilities of GIS, such as the definition by Burrough (1986) that GIS are a set of tools for collecting, storing, retrieving, transforming and displaying geographical data; the definition by Devine & Field (1986) that GIS are systems that promote the use of maps; or the definition by Tomlinson (1987) that GIS are simply digital systems that capture, manipulate and display geographical data for various analyses. Earlier GIS, therefore, were mostly viewed as systems for improving the management of geographical information, and contributions of earlier GIS as decision-making aids were limited to geographical representations or maps.

After GIS were integrated with more complex spatial analysis tools, various disciplines, including business, found multiple applications and uses for such systems. Cowen (1988) defined GIS as a decision support system that integrates location-referenced data into problem-solving space. Over the years, the definitions of GIS have grown to include functionalities for possible users in various fields. As a result, different users are likely to have different perspectives on GIS. For example, at the operational level, workers automatically generate maps using GIS in their day-to-day business. In this context, GIS can be viewed more as systems intended for improving operational performance. On the other hand, top-level managers can use GIS for planning and decision-making, giving such systems a DSS perspective. Thus, the classification of GIS depends on their intended use. Applications of GIS in business disciplines are due largely to the belief that the use of GIS would improve decision-making and give an edge over competitors. Currently, the development of GIS is focused on increasing the analytical capabilities of GIS that support the requirements of complex decision tasks (Rogerson & Fotheringham, 1994; Malczewski, 1999).

For this chapter, we define GIS as a technology for the storage, processing, retrieval and display of information with spatial dimensions (information about location of entities which can be points, lines or polygons), and with attribute dimensions (information about entities and objects), with capabilities for manipulating data into different forms, extracting additional meaning, and presenting the information in various forms (map, table, graph, etc.). This definition of GIS as a shell embodies various applications of GIS in different fields and areas.

GIS Applications for Decision Support

Scott Morton (1971) categorized decisions as unstructured, semi-structured, and structured; and business controls as operational, tactical/managerial, and strategic. GIS-based SDSS have been applied in many fields for enhancing various types of decisions and business controls.

In marketing, Thrall & Fandre (2003) demonstrate the use of GIS-based spatial analysis to define the trade areas for a retail business in Florida. They used CACI/Coder Plus to perform geocoding the point of sale (POS) data and used ArcView with Business Analyst to calculate a trade area of the business. Together with marketing models for characterizing customers, a manager can identify customer patterns that are useful to develop a strategic business plan. A similar GIS-based decision support has been used in Beijing to study the retail accessibility. The results help business managers in Beijing gain better insights of their marketing potential and aid city planners with their downtown revitalization (Yongling, 2002).

GIS-based SDSS has been used in court systems to settle disputes. For example, in Maryland, a GIS-based DSS has been developed to help settle trade disputes in court (Brodsky, 2002).

Government agencies have used GIS-based SDSS for various types of decisions. In cities and municipalities, GIS-based decision support technology has been long used to improve the efficiency and effectiveness of operation and planning. For example, implementing distributed GIS (including Internet GIS and Mobile GIS), Taiwan's capital city Taipei allows their residents to report needed repair and maintenance of public facilities via the Internet, and allows city workers to receive and close the work orders via mobile devices. Another case is the village of Gurnee, Illinois, which is located about an hour north of Chicago. It has customized GIS-based technology to provide a graphic user interface similar to that of standard Microsoft software (Excel and Word) for the notification process to the Gurnee residents (Venden & Horbinski, 2003). The customized GIS application allows village staff, who have few skills in using GIS but are familiar with the MS software platform, to efficiently perform operational tasks, such as notification for zoning, violation, construction, and repair.

In addition to the use of GIS for operation decisions by government agencies, GIS technology has been used for strategic planning, such as modeling water demand, pest management, simulating urban growth, and simulating dispatch responses during big events and catastrophes (Barnes, 2002; Brewster et al., 2002; Lee et al., 2002; Pimpler & Zhan, 2003; Price & Schweitzer, 2002). For example, expanding cities and metropolitan areas require national and local government agencies to plan for the future resources to support residents and businesses. For instance, the state of Texas has developed the GIS-based water-demand model to support strategic decision-making about potential water shortage problems (Pimpler & Zhan, 2003). Similarly, environmental agencies have developed specific applications to simulate the growth of urban areas, examine its impact on the environment, and develop strategic plans and policies for upcoming problems (Lee et al., 2002).

GIS-based SDSS with proper tools have been used to develop a floodplain model in order to identify the risk areas (Price & Schweitzer, 2002). Several interested parties can use this model and information it produces to make decisions in flood emergencies. For example, government agencies use the model to develop the evacuation plans in the case of floods; insurance agencies use the model to assess the risk and set the insurance premium; and individuals use it to help them decide whether to buy a property.

Developing countries use GIS-based SDSS for strategic planning as well. For example, Nepal uses GIS-based SDSS to develop a spatial energy information system model to assess the need for energy (Pokharel, 2000).

GIS-based SDSS have applications for decision-making in security threats and emergency responses. After 9-11, GIS-based SDSS have been used to address homeland security concerns. Vexcel Corporation (www.vexcel.com) has developed GIS-based SDSS that include emergency preparation and emergency response simulation functions with a 3-D visualization capability. Similarly, Science Applications International Corporation has combined GIS-based SDSS with other IS technology, including Internet network, WAN, and GPS, to develop strategic planning for various emergency situations and to run operations during the 2002 Winter Olympics at Salt Lake City, Utah (Barnes, 2002).

Another example of the critical role of GIS-based SDSS is their use by the United States in the Iraqi Freedom operation. As shown in CNN coverage of the operation, GIS-based SDSS were used to integrate intelligence information from various sources in real time for strategic planning at the United States Central Command (USCENTCOM) room. Meanwhile, the intelligent geospatial information was downloaded by U.S. soldiers and their allies in the field for their day-to-day operations. Similar information and GIS technology were used by U.S. news media to give public an understanding of the events in Iraq and were used by U.S. State Department for rebuilding the Iraq after the war (Barnes, 2003).

In the healthcare industry, GIS-based SDSS technology using proper functionalities (such as network analysis, and spatial statistical analysis) and other technologies (e.g., wireless devices and GPS) can be used to support healthcare providers in making operational, tactical, and strategic decisions. At the operation level, GIS technology with network analysis module, GPS and wireless devices can help a dispatch center identify the best route to pick up patients and take them to a hospital. At the strategic level, GIS-based SDSS with spatial statistical analysis and network analysis modules has been used for modeling the location of healthcare clinics to determine whether additional clinics are needed or existing clinics should be relocated to improve their accessibility for the target population (Hyndman & Holman, 2000).

There are a number of research studies in designing and using SDSS for solving specific complex problems. For example, Ioannou et al. (2002) and Tarantilis & Kiranoudis (2002) report on the design and development of GIS-based SDSS for solving vehicle routing problems. Ghinea et al. (2002) perform a comparative study of various solutions for visualizing and managing back-pain problems and conclude that a GIS-based solution is one of the most appropriate approaches for analyzing back-pain data.

The synthesis of GIS-based SDSS with other technologies for improving decisions has also been advocated in the literature. Weber (2001) suggests using GIS and online analytical processing (OLAP) for the accurate evaluation of lands for development. De Silva & Eglese (2000) propose the design of a GIS-based decision support for planning emergency evacuations, which combines GIS and simulation techniques. Seffino et al. (1999) report on using a synthesis of workflow spatial decision support system and GIS to provide spatial decision support for environmental planning, and report on its use in agri-environmental planning activities. Talen (2000) has proposed a GIS-based group support system, which includes the concepts and functionalities of group support systems. He reports the applications of this system for involving residents in city planning. West & Hess (2002) propose the synthesis of GIS-based SDSS with knowledge management tools in order to facilitate and enhance the SDSS usability.

Such designs and applications use various capabilities and functionalities of SDSS to accomplish their intended objectives. Therefore, it is important to briefly review the capabilities available in the GIS technology.

GIS Capabilities

Scholten & van der Vlugt (1990) classified information systems into non-spatial information systems and spatial information systems. Non-spatial information systems are traditional systems that do not store and display spatial information. Information systems classified as non-spatial information systems are, for example, transaction processing systems (TPS) or management information systems (MIS) in business. The outputs of these systems are usually either tables or graphs. The outputs of spatial information systems, however, are not limited to tables or graphs but also include the dynamic presentation of maps.

Based on capabilities, spatial information systems can be subcategorized into spatial design systems (CAD), land-use information systems (LIS), and geographic information systems (GIS), *Table 1*.

Spatial information systems, including CAD, LIS and GIS, allow users to accurately and effectively create and maintain spatial data and maps. While both LIS and GIS can store and manage spatial and attribute information, only GIS have capabilities to perform spatial analysis. The functionalities for management, manipulation, and analysis of attribute and spatial data distinguish GIS from most map-drawing systems, even though they share similar capabilities for displaying spatial maps (Huxhold, 1991). On the other hand, MIS and GIS are similar in their functionality of storing, analyzing, and retrieving attribute data. Only GIS, however, have analytical functionalities that use both the spatial and attribute data (Malczewski, 1999). Hence, GIS could provide distinct tools — map visualization and spatial analysis — for making spatial decisions not found in MIS.

Analytical Tools in GIS

Analytical tools in the context of our discussion are functionalities for processing information in supporting decision-making. Given an appropriate model, procedure or rule, information systems with analytical tools can provide the optimal solutions, appropriate suggestions, or decision-enhancing informational intelligence. Decision support systems are built on this premise and have database, model base, and interface components. GIS could support spatial decision-making with a wide array of capabilities during decision-making processes (Crossland et al., 1995). The capability of GIS to manipulate and analyze the spatial and attribute data using various statistical, mathematical, geometric, and cartographic methods can be referred to as spatial analytical tools, which make GIS unique (Huxhold, 1991; Malczewski, 1999). GIS provide interactive map presentation that, in conjunction with analytical tools, could be used to probe maps at various levels of specificity, a feature that is missing in paper maps (Crossland et al., 1995). The spatial analysis tools can be used for many different forms to answer questions or issues related to location. These questions can range from simple calculative questions, such as the distance between two locations, to more complex quantitative problems, such as the most suitable location of a new retailer.

Table 1. Classification of Information Systems and Capabilities

Information Systems	Visualization		Analysis	
	Spatial	Attribute	Spatial	Attribute
Non-spatial Information Systems				
- Transaction Processing Systems		✔		
- Management Information Systems		✔		✔ *
- Decision Support Systems		✔		✔
Spatial Information Systems				
- Spatial Design Systems	✔	✔		
- Land-Use Information Systems	✔	✔		
- Geographic Information Systems	✔	✔	✔	✔

*Note: * the analysis capabilities are limited when compared to those of decision support systems*

We identify two broad categories of GIS functionalities: standard and advanced (Malczewski, 1999). First, standard functions are tools available to users in standard GIS systems and are used to either assist dynamic visualization or perform basic analysis. For example, ZOOM IN and ZOOM OUT functions are used to visualize geographic information in the form of maps in different scales, while PAN function is used to visualize a map in different angles. Measurement functions calculate the distance, area or volume of features (e.g., point, line, or polygon), while proximity or buffer functions define a zone or region around given features (*Figure 1*). Map overlay functions combine two or more maps to synthesize maps into one (*Figure 2*), or subtract two or more maps to simplify the map to include only features of interest. In general, these functions are used to provide dynamic manipulation of maps for visual thinking and visual spatial analysis.

Second, advanced (or specific) functions are tools that are designed for specific tasks and thus useful for special-purpose spatial decision-making. Such tools include 3-D presentations, statistical modeling, or mathematical modeling for spatial analyses. Examples of analytical tools used in mathematical modeling include network analysis and shortest path. According to Malczewski (1999), conventional statistics can be used to analyze spatial data, but their use in geographical data is questionable because geographical data by nature are statistically dependent — adjoining features are spatially related to each other. There exist spatial statistical tools that could be used in GIS. To take full advantage of GIS capabilities, users should understand spatial statistics well, which is not common knowledge in business disciplines.

In short, most of basic spatial analysis can be done using GIS with standard functions, which have use in multiple disciplines. However, the use of advanced functionalities is task-dependent. In this respect, GIS could enhance decision performance only if their functionalities support task requirements.

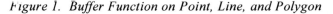

Figure 1. Buffer Function on Point, Line, and Polygon

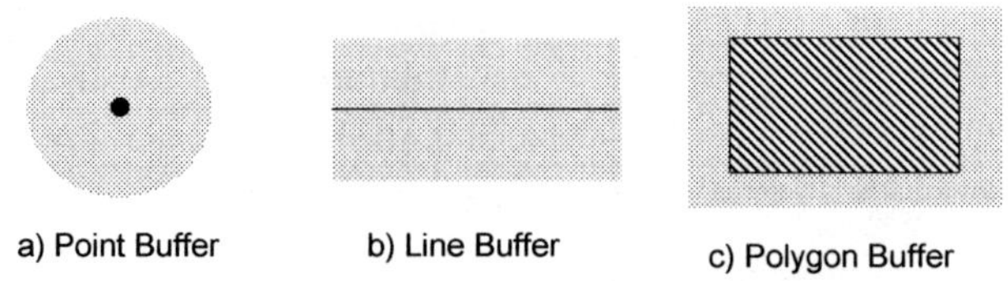

Figure 2. Map Overlay Function

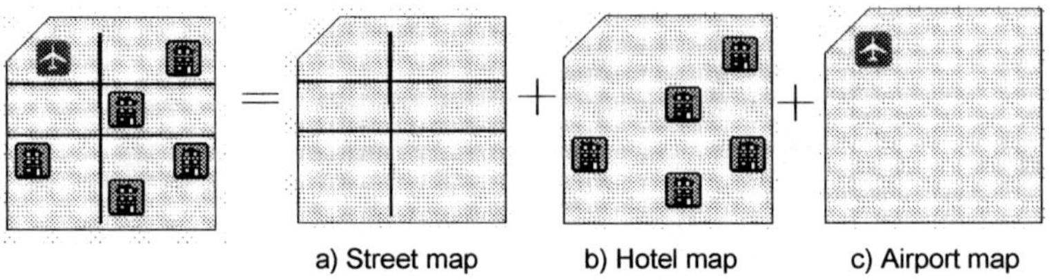

GIS + DSS = SDSS

The stages of GIS development could be compared to those in IS and DSS, as shown *Figure 3*.

In IS, transaction processing systems (TPS) used file systems to generate reports and process transactional data. Similarly, early GIS were used mostly for generating off-line maps, replacing the need to draw paper maps. In the 1970s, IS advanced to include database techniques. In the 1980s, IS gained interactive online capabilities, and evolved to include decision support systems (DSS), in which models and databases were used for supporting complex decision making tasks (Marakas, 2003). At the same time, GIS also evolved into interactive systems that included advanced database and statistical modeling tools for complex visualization of locational data. As the Internet gained popularity in the 1990s, DSS moved on to the Web, as did GIS, both with the purpose of helping web-users in their search for information and decision support. There are efforts under way to make both traditional DSS and GIS available to mobile users.

Yoon et al. (1995) differentiate between the two systems in that TPS are used for routine data processing and emphasize business procedures, while DSS are used for supporting decision-making and focus on statistical and decision models. While effective in their specific objectives, both TPS and DSS are deficient in dealing with location information and decisions (Scholten & van der Vlugt, 1990). On the other hand, GIS at present are viewed to be deficient as decision support systems (Harris & Batty, 2001; Klosterman, 2001). Hence, GIS are natural complements for DSS in providing visualization support in areas where traditional DSS lack the appropriate tools. Combining GIS capabilities with

Figure 3. Parallel Stages of Development in IS-DSS and GIS

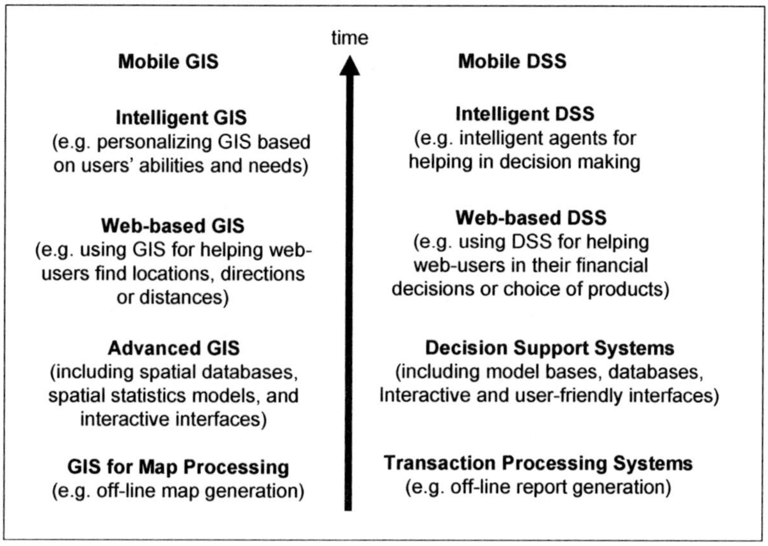

DSS modeling and data support results in "spatial DSS," or SDSS. Given the fact that many decision tasks and problems have some spatial or locational components, the development of SDSS could enhance the quality of support for decision-making.

The need for SDSS has already been recognized. Malczewski (1999) categorize GIS into four types: spatial data processing systems, SDSS, spatial expert system, and spatial expert support systems. "Spatial data processing systems" generate map reports and solve structured problems automatically. SDSS are equipped with more advanced functionalities and are used for dealing with semi-structured problems. "Spatial expert systems" combine expert systems and GIS to provide support for dealing with semi-structured problems that require encoded knowledge in the domain. Lastly, Malczewski (1999) integrated the concept of DSS and ES and referred to them as "spatial expert support systems" that are used for semi-structured problems, for which not all knowledge for solving can be encoded. In short, GIS can be integrated with different types of IS functionalities to use on the continuum of improving quality of information processing and decision-making. In all such syntheses, visualization plays a critical role.

SDSS and Visualization

Visual information processing has been the critical key for human development and human cognition (Chase, 1972). Humans use visualization to comprehend new knowledge and make every day decisions. According to cognitive psychology, humans construct a mental model of reality, which does not have to be the exact image but rather an abstract form of problems or simplified versions of real systems. In the study of nonroutine problem solving, Presmeg & Balderas-Canas (2001) found that when people use analytical strategies for solving mathematical problems, they use visual methods in helping them to understand problem tasks and to validate solutions. Furthermore, visual representations are found to influence people to better assimilate and understand new concepts (Gershon & Page, 2001; Kraidy, 2002). According to the DiBiase model, during the problem solving process, humans use visual thinking to reveal unknown relationships, and use visual communication to present known relationships (DiBiase, 1990). In the visual thinking stage, people use visualization aids to explore problems and to verify assumptions and hypotheses. In the visual communication stage, individuals use visualization aids to focus attention on the information needed to convince others (DiBiase, 1990; Dransch, 2000). Dransch (2000) suggested that in visual thinking and visual communications, visualization should help individuals in constructing complex mental models, putting information into a greater context, preventing information overload, getting the correct understanding of problems and solutions, supporting double encoding of information, and highlighting important information (Dransch, 2000). Hence, we conclude that visualization plays a critical role in decision-making and communication.

Visualization in Non-Spatial Information Systems. In IS literature, visualization tools have been shown to be useful for decision-making (Vessey, 1991; Meyer, 2000) and data mining (Mitchell, 1999; Witten & Frank, 2000). Representation tools include decision trees, charts, graphs, diagrams, and tables. Among those, tables and graphs are tools that have been rigorously examined (e.g., Jarvenpaa, 1989; Meyer, 2000; Vessey, 1991). Vessey (1991) proposed the cognitive fit theory to show that individuals' performances depend on the "fit" between characteristics of representation, characteristics of tasks, and individuals' abilities (Umanath & Vessey, 1994). Meyer (2000) found that people perform better when they gain more experience in using graphs and tables. In general, the use of visualization tools could enhance performance when they are congruent with task characteristics and individuals' abilities to use the tools in performing the tasks.

Since as much as 80% of business data has a geographical component, many tasks require decision makers to comprehend spatial relationships of business phenomena. Unfortunately, visualization tools available in traditional IS are not designed to represent geographical and spatial information visually.

Visualization in Geographic Information Systems. While GIS provide different types of presentation, the most dominant visualizations in GIS are in the form of maps. The capability of GIS technology to visualize the data in the form of interactive maps is one of the key features (Smelcer & Carmel, 1997) that distinguishes GIS from other IS technology (Huxhold, 1991).

In the cartography literature, map visualization is considered a powerful tool for exploration, analysis, and communication of spatial context (Wood, 1994). Furthermore, according to human problem solving theory, the human mind has a limited capacity for storing, retrieving, and processing information (Newell & Simon, 1972). Due to the limitation of human cognition, filtering schemes are needed to filter out unrelated or unimportant information and to focus only relevant and important information for decision-making (Cooper, 1988). As tasks increase in complexity, additional aid is needed to reduce information load in working memory. Such aids are considered as "external memory" (Cooper, 1988; Newell & Simon, 1972). In this context, maps can be viewed as external memories that help decision makers overcome their limited spatial information processing, particularly in dealing with large volumes of multi-dimensional information. The dynamic nature of map presentations in GIS makes such systems useful for complex decision tasks.

Hence, an important difference between the map visualization in GIS and non-spatial visualization is the built-in dynamic nature of map visualization. For example, decision makers could use functions like ZOOM IN, ZOOM OUT, PAN and MAP OVERLAY to dynamically explore a host of complex data, whereas such capabilities are not present in mostly static graphical visualization. Furthermore, static non-spatial visualization could not easily accommodate large and multi-dimensional datasets, and have no built-in dynamic functionality for online exploration of data. Hence, the dynamic map visualization in GIS provides superior capabilities for building SDSS in dealing with complex and multi-dimensional decisions.

A number of studies in GIS support the importance of map visualization in decision-making, as summarized in *Table 2*.

Table 2. Summary of Studies of GIS in IS Literature

Studies	Findings
Map Representation	-Map users make faster decisions than those who use tables (Smelcer & Carmel, 1997). -For a geographic task that does not require examining spatial relationships, using maps results in less accurate but faster decisions than using tables (Dennis & Carte, 1998). -Performance deteriorates as problem size increases, data aggregation is reduced, and data dispersion is increased (Swink & Speier, 1999).
Visual analysis tools	-GIS map performs better than paper maps because GIS tools reduce the load on the human cognitive information process (Crossland et al., 1995). - Experts are more accurate than novice when using GIS technology to perform geographical tasks (Mennecke et al., 2000)
Implementation	- Education and training are important for the success of implementing GIS (Walsham & Sahay, 1999; Robey & Sahay, 1996)

Wood (1994) observed that visualization is used for visual thinking, which helps humans construct mental models for understanding and gaining insight about spatial relationships (Wood, 1994). In his review of literature, Wood (1994) found that maps can be used effectively for scientific visualizations — visual communication or visual thinking.

Crampton (2001) argued that people use map representation more for visual thinking than for visual communication, and observed that maps with low interactivity are appropriate for presenting "known" information and for public use, whereas highly interactive maps are used for revealing "unknown" information and for private use. In other words, when maps are used for visualization or data exploration, users need to engage in a high level of interaction with maps. On the other hand, when maps are used for representation or communication, the need for map interactivity is not as high. With the emergence of web-based GIS, many end users can also be their own map-makers, which leads to a number of issues concerning web-users' levels of skills and knowledge, data availability, and privacy and security issues related to making data available over the Web.

Smelcer & Carmel (1997) compared the effectiveness of map and table representations in different geographic relationships (such as proximity, adjacency, and containment), with different levels of task difficulty (simple, moderate, and complex), and cognitive skills (low spatial ability and high spatial ability). Geographic proximity refers to relationships of features that are at a distance; geographic adjacency refers to relationships of features that share borders; and geographic containment refers to relationships of features where one feature contains others. Using cognitive fit theory (Vessey, 1991) and the proximity compatibility principle (Wickens, 1992), Smelcer & Carmel (1997) argued that in geographic relationships, decision makers who used maps would have better performance than those who used tables. (The proximity compatibility principle is somewhat similar to cognitive fit theory in that the task requirements and information representation should be compatible.) They tested their propositions through lab experimentation in which decision makers had access to different representations in dealing with different level of difficulty of decision tasks. They found that decision makers (in this case undergraduate students) who used maps spent less time solving decision problems than those who used tables. Also, as tasks increased in complexity, the time required to perform the tasks increased much more for those who used tables than for those who used maps. However, decisions were more accurate using tables than using maps only when tasks involved geographic containment relationships.

Dennis & Carte (1998) examined the impact of map and tabular representations on decision processes (analytical and perceptual) and on decision outcomes (time and accuracy). They argued that maps are spatial representations, which are suitable for spatial analysis such as geographic adjacency tasks, whereas tables are symbolic representations, suitable for symbolic analysis such as geographic containment tasks. They argued that map representation would induce decision makers to use perceptual processes in decision-making, while table representation would induce decision makers to use analytical processes. Using the cognitive fit theory, they posited that the match between task and representation would lead to good decision outcomes. They performed a lab experiment in which undergraduate participants used maps (perceptual inducing) and tables (analytical inducing) to perform adjacency and containment tasks. They found that using maps, compared to tables, led to faster and more accurate decisions in geographic adjacency tasks, whereas using maps in geographic containment tasks resulted in less accurate but faster decisions.

Swink & Speier (1999) examined the impact of different characteristics of geographic information (problem size, data dispersion, and data aggregation) on decision outcome when using maps for a facility network design problem. The result from the lab experiments indicated that all three characteristics have significant impacts on decision performance, in that performance is deteriorated when problem size is large, data is less aggregated, and data is more dispersed. Problem size increases the complexity of the decision process, which results in lengthier decision time and less accurate decisions. Data aggregation effects decision time but does not influence the accuracy of decisions. Data dispersion, on the other hand, has a significant impact on the accuracy of decisions but does not change decision time. The study also revealed that the decision time is impacted by the interaction between problem size and data dispersion and the interaction between problem size and data aggregation. However, only the interaction between data dispersion and data aggregation has a significant impact on decision quality.

Based on the studies in GIS, we can conclude that map visualization has good potential for improving decisions. However, there is inadequate research documenting the efficacy of SDSS in enhancing decisions.

Toward Testing the Efficacy of SDSS

GIS in business has been popular for almost twenty years but, as shown above, there are only a handful of studies in IS literature evaluating different facets of the technology (e.g., Crossland et al., 1995; Dennis & Carte, 1998; Smelcer & Carmel, 1997; Swink & Speier, 1999). Most of the empirical studies have focused on the impacts of map representation. Only the studies by Crossland et al. (1995) and Mennecke et al. (2000) examined the effects of spatial analysis tools — map overlay and buffer functions. None of the studies has examined the advanced spatial analytical tools used in GIS.

To evaluate the effectiveness of GIS as spatial decision support systems, we argue that both map representation and advanced spatial analytical tools should be evaluated for their efficacy in decisions that involve multi-dimension and spatial features (Jarupathirun & Zahedi, 2001). In doing so, we draw from a number of theories to build a conceptual model for examining the efficacy of SDSS (*Figure 4*).

Task-Technology Fit. Since SDSS in most cases would involve using advanced analytical tools in GIS, we need to evaluate the fit between given decision tasks and the features of SDSS that support them. Thus, we propose the use of task-technology fit (TTF) theory to evaluate the impact of GIS-based SDSS. According to TTF theory (Goodhue & Thompson, 1995), each function should fit the task requirements in order to have a positive impact on performance. In this context, spatial analysis tools with map representation are used to overcome the limitation of humans' analytical and visualization abilities for spatial decision-making. To achieve better decision performance and

Figure 4. Conceptual Model for Examining the Efficacy of SDSS

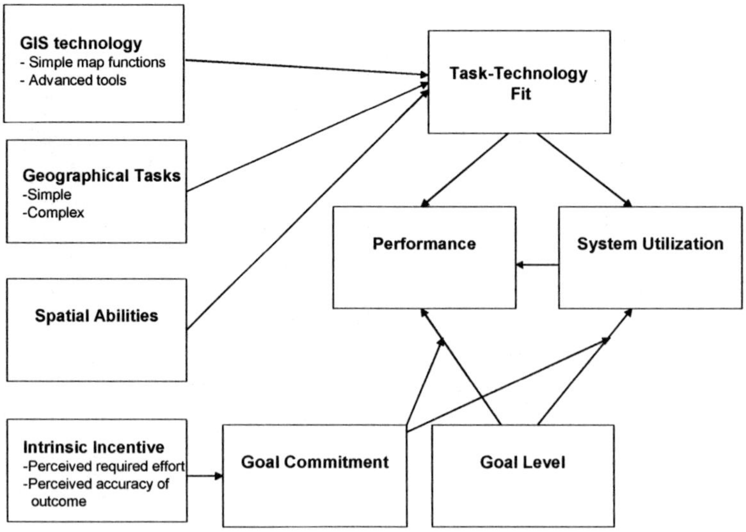

higher use of SDSS, we posit that the higher complexity of decision tasks in the form of higher dimensions and a larger set of alternatives generate greater needs for advanced spatial analytical tools and increased interactivity between users and dynamic maps.

Abilities. Marcolin et al. (2000) argued that abilities play a major role in TTF. Successful visualization using GIS-based SDSS requires not only the external stimuli for using such systems, but also the knowledge about using interactive maps as well as about task domain. In addition, to perform spatial analysis, decision makers have to understand how to model spatial problems and how to operate available spatial analytical tools. In the survey of background coursework for GIS, Wikle (1994) found that more than 60% of respondents agree that map reading, database management and spatial analysis knowledge are extremely important to the effective use GIS.

Following these findings, we argue that higher levels of decision makers' abilities are associated with effectiveness of SDSS, and this relationship is mediated by the higher level of task-technology fit. Since the use of GIS-SDSS requires maps for visual thinking, we argue that salient abilities in the context of using SDSS are users' spatial abilities. Spatial abilities could be measured in a number of ways. However, two dominant spatial ability dimensions are viewed as important. They are "spatial orientation" and "visualization" (e.g., Cooper, 1988; Golledge & Stimson, 1997). The visualization ability is defined as "*the ability to manipulate or transform the image of spatial pattern into other arrangements*" (Ekstrom et al., 1976, p. 173). The spatial-orientation ability is defined as "*the ability to perceive spatial pattern or to maintain orientation with respect to objects in space"* (Ekstrom et al., 1976, p. 149). Both visualization and spatial orientation dimensions are used as measures of spatial abilities. Smelcer & Carmel (1997) suggested that spatial visualization ability is needed for navigational tasks. Swink & Speier (1999) found that when using GIS, decision makers with high spatial orientation ability significantly provided better decision quality for both small and large problems than decision makers with low spatial orientation ability. In making decisions, only a single map may not be sufficient for visual communication and visual thinking, particularly when tasks are increasing in complexity. As decision complexity increases, decision makers using GIS may need to manipulate and analyze various maps to identify the problems or discover patterns. Hence, the effective use of GIS for dealing with complex decisions requires spatial abilities. Decision makers with low spatial ability may feel less comfortable using visualization to solve or recognize the problems. Therefore, we argue that the performance of decision makers with low levels of spatial abilities would be associated with low levels of TTF, leading to a lower performance.

Goal Setting and Commitment. One of the well-established theories in organizational behavior is the theory of goal setting, described as a positive relationship between goal level and task performance (Locke et al., 1988). A higher goal level motivates individuals to spend more effort to achieve the desired decision performance. Hence, we posit that a higher level of goal setting is positively associated with decision performance and system utilization.

However, goal setting does not work if goal commitment is not maintained over the entire period in which the decision task is performed. Goal commitment has been found to have a moderating effect on goal level and performance (e.g., Hollenbeck & Klein, 1987). Commitment is influenced by external factors (e.g., external reward), interactive factors (e.g., competition), and internal factors (e.g., internal satisfaction). Thus, we posit that goal commitment moderates the impact of goal setting on performance and system utilization, in that a higher level of goal commitment makes the impact of goal setting on performance and system utilization more pronounced.

Intrinsic Incentive. According to the cost/benefit theory in DSS, using a system for decision-making involves costs and benefits for decision makers (Todd & Benbasat, 1992). The cost involves the efforts required in learning and using the system and the benefits are the increased accuracy in decision outcome. In this theory, decision makers evaluate the trade-off between effort and accuracy in selecting a problem-solving strategy. We define intrinsic incentive as the difference between cost and benefit, where cost is measured in terms of perceived effort and benefit is measured in terms of perceived accuracy of results in using the SDSS. Our model builds on and modifies the role of incentive in Todd-Benbasat's model (1999) by introducing users' goal setting and goal commitment. We posit that the impact of incentive for utilization and performance is mediated by the level of users' goal commitment.

In sum, the proposed conceptual model could be used to measure the efficacy of SDSS in terms of decision performance (quality of decisions) and extent of system utilization. However, we argued that a number of factors come into play in such evaluations, which would influence the efficacy of SDSS. These factors are: the fit between decision tasks and the SDSS functionalities, users' spatial abilities, users' net cost/benefit in terms of expenditure of efforts and expected accuracy of decisions, as well as their goal commitments and goal settings. We drew on the theories in TTF, goal setting, and cost/benefit in DSS in constructing the relationships between these factors and the outcome of using SDSS in terms of decision performance and system utilization. We believe that testing this model could enhance our insight regarding the efficacy of using GIS-based SDSS.

Discussion

In this chapter, we argued that the unique visualization capabilities of GIS combined with their advanced analytical tools make them appropriate for supporting complex decisions that involve spatial features, multiple dimensions, and a large alternative set. We explored the existing literature in search of studies evaluating the efficacy of SDSS, and found few studies that include salient factors for such evaluation. This gap was the motivation for proposing a conceptual model for exploring the efficacy of GIS-based SDSS that includes a theoretically based set of salient factors in SDSS evaluation.

Although we have not discussed the evaluation of the conceptual model presented in this chapter, the arguments for the relationships among factors in the model are based on well-grounded theories and empirical studies in IS literature, organization literature and psychology literature. The key idea that IT managers and business managers can draw from this chapter is that there may not be a universal SDSS for all decision types and business applications. The success of adapting GIS as SDSS may involve the right mix of human input and technology. For the effective use of GIS as SDSS, the GIS need to be customized and integrated with other IS technology. In doing so, the selection of analytical tools should fit with decision tasks. Furthermore, decision makers' abilities and motivations to use SDSS may contribute to the successful use of SDSS.

This conclusion is in line with the Scott Morton (1971) categorizations of decisions for using DSS. The use of SDSS in general can be classified based on the type of decision: structured, semistructured, and unstructured; and the type of business control: operational control, management control, and strategic planning. The tools and functionalities of SDSS should match decision and business control types.

For example, GIS-based SDSS could help customer services at the operational level to determine the nearest service location, such as finding the nearest ATM, and in an emergency dispatch center to help paramedics find the shortest route to carry a patient to a hospital or locate alternative routes (the second best), when the first choice is not practical at that moment. The VISA website uses Internet-based GIS to provide an interactive map for customers to locate the nearest ATM and to display the directions for getting to it. The needed tools for the VISA website may be only a map engine to display maps, a simple spatial query to answer the "where is located" question, and an overlay function, whereas the required SDSS tools for the emergency dispatch service center may be not only a map engine and spatial query engine but also the network analysis engine that ranks the shortest routes. In both applications in the ATM locator and the emergency route analysis, the map visualization is used mainly for communication rather than for the identification and solution of unstructured decision problems.

On the other end of the spectrum of SDSS application, SDSS could be applied for dealing with unstructured decisions and strategic planning. Such applications use visualization to aid the cognitive process that involves identifying the problem, setting up hypotheses or confirming the analysis, and finding solutions. The decision process is more complex in these cases and the SDSS would require advanced tools and functionalities, such as spatial statistical analysis.

Hence, in implementing SDSS for increasing decision-making effectiveness, IT managers and business managers may need to ask the following questions. Does the SDSS have the capabilities and functions that support the task requirements? Who is going to use a particular SDSS? Do SDSS users have adequate knowledge about the problem domain to ask the right questions, skill to use the technology, and spatial abilities to understand the visual outputs?

Future Trends

The present perspectives on GIS-based SDSS could be extended to go beyond visualization through geographical maps. Dransch (2000) proposed the use of other media such as 2D and 3D pictures, animation, sound, and video together with maps to enhance the process of scientific visualization — visual thinking and visual communication. Instead of separately investigating the effectiveness of each media, the development of SDSS should also focus on the combined use of different representations and media, each addressing a given set of decision makers' needs in dealing with complex decisions. For example, animation could be used to visualize time series data (such as the presentation of weather maps for different time periods), whereas video could be used to visualize highly interrelated data, such as information regarding the damages caused by a fire. The examination of such synergy requires extending the existing theories of task-technology fit, cognitive fit, or proximity compatibility principle that could help in developing conceptual models for exploring the efficacy of integrated SDSS.

Another direction of enhancement in GIS-based SDSS is the intelligent interface. Users' abilities have been identified to be critical in information system's success (Marcolin et al., 2000). The use of GIS equipped with intelligent components such as a knowledge base or an intelligent agent may help business users to overcome the lack of knowledge in cartography or spatial analysis.

In addition, an intelligent agent could be used to complement the lack of decision makers' spatial abilities. An intelligent agent could also guide decision makers to use a "divide and conquer" strategy to reduce the complexity of a problem when it recognizes a lack of spatial abilities in a user profile. Teng & Fairbairn (2002) used a fuzzy expert system and an adaptive neuro-fuzzy system so that the computers can recognize the objects and pattern shown on a map.

All in all, in creating intelligent SDSS, it is important to investigate the issues related to the use of intelligent technology for increasing the efficacy of SDSS. For example, one needs to investigate the level of required knowledge for efficiently and effectively using SDSS at different phases of a decision making process. Or, whether the loss of control due to the presence of intelligent agents could possibly cause reluctance in using SDSS.

Yet another interesting direction in the development of GIS as SDSS is in making it available on the Web. Putting the GIS-based DSS on the Web for use by the globally connected community of web users has its own challenges and issues. As shown in *Figure 5*, as the GIS technology has moved from mainframe to desktop, then on to the Internet and mobile tools, it has gained a wider audience while losing some of its functionalities due to limited hardware and communication capabilities, although the gap between functionalities of mainframe and desktop GIS is narrowing. The mass use of GIS-based SDSS over the Internet introduces cultural factors in the successful use of the technology.

One of the key factors that influence the diffusion of technology globally is culture (Burn, 1995; Walsham & Sahay, 1999). Walsham & Sahay (1999) studied the implementation of GIS technology in India and had problems in communicating with typical Indians using maps, since maps are not used in their daily lives. Therefore, in addition to the impact of culture on diffusion, it is important to understand how culture influences the perceived task-technology fit and other perceptual constructs of people from different cultures. For example, some cultures rely more heavily than others on consensus building and collective decision-making. In such cases, the SDSS would require the capability for group decision-making or multiple inputs. We are just starting to explore the impact of

Figure 5. GIS Platforms

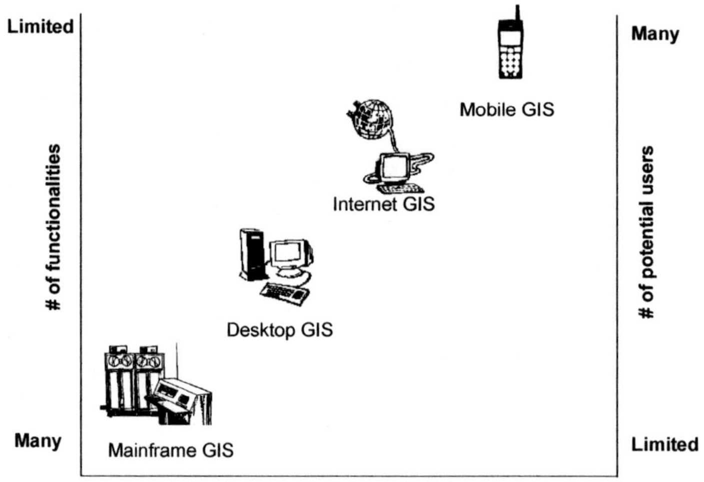

culture in the use of technology, and GIS-based SDSS will be among technologies on which culture may have significant influence.

Summary

With the rapid growth and increased availability of spatial business data, the significant advances in GIS technology, and the considerable decline in technology cost, the GIS-based decision systems have become increasingly more popular in public and private organizations and over the Internet. In this chapter, we discussed the nature of GIS for building spatial decision support systems and reviewed its applications in various areas and industries. We discussed the capabilities that distinguish GIS from traditional information systems and provided a historical perspective for the parallel development of IS-DSS and GIS technologies, which provided a foundation for the important role of GIS-based decision support systems in decision-making. We argued that visualization plays an important role in decision-making, particularly in dealing with complex mental models. Since the important features of GIS-based SDSS are the visualization capabilities for representing spatial and attribute variables in the form of maps and advanced analytical tools, GIS-based SDSS are uniquely positioned to aid decision makers in a wide spectrum of decision problems, from spatial data inquiry to unstructured strategic decisions.

Given the significant potential of GIS-based SDSS in improving decision-making, we observed the inadequacy of research documenting the effectiveness of SDSS. In an attempt to reduce this gap, this chapter developed a model for measuring the efficacy of SDSS by integrating theories across multiple disciplines, including psychology, organizational behavior and IS. Based on this conceptual model, we alerted practitioners to the potential importance of including functionalities that match with the intended decision tasks as well as the abilities and motivations of the target decision makers. Finally, observing the increasing need for visualization, particularly on the Web, we discussed future directions for possible enhancements of GIS-based SDSS.

References

Attenucci. J.C., Brown, J., Croswell, P.L., Kevany, M.J., & Archer, H. (1991). *Geographic information system: A guide to technology*. New York: Van Nostrand Reinhold.

Barnes, S. (2002). Salt Lake hosts: Spatial olympics. *Geospatial Solutions*, *12* (4), 26-33.

Barnes, S. (2003). Spatial intelligence and Iraqi freedom. *Geospatial Solutions*, *13* (5), 28-33.

Bracken, I., & Webster, C. (1989). Toward a new typology of geographical information systems. *International Journal of Geographic Information Systems*, *3* (2), 137-152.

Brewster, C.C., Holden, E.L., & Allen, J.C. (2002). Spatial tools for pest management. *Geospatial Solutions, 12* (6), 26-31.

Brodsky, H. (2002). GIS goes to court: Making a case for forensic mapping. *Geospatial Solutions, 12* (3), 48-51.

Burn, J.M. (1995). The new cultural revolution: The impact of EDI on Asia. *Journal of Global Information Management, 3* (4), 16-23.

Burrough, P.A. (1986). *Principles of geographical information systems for land resources assessment.* Oxford: Clarendon Press.

Chase, W.G. (1972). *Visual information processing.* New York: Academic Press.

Cooper, L.A. (1988). The role of spatial representations in complex problem solving. In S. Schiffer & S. Steele (Eds.), *Cognition and Representation* (pp. 53-86). CO: Westview Press.

Cowen, D.J. (1988). GIS versus CAD versus DBMS: What are the differences? *Photogrammetric Engineering & Remote Sensing, 54* (11), 1551-1555.

Crampton, J.W. (2001). Maps as social constructions: Power, communication and visualization. *Progress in Human Geography, 25* (2), 235-252.

Crossland, M.D., Perkins, W.C., & Wynne, B.E. (1995). Spatial decision support systems: An overview of technology and test of efficacy. *Decision Support Systems, 14* (3), 219-235.

Dawes, S.S., & Oskam, S. (1999) The Internet, the state library and the implementation of statewide information policy: The case of the NYS GIS clearinghouse. *Journal of Global Information Management, 7* (4), 27-33.

Dennis, A.R., & Carte, T.A. (1998). Using geographical information systems for decision making: Extending cognitive fit theory to map-based presentations. *Information Systems Research, 9* (2), 194-203.

De Silva, F.N., & Eglese, R.W. (2000). Integrating simulation modelling and GIS: Spatial decision support systems for evacuation planning. *Journal of Operational Research, 41* (4), 423-430.

Devine, H.A., & Field, R.C. (1986). The gist of GIS. *Journal of Forestry, 84* (8), 17-22.

DiBiase, D. (1990). Visualization in the earth sciences. *Earth and Mineral Sciences. Bulletin of the College of Earth and Mineral Sciences, PSU, 59* (2), 13-18.

Dransch, D. (2000). The use of different media in visualizing spatial data. *Computers and Geosciences, 26* (1), 5-9.

Ekstrom, R. B., French, J.W., & Harman, H.H. (1976). *Cognitive factors: Their identification and replication.* NJ: Princeton.

Gagne, C. (1999). Geographic information systems. *Computerworld, 33* (16), 90.

Gershon, N., & Page, W. (2001). What storytelling can do for information visualization. *Communications of the ACM, 44* (8), 31-37.

Ghinea, G., Gill, G., Frank, L., & de Souza, L. H. (2002). Using geographical information systems for management of back-pain data. *Journal of Management in Medicine, 16* (2/3), 219-237.

Golledge, R.G., & Stimson, R.J. (1997). *Spatial behavior: A geographical perspective.* New York: The Guilford Press.

Goodchild, M.F. (2001). A geographer looks at spatial information theory. In D.R. Montello (Ed.), *Spatial Information Theory: Foundations of Geographic Information Science* (pp. 1-13). New York: Springer.

Goodhue, D.L., & Thompson, R.L. (1995). Task-technology fit and individual performance. *MIS Quarterly, 19* (2), 213-236.

Groupe, F.H. (1990). Geographic information systems: An emerging component of decision support. *Journal of Information Systems Management, 7* (3), 74-78.

Harris, B., & Batty, M. (2001). Location models, geographic information, and planning support system. In R. K. Brail & R. E. Klosterman (Eds.), *Planning support systems: Integrating geographic information systems, models and visualization tools* (pp. 25-58). CA: ESRI Press.

Heikkila, E.J. (1998). GIS is dead: long live GIS! *Journal of the American Planning Association, 64* (3), 350-360.

Hollenbeck, J.R., & Klein, H.J. (1987). Goal commitment and the goal-setting process: Problems, prospects, and proposals for future research. *Journal of Applied Psychology, 72* (2), 212-220.

Huxhold, W. (1991). *An introduction to urban geographic information systems.* New York: Oxford University Press.

Hyndman, J., & Holman, C. (2000). Differential effects on socioeconomic groups of modeling the location of mammography screening clinics using geographic information systems. *Australian and New Zealand Journal of Public Health, 24* (3), 281-286.

Ioannou, G., Kritikos, M.N., & Prastacos, G.P. (2002). Map-route: A GIS-based decision support system for intra-city vehicle routing with time windows. *Journal of Operational Research, 53* (8), 842-854.

Jarupathirun, S., & Zahedi, F.M. (2001). A theoretical framework for GIS-based spatial decision support systems: Utilization and performance evaluation. *Proceedings of the Seventh Americas Conference on Information Systems* (pp. 245-248). Boston, August 2001.

Jarvenpaa, S.L. (1989). The effect of task demands and graphical format on information processing strategies. *Management Science, 35* (3), 285-303.

Johnson, M.L. (1996). GIS in business: Issues to consider in curriculum decision-making. *Journal of Geography, 95* (3), 98-105.

Kitchin, R., & Tate, N.J. (2000). *Conducting research in human geography: Theory, methodology and practice.* New York: Prentice Hall.

Klosterman, R.E. (2001) Planning support systems: A new perspective on computer-aided planning. In R. K. Brail & R. E. Klosterman (Eds.), *Planning support systems: Integrating geographic information systems, models, and visualization tools* (pp. 1-24). CA: ESRI Press.

Kraidy, U. (2002). Digital media and education: Cognitive impact of information visualization. *Journal of Educational Media, 27* (3), 95-106.

Lee, J., Brody, T.M., Zhang, R., Kim, H. J., & Bradac, M. (2002) Simulating urban growth on the Web. *Geospatial Solutions, 12* (11), 42- 45.

Locke, E.A., Latham, G.P., & Erez, M. (1988). The determinants of goal commitment. *Academy of Management Review, 13* (1), 23-39.

Malczewski, J. (1999). *GIS and multi-criteria decision analysis.* Toronto, Canada: John Wiley & Sons.

Marakas, G.M. (2003). *Decision support systems in 21st century.* NJ: Prentice Hall.

Marcolin, B.L., Compeau, D.R., & Munro, M.C. (2000). Assessing user competence: Conceptualization and measurement. *Information Systems Research, 11* (1), 37-60.

Mennecke, B.E., Crossland, M.D., & Killingsworth, B.L. (2000). Is a map more than a picture? The role of SDSS technology, subject characteristics, and problem complexity on map reading and problem solving. *MIS Quarterly, 24* (4), 601-629.

Meyer, J. (2000). Performance with tables and graphs: Effects of training and a visual search model. *Ergonomic, 43* (11), 1840-1865.

Mitchell, T.M. (1999). Machine learning and data mining. *Communications of the ACM, 42* (11), 30-36.

Murphy, L.D. (1995). Geographic information systems: Are they decision support systems? *Proceedings of 28th Annual Hawaiian International Conference on Systems Sciences* (pp. 131-140). Hawaii.

Newell, A., & Simon, H.A. (1972). *Human problem solving.* NJ: Prentice-Hall.

Pimpler, E., & Zhan, F.B. (2003). Drop by drop: Modeling water demand on the lower Colorado. *Geospatial Solution, 13* (5), 38-41.

Pokharel, S. (2000). Spatial analysis of rural energy system. *International Journal Geographical Information Science, 14* (8), 855-873.

Presmeg, N.C., & Balderas-Canas, P.E. (2001). Visualization and affect in non-routine problem solving. *Mathematical Thinking and Learning, 3* (4), 289-313.

Price, J., & Schweitzer, T. (2002). Before the deluge: FEMA, floodplains, and GIS. *Geospatial Solutions, 12* (10), 38-42.

Robey, D., & Sahay, S. (1996). Transforming work through information technology: A comparative case study of Geographic Information Systems in country government. *Information Systems Research, 7* (1), 93-110.

Rogerson, P.A., & Fotheringham, S. (1994). GIS and spatial analysis: Introduction and overview. In S. Fotheringham & P. Rogerson (Eds.), *Spatial Analysis and GIS* (pp. 1-10). London: Taylor & Francis.

Scholten, H., & van der Vlugt, M. (1990). A review of geographic information systems application in Europe. In Les Worrall (Ed.), *Geographic Information Systems: Developments and Application* (pp. 13-40). New York: Belhaven Press.

Scott Morton, M.S. (1971). *Management Decision Systems: Computer-based support for decision making.* MA: Harvard University Press.

Seffino, L.A., Medeiros, C.B., Rocha, J.V., & Yi, B. (1999). WOODSS - A spatial decision support system based on workflows. *Decision Support Systems*, *27* (1/2), 105-123.

Smelcer, J.B., & Carmel, E. (1997). The effectiveness of different representations for managerial problem solving: Comparing tables and maps. *Decision Sciences*, *28* (2), 391-420.

Swink, M., & Speier, C. (1999). Presenting geographic information: Effects of data aggregation, dispersion, and users' spatial orientation. *Decision Sciences*, *30* (1), 169-195.

Talen, E. (2000). Bottom-up GIS: A new tool for individual and group expression in participatory planning. *Journal of the American Planning Association*, *66* (3), 279-294.

Tarantilis, C.D., & Kiranoudis, C.T. (2002). Using a spatial decision support system for solving the vehicle routing problem. *Information & Management*, *39* (5), 359-375.

Teng, C.H., & Fairbairn, D. (2002). Comparing expert systems and neural fuzzy systems for object recognition in map dataset revision. *International Journal of Remote Sensing*, *23* (3), 555-567.

Thrall, G.I., & Fandre, M. (2003). Trade areas and LSPs: A map for business growth. *Geospatial Solutions*, *13* (4), 48-51.

Todd, P., & Benbasat, I. (1992). The use of information in decision making. *MIS Quarterly*, *16* (3), 373-393.

Todd, P., & Benbasat, I. (1999). Evaluating the impact of DSS, cognitive effort, and incentives on strategy selection. *Information System Research*, *10* (4), 356-374.

Tomlinson, R.F. (1987). Current and potential uses of geographical information systems: The North American experience. *International Journal of Geographical Information Systems*, *1* (3), 203-218.

Umanath, N.S., & Vessey, I. (1994). Multi-attribute data presentation and human judgment: A cognitive fit perspective. *Decision Sciences*, *25* (5/6), 795-824.

Venden, E., & Horbinski, T. (2003). Gurnee streamlines notification process. *Geospatial Solutions*, *13* (1), 26-27.

Vessey, I. (1991). Cognitive fit: A theory-based analysis of the graphs versus table literature. *Decision Sciences*, *22* (2), 219-241.

Walsham, G., & Sahay, S. (1999). GIS for district-level administration in India: Problems and opportunities. *MIS Quarterly*, *23* (1), 39-66.

Weber, W. (2001). The use of GIS & OLAP for accurate valuation of developable land. *Journal of Real Estate Portfolio Management*, *7* (3), 253-310.

West, L.A., Jr., & Hess, T.J. (2002). Metadata as a knowledge management tool: Supporting intelligent agent and end user access to spatial data. *Decision Support Systems*, *32* (3), 247-264.

Wickens, C.D. (1992). *The proximity compatibility principle: Its psychological foundation and its relevance to display design* (Technical Report ARL-92/NASA-923). Savoy. IL: Aviation Research Laboratory, Institute of Aviation, University of Illinois at Urbana-Champaign.

Wikle, T.A. (1994). Survey defines background coursework for GIS education. *GIS World, 7* (6), 53-55.

Wilson, R.D. (1994). GIS and decision support systems. *Journal of Systems Management, 45* (11), 36-40.

Witten, I.H., & Frank, E. (2000). *Data mining: Practical machine learning tools and techniques with Java implementations*. New York: Morgan Kaufmann Publishers.

Wood, M. (1994). The traditional map as a visualization technique. In H.M. Hearnshaw & D.J. Unwin (Eds.), *Visualization in geographical information systems* (pp. 9-17). New York: Wiley & Sons.

Worrall, L. (1991). *Spatial analysis and spatial policy using geographic information systems*. London: Belhaven Press.

Yongling, Y. (2002). Beijing downtown: Mapping customer reach in an urban core. *Geospatial Solutions, 12* (4), 40-43.

Yoon, Y., Guimaraes, T., & O'Neal, Q. (1995). Exploring the factors associated with expert systems success. *MIS Quarterly, 19* (1), 83-106.

Chapter VIII

The Value of Using GIS and Geospatial Data to Support Organizational Decision Making

W. Lee Meeks, George Washington University, USA

Subhasish Dasgupta, George Washington University, USA

Abstract

For several years GIS has been expanding beyond its niche of analyzing earth science data for earth science purposes. As GIS continues to migrate into business applications and support operational decision-making, GIS will become a standard part of the portfolio that information systems organizations rely on to support and guide operations. There are several ways in which GIS can support a transformation in organizational decision-making. One of these is to inculcate a geospatial "mindset" among managers, analysts, and decision makers so that alternative sources of data are considered and alternative decision-making processes are employed.

"If a man does not know to what port he is steering, no wind is favorable."

- Seneca, 4 B.C.-65 A.D.

"I am told there are people who do not care for maps, and I find it hard to believe."
- Robert Louis Stevenson, 1850-1894

Chapter Organization

This chapter presents a data-centric view of the ways that GIS and geospatial data support organizational decision-making. As such, the chapter is organized to cover the following topics:

- *Introduction*
  - Non-traditional uses of GIS
  - GISs are descriptive and can be prescriptive
  - Having geospatial and spatiotemporal mindsets are important
  - Three themes to carry away
- *Background*
  - GIS fit the general systems model
  - Three relevant geosciences functions
  - Spatial and spatiotemporal data and information
- *Issues in Decision-Making*
  - Decision models fit general systems models
  - Decision inputs can include geospatial data
  - Understanding errors in modeling organizations
- *GIS Support to Decision-Making*
  - Age of the spatial economy
  - Integrating business, geospatial, and remotely sensed data
  - Incorporating geospatial and spatiotemporal contexts
  - Data aspects of GIS in decision-making
- *Geospatial Data Issues to Improve Decision-Making*
  - Drilling into geospatial data types and uses
  - Considering evaluation paradigms for geospatial data in GIS
- *Evaluating the Value of Geospatial Information in GIS*
  - Considering the value of information
  - The status quo for evaluating geospatial data

- An alternative: Geospatial Information Utility
- *Looking into the Future*
  - Advances in remote sensing and opportunities
- *Summary*
  - Recapping the three themes

Introduction

This chapter explores themes that are based on some non-traditional ways geographic information systems (GIS) are able to support businesses through transforming organizational decision-making at operational, tactical, and strategic levels. The starting point is the recognition that GIS and the geospatial sciences are mostly gaining prominence and popularity by providing answers to analysts, researchers, practitioners and decision makers for those problems — of different types and levels of complexity — that have a recognized spatial or spatiotemporal component. This, in fact, is what most organizations come to GIS for: to find descriptive and prescriptive answers to space and time problems. Within the geospatial sciences, and considering the use of GIS, descriptive answers are provided through analysis of collected sampling data. An entire field of statistical analysis, called geostatistics (Isaaks & Srivastava, 1989), exists to guide and improve the quality of statistical analysis of spatially oriented data. Just as other fields combine expert domain knowledge and inferential statistical analyses to make probabilistic predictions about future operating environments or activities, GIS is also used in prescriptive ways to support decision-making. For managers in different industries and in firms of different sizes, using GIS to provide both descriptive and prescriptive answers means being able to adopt a spatiotemporal "mindset" that automatically presumes business data have space and time components that can be mined and analyzed to improve decision-making.

We consider the distinction between spatial and geospatial (or geographic) data important: spatial roughly means "place" or "space" (e.g., answers where, how far, and how long or wide kinds of questions) whereas geospatial, which is properly a subset of spatial, means "place or space tied to a geographic reference." We also consider the distinction between spatial (or geospatial, depending on the context or frame of reference) and spatiotemporal to be important because of the exclusion or inclusion of a temporal frame of reference, which includes place or space changes over time.

A geospatial mindset means having a pre-disposition towards considering the analysis of business problems from a spatial or spatiotemporal perspective. Thus both the spatiotemporal and geospatial mindsets are important and worth mentioning separately. For example, a manager uses her spatiotemporal mindset to examine her continental transportation and logistics operations as occurring in "4-D" (e.g., latitude, longitude, elevation, and time) or her intra-factory materials movement as occurring in "4-D" (e.g., length, width, height, and time) by including consideration of the space/place relationships that change over time. Hence, space-time or spatiotemporal issues matter. Another manager seeks to optimize the expansion of a cellular telephone transmission network through the mini-max decision of how few new cell phone towers should be built (i.e., fewer towers, lower cost) vs. how many are needed (i.e., more towers, better signal strength) — the mini-max decision is about minimizing cost while maximizing signal coverage to users. The use of a GIS — which requires geospatial and other cost and performance data — allows managers to perform better analyses for these kind of problems.

To extend and employ the geospatial and spatiotemporal mindsets, the biblical proverb about the difference between giving a person a fish to feed them for a day or teaching them to fish so they can feed themselves for all days is useful. The real question the business community should be asking of GIS is not "what is the (spatial) answer to my current problem?" But rather, "in what ways should I be thinking about my current sources of business data within a spatial context?" And, "what other sources and types of data support a spatiotemporal 'mindset' useful in improving the accuracy and speed of my organizational decision-making?" This is a key opportunity for GIS to support business in innovative ways. Three themes running through this chapter are:

- GIS can improve organizational decision-making through the awareness that all business decisions include space and time components. The benefit is that thinking spatiotemporally provides additional analytical approaches and methods.
- GISs use both business data and remotely sensed data. An awareness of the power of the different forms and sources of remotely sensed data and the ways their integration can transform organizational notions about how and where to collect business data helps improve both GIS-based and non-GIS based decision making.
- Accessing many different data sources and types imply challenges with using these data; these challenges include determining the quality of the data ingested, manipulated and outputted; and, equally as importantly, determining the utility and relevance of the ingested and outputted data and information as they pertain to the result of the final decision or action.

Another definition is useful: from Lillesand & Kiefer (2000, p. 1), "Remote sensing is the science and art of obtaining information about an object, area, or phenomenon through the analysis of data acquired by a device that is not in contact with the object, area, or phenomenon under investigation." The terms remotely sensed data (noun) and remote sensing (verb) are different forms of the same concept: to collect data on objects of interest from afar. Far and close are immaterial, e.g., imaging is becoming very prominent in the medical world where the distances are very small when compared to the altitude of a satellite orbiting hundreds of miles in space. Our thesis includes remote sensing (or remotely sensed data) as fact and as metaphor.

In the geospatial sciences, normally an actual sensor (e.g., electro-optical, radar, laser, radiometer) is employed to passively or actively perform the remote sensing. In both the geosciences and business operations, the distinction between passive and active data collection is important. Passive sensors rely on emitted radiance or other phenomena from the object of interest to perform the data collection. Active sensors emit electromagnetic energy to excite or illuminate the object of interest so that the sensor detects reflected energy from the object of interest. Both have their advantages and disadvantages. In our chapter, this has a normative meaning — remote sensing taken at its face — and a metaphorical meaning that means managers and decision makers should consider the act of remote sensing as a guide to finding alternative means for collecting and processing the data and information they need to make competitive business decisions at all levels of the operating spectrum.

Considering our central theme of opening business managers' minds to alternative forms of data, alternative sources of data, and alternative concepts for tying data, sources, and decision models together, other means might be considered as the vehicles that perform remote sensing. For example, agent-based queries on a firm's operating network retrieve local sales updates from distributed databases and then autonomously feed those changes to a central decision support system for analysis; this could be considered as remote sensing in a non-traditional context.

Background

Any discussion of GIS should begin with recognition of that GIS represents a holistic system of the systems with several complex yet easy to use components, such that:

GIS = *f*{Hardware, Software, Data, Connectivity, Procedures, Operators}

Where *Hardware* represents all systems hardware; *Software* represents operating software and other applications and tools; *Data* are primary and supporting data received from many sources, ingested into, manipulated by, and outputted from GIS systems; *Connectivity* represents system inter-networked connectivity linking GIS to remote data sources and other supporting applications; *Procedures* are the automated and manual processes, methods, and other algorithms necessary to use the GIS system; and *Operators* are the operators, analysts, researchers and others who use GIS hardware, software, data, connectivity, and procedures in order to support spatial analyses and other organizational decision-making. To further ground this view of the value of GIS to organizational decision-making and performance, it is also important to know that GIS operate as any other system according to a general system model incorporating inputs, processes, outputs, and feedback as shown in *Figure 1*.

In its niche, GIS evolved by analyzing earth science data primarily for earth science reasons. Much of this data is collected from remote sensing devices and specialists' fieldwork. Bossler (2002) and others point out the three main components, functions, or fields in the geospatial sciences: remote sensing, global positioning systems (GPS) and GIS, which functionally translate into: *collecting data* (i.e., remote sensing), *locating objects* (i.e., through the use of global positioning system or other survey or locational technique), and *analyzing data* and information (i.e., through the use of GIS). Many sub-fields and applications exist to develop and hone these functions. The "sub-fields" are

Figure 1. A General Systems Model

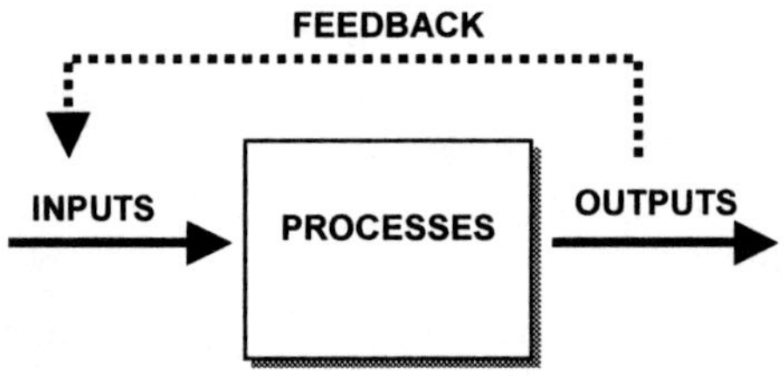

the decomposition of these three main functions or fields. For example, collecting data can be decomposed into passive versus active sensors; or sensors based on placement relative to the surface of the earth, e.g., space-, air-, surface-, or sub-surface-based sensors; or by their phenomenologies, e.g., imagery-, acoustic-, magnetic-, radio-electronic-, signal-, olfactory-, thermal-, seismic-based sensors. Therefore, there are many different ways to decompose each of collect, locate, and analyze functions and each of these decompositions constitutes the "sub-fields" to us. The point is relevant because each of these fields and sub-fields are developed and advanced relatively independently of the others. Each independent evolution includes hardware capabilities, software applications, data structures and forms, means for data transmission and sharing, operator training, and process improvements. All of these fields and sub-fields will continue to grow at a rapid rate as each uncovers faster, more versatile, and more accurate tools and ways to collect, locate, analyze and present GIS input and output data. This growth in capability, performance and versatility will be in part driven by external demands placed by communities of users as GIS penetrates into mainstream business operations.

Spatiotemporal information is comprised of spatial and temporal information. Spatial information is a component of organizational information that links to a place without respect to any specific geographic reference orientation (Longley et al., 2001). As mentioned, spatial information addresses "where" and "how far" kinds of questions. For example, sales figures for a region, inventory at distributed locations, machinery laid out on a shop floor, and even the swirls and whorls on a fingerprint are all spatial information. Geospatial information is a subset of spatial information, which includes an absolute or relative geographic or relative geographic basis, called geo-referencing. We will mostly use the term geospatial data as the term of preference in the latter part of the chapter. References to spatial data (versus geospatial data) are made to form the context for the focus on geospatial data. Detailing the issues and potential applications for non-geographic spatial data (e.g., particularly in the case of the geospatial information utility) is beyond the scope of our interest and work for now. Also, geospatial data occur in GIS far more frequently. There are, however, classes of systems such as automated computer-aided drawing (CAD) and computer-aided manufacturing (CAM) that specialize in the use, analysis and output of non-geographic spatial data and information. One caveat: GISs predominantly use geospatial data, but also use other forms of data, CAD/CAM applications predominantly use non-geographic spatial data, but some can also use geospatial data.

Planned locations for new cell phone towers and the real-time dynamic location information of en-route FedEx delivery trucks are examples of geospatial information. Temporal information is another component of organizational information, which is linked to particular events or places, which can also be specific as occurring at a specific time. Temporal information addresses "when" and "how often" questions, including start and stop times, or alternatively, duration intervals, and irrespective of the type of system (mechanical, electronic, or organizational), networks' latency for node-to-node processing and delay times. Temporal information is particularly useful in longitudinal analyses; that is, making assessments of changes in events or places over time, and in forecasting future changes in events or places over time. Just-in-time techniques widely-used in manufacturing today aim to deliver raw materials to factories or finished goods inventory to distributors at tightly specified intervals. In the previously mentioned case of the FedEx truck on delivery, it is important to consider not only where the truck is located, but also when it is located there. These are examples of using temporal information. By convention, the term spatiotemporal information includes information having a spatial orientation, a temporal orientation, or both. Considering the value of dynamic and static data, both spatial and temporal data can be static or dynamic. Cognitively speaking, spatial and temporal reasoning are common forms of reasoning; so much so they are not commonly thought of in any determined way. However, the value of reasoning and problem solving in spatiotemporal terms is gaining attention. Organizations like the National Center for Geographic Information and Analysis (NCGIA) are pursing spatiotemporal reasoning and analysis (Frank et al., 1992).

There is nothing inherently transformational about spatiotemporal information per se. The types of information provided as examples above are already being collected and analyzed in organizations. Today, however, decision timelines, like other operational aspects of organizational life, are becoming highly compressed. Advances in remote sensing and information systems provide the means to collect, analyze, exploit and disseminate questions, decisions, actions, and their results with ever-greater fidelity and robustness with ever-shorter timelines (Johnson et al., 2001). This is how a spatiotemporal mindset will be transformational: using information collected from a broader range of sources in more innovative ways to solve complex analytical and decision problems — the "faster, better, cheaper" paradigm.

Decision Making

Once thinking is focused in terms of a systems process model, or better yet, in systems-of-systems terms, it is natural to consider the extension of the general systems model to the art and science of decision-making. To do this, it is necessary to consider the nature of decision inputs, decision processes, and decision outputs. One current focus in decision sciences is on developing systematic methods to improve decision-making, because "... interest in decision making is as old as human history" (Hoch & Kunreuther, 2001, p. 8). The recognition that up to 80% of an organization's data are spatial (Bossler, 2002) is forcing, in part, this transformation to further systematize decision-making.

Figure 2. General Systems Model Applied to Decision Making

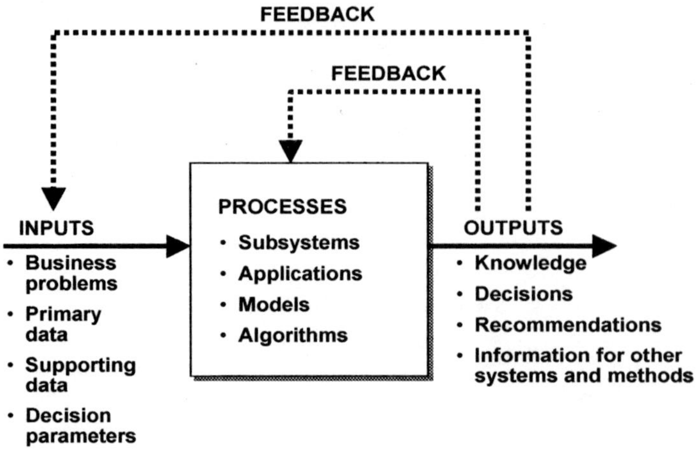

Christakos et al. (2002) describe how spatiotemporal information associates events with their spatial and temporal ordering, and that by using these data in new decision models, managers are able to achieve improved fidelity and quality in their decision-making. The future of this field is to encourage a mindset of spatial thinking in managers of all disciplines (DeMers, 2000). Broadly speaking, consistent with the general systems model, *Figure 2* depicts how systematized decision-making is comprised of decision inputs, decision processes, and decision outputs. Also essential are feedback loops to evaluate decision inputs and processes. Decision analysis is a related science, which is also being systematized (Clemen & Reilly, 2001).

Decision inputs include the decision requirements (e.g., what needs to be decided and other parameters), primary and other supporting data and metrics, and the degree of uncertainty or risk that is present or can be tolerated in the decision. Decision processes include the phases of the decision (e.g., generating and evaluation options), roles of actors involved in the decision, and decision support tools or systems. Specifically included are the internal systems, subsystems and processes found within the organization (labeled as "subsystems" on the figure above); and the applications, models and other domain specific algorithms affecting decision processes that determine how decision inputs are manipulated and otherwise analyzed in order to arrive at decision outputs. Finally, decision outputs include the decision in a form that can be communicated clearly to those who must act on it and may include a statement of the level of confidence associated with the decision result. A critical area is the impact errors have on analyzing organizational systems' behavior (i.e., with their associated operating activities), and how they are accounted for or dealt with within decision-making paradigms. *Figure 3* depicts key issues that must be considered in determining, accounting for, and ultimately eliminating errors found within organizational systems as a result of operating activities and managerial decision-making. It should be the goal of analysts, researchers, managers, and decision makers to be able to account for random errors and to eliminate bias errors. This model is particularly useful in considering the "nouns and verbs" that go into an assessment of what makes an organizational system

Figure 3. Understanding Sources of Errors in Organizational Systems

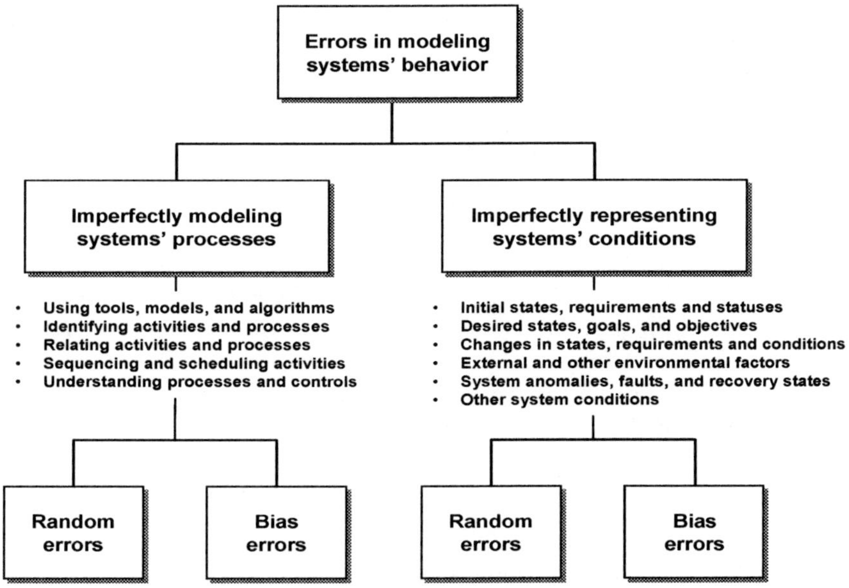

work. It represents a broad systems view of how errors are induced into analyses of businesses' operational activities and processes. The reason this view of errors is relevant to considering how GIS support decision-making in business is that the analytical structure resembles the analytical process in GIS analyses. Later in the chapter we will focus on errors within a more narrow data quality perspective, with respect to the geospatial data ingested into GIS to support decision-making.

GIS Support to Decision-Making

Fujita, Krugman, & Venables (1999) call this the age of the spatial economy and describe "economic geography," which considers where and why economic activity occurs. Business leaders must accommodate themselves to the distributed, geographic nature of their industries and, significantly, also to the growing use of spatial, geographic, and temporal information within decision-making paradigms. The increased use of these data is forcing new decision-making methods and tools. GIS, as an aid to decision making, provides crucial support to decision makers in many fields. However, the effective use of GIS requires high quality, accurate information as inputs. And, effectively used GIS outputs high quality, accurate, relevant, and "actionable" information to decision makers. Later, we will address how to value the spatiotemporal information used in GIS and how to value the output of GIS to decision makers within a spatiotemporal reasoning mindset, for as Albert Einstein's contemporary, Hermann Minkowski, lectured in 1905: "Henceforth space by itself and time by itself, are doomed to fade away to mere shadows, and only a kind of union between the two will preserve an independent reality" (Raper, 2000).

Incorporating the geospatial sciences into organizational decision-making complements and improves the quality of decisions (Malczewski, 1999). Further, new decision constructs supported by GIS are also supporting participatory group decision-making (Jankowski & Nyerges, 2001): generating creative options, identifying and quantifying multiple evaluation criteria, developing the causal linkage between options and evaluation criteria, and analyzing the uncertainty associated with the options-evaluations criteria. GIS and spatiotemporal data can help with all.

Readers should consider the lessons found in the history of information systems (IS) and management information systems (MIS). From the early, very limited beginnings of automated data processing on punch cards to vastly complex and responsive IS and MIS today, the inputs, processes, outputs, and uses of IS and MIS have evolved; sometimes incrementally and predictably, and sometimes dramatically and innovatively. GIS as a subclass of management information systems is gaining mainstream use and acceptance. As with the evolution of other forms of MIS, new and innovative uses and applications of GIS can spawn competitive advantage for managers and decision makers who grasp their potential significance and who are willing to experiment with the traditional and non-traditional use of these systems to solve many different types of problems.

The value of GIS, to our data-centric view, is:

- To allow decision makers to analyze and correlate their organization's operating activities in spatial and spatiotemporal terms, in ways probably not employed before.
- To inculcate in managers and decision makers the mindset that spatial, geospatial, and spatiotemporal factors are important. Specifically:
  - To guide decision makers to consider alternative forms of data.
  - To guide decision makers to consider alternative sources of data, where this means considering traditional remotely sensed data.
  - To guide decision makers to consider using geoscience remote sensing as a metaphor for innovations in business data collection and processing.
  - To guide decision makers to reevaluate their decision-making processes to incorporate GIS-provided alternatives.

As GIS breaks into the organizational mainstream, geospatial and spatiotemporal analyses allow managers and decision makers to incorporate additional data types and sources. Specifically, traditional "business data," which all along had space and time components, but which often went un-analyzed (or at the very best, under-analyzed in not-so-sophisticated ways considering today's computational tools and processing power), can also now be incorporated into GIS analyses for improved organizational decision-making. Thinking back to when business data was not thought of in robust spatiotemporal terms, it is not hard to imagine the manager of Sears Catalog division from 100 years ago trying to collect sales data for goods shipped across a growing country. What decisions did that long-ago manager face? What could he influence to improve Sears' profitability through managing the goods sold for shipment across the country? More importantly, what could he have done differently if he had the analytical tools of

GIS to help more precisely analyze his (textual) sales data within a place and time construct overlain on a map of production sites, distribution sites and networks, transportation nodes and links, and customers sorted by demand preferences and purchasing patterns?

Of longtime interest in GIS analyses are incorporation of remotely sensed data, which include environmental, intelligence, scientific, and other data from space-, air-, surface-, and subsurface-borne sensors. Beyond the use of remotely sensed data, in increasing complexity and utility are the collection and use of various types of structured and unstructured organizational data, which are analyzed in GIS within spatial and spatiotemporal contexts, or are correlated, with the available remotely sensed data in industry- and organization-unique ways. While many types of organizations need these data for operational purposes, managers and decision makers know that after the geospatial function "locate it," a decision must be made to act on "it".

Considering the breadth of topics in the use of GIS in business today, including the sentence, "GIS has been built up based on a combination of theories and concepts from IS and geography," we consider that many technical GIS readers will be interested in the "manipulate it" and "act on it" functions, which are most fundamental to GIS users, developers, and data providers. However, it is not just environmental data directly resulting from remote sensing — and remote sensing can take on many non-traditional forms — but also the spatiotemporal associations being addressed within business data that are important for business leaders and managers. Therefore, a new imperative in the field centers on the issue of the value of spatiotemporal data and information ingested into and outputted from GIS, specifically as they pertain to improving organizational decision-making. One focus considers both the state of today's GIS analysis technologies and also the technology and process advances that will shape tomorrow's convergence of GIS users and organizational decision makers.

Figure 4 shows how data inputs are moved into the system towards data uses, passing through several filters that consider data form and type, data resolution and data accuracy.

Figure 4. Considering Different Aspects of Geospatial Data in GIS

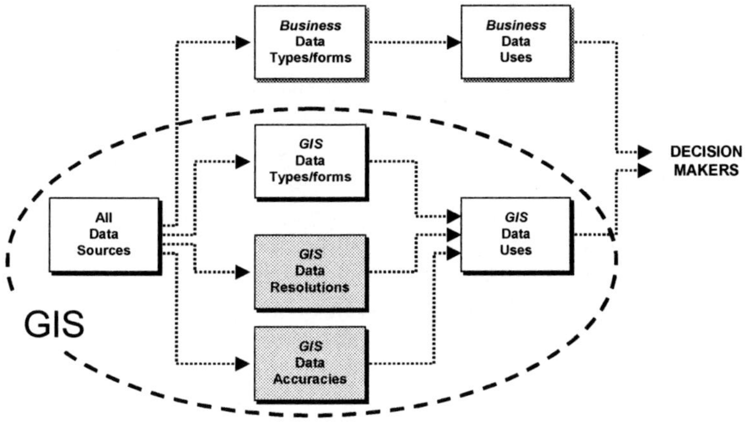

Figure 5. Considering Data Issues in Business Analyses

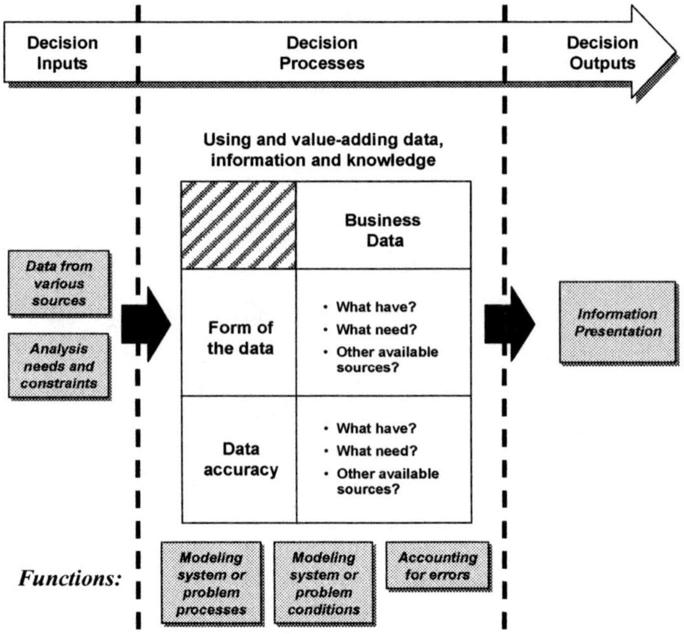

Both GIS and other business data are provided in several forms and types. For GIS, these data are typically raster, vector, elevation, spectral, textual, or structured (i.e., numerical data in tables, etc.) types. Business data can also take many forms; however, the most common seem to be the forms found in typical office productivity software suites: e.g., structured data found in databases and spreadsheet programs and non-structured data found in textual documents and presentation graphics. These data are often of interest to GIS analysts and can be integrated into GIS-supported decision-making.

As mentioned, data comes in many forms: vector, raster, gridded elevation, textual, and spectral are most common. Form is important because, as with application file types that most managers are comfortable with (e.g., using spreadsheet files, database files, word processing files, etc.), only allowed file types operate within specific applications. Data accuracy refers to getting things right where "right" means correct in many different dimensions, as with resolution. Data accuracy most often refers to content correctness and locational correctness. Accuracy can be: content, horizontal, vertical, spectral, radiometric, and temporal. Another consideration is data resolution, which refers to the ability to discretely discriminate between objects in the area of view. This can be thought of as how well something can be seen. It is important to note that there are many different types of resolution: spatial, spectral, radiometric, and temporal to name the most common.

Figure 6. Combining Geospatial and Business Data

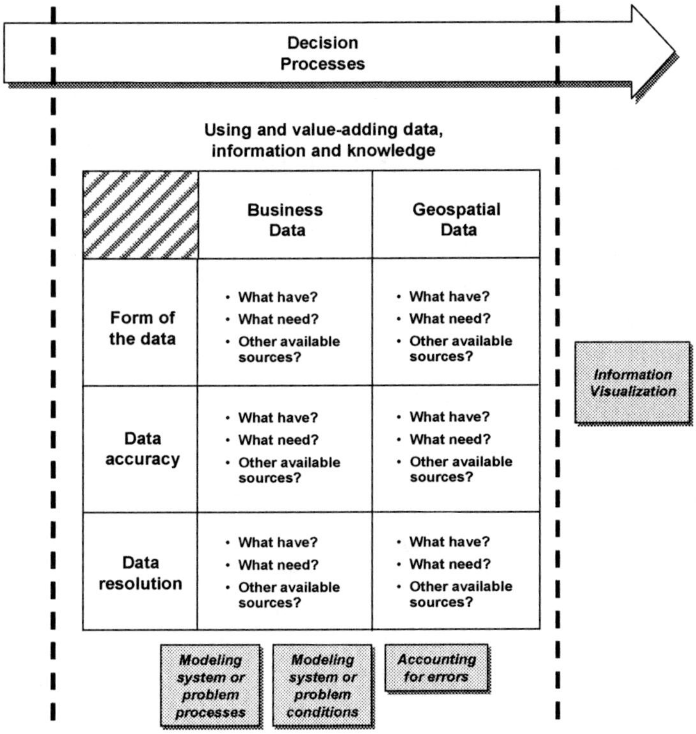

Data sources are varied (Decker, 2001). One way to classify data ingested into GIS and used for organizations' operational activities is to consider the remotely or environmentally sensed data, which are often purchased from commercial geospatial data collection vendors (Althausen, 2002) or are found in government, public, and other databases. Next are firm-specific operational data collected internally and finally industry-specific data procured from industry analysts and associations. Many of the data type, data form, and data accuracy issues are inimical to traditional business data, though we have considered these primarily on geospatial data terms. *Figures 5* and *6* link the decision process model with some of the data issues encountered within "decision processes." Managers and analysts must always ask themselves, what data do I have already available? What data do I need to perform a certain analysis? And, what other sources of data exist so that I can augment what I have to satisfy what I need? In *Figure 6*, the inclusion of geospatial data in the decision process increases the sophistication of what can be accomplished, however, it also increases the complexity of issues, such as also being forced to consider data resolution.

Geospatial Data Issues for Improved Decision-Making

Thinking specifically about GIS-oriented data types, raster data are organized arrays of data structured into regular cells such as pixels making up a satellite image or orthophoto (Decker, 2001). Rasters are used to classify some set of one or more "real world" areas of features found within the area encompassed by the pixel boundaries. As each cell represents a unit of surface area (Ormsby et al., 2001), this requires that each cell or pixel assume the measured or estimated value of the most dominant soil, terrain, or vegetation feature found within it. The benefit of raster data is that the use of regions of pixels can easily be formed to represent regions of common characterization (e.g., soil or vegetation type) or activity (e.g., changes in land cover type, such as through construction of buildings, roads, etc., or in dynamic activities such fires, floods, etc.) Pixels and grid arrays come in many different sizes, known as resolution. If the pixel size is small, e.g., representing inches by inches of ground surface, this is not a problem; however, in the case of French SPOT imagery of 10-meter resolution or early Landsat imagery of 30-meter resolution, pixels equal cells of 10x10 meters or 30x30 meters in size, respectively, which represent simplification of larger amounts of earth surface or surface-based activity that must nominally be represented as a single constant value within this relatively large space. Thus, depending on the ground sample distance, which translates to pixel size, a possible disadvantage of raster data is that of representing too coarse a degree of "mapping" from the real world to the represented world. Another disadvantage is that the pixel is what the pixel is; what information one gets from viewing the raster pixel (or more realistically, the grouping of several raster pixels clustered together) is interpretive. That is, how well the GIS or other system codes the pixel and how well the analyst understands the coding of the pixel determine the limit of information that can be gleaned from the pixel. Common uses of rasters include map background displays upon which other data are overlaid. A principal benefit of rasters is to create various theme-based data layers, which can be stacked upon one another to integrate together different pieces of information. For example, raster displays often provide photo- or map-based backgrounds that analysts can easily identify with while more easily manipulated data such as vector data are overlaid to aid the analysis.

Vector data are mathematical representations of geographic objects, or other business-content objects having some geospatial meaning. Vector data are normally thought of as points, lines, and polygons. These data can be maintained in relational data tables and, through the tools and applications commonly found in GIS, can be used to create stand-along vector displays or overlaid upon raster-based displays. Couclelis (1992) in Burrough & McDonnell (1998) said, "Objects in vector GIS may be counted, moved about, stacked, rotated, colored, labeled, stuck together, viewed from different angles, shaded, inflated, shrunk, stored and retrieved, and in general, handled like a variety of everyday objects that bear no particular relationship to geography." Vector data, as objects, can be stored in databases with various attributes, properties, and behaviors. For example, a stream (or a line segment in vector terms), drawn on a map display as a blue line, could be clicked to reveal an information table of stream attributes, such as stream width, depth, bottom type, bank slope, bank height, current speed, average temperature, etc., which delivers far more information than could displayed with a series of blue pixels or the photographed image of the stream in a raster display. An example of a vector object in a business context might be a wireless telephone company's grid of points depicting each cell phone transmitter tower, such that each point that represented a single cell phone tower could be linked to an attribute table with technical information about the tower and its current status in the operation of the wireless telephone network. Ingested into an application linked to the GIS that models network performance, the phone company could model different hypothetical locations for the towers to increase network "up" time and signal strength to users who depend on a certain quality of service when selecting their wireless phone service provider.

Other GIS data include elevation data, which is normally represented as "z" values representing height or elevation in an x, y grid that locates these points on the surface of the earth, where "x" and "y" could represent longitude and latitude, or some other referencing scheme. Elevation models have been used extensively to add relief or 3-dimensionality to normal 2-dimensional map displays. This is often accomplished by "draping" raster or vector layers over the top of wire-frame grid of x, y, and z points so that photos and maps are seen in their dimensional relief. This allows whole new classes of analyses and uses to emerge, including anything from airline pilot training simulations to the highly technical computer gaming and simulation industry. It is important to note that there is a big difference in computational processing power required and application complexity found between viewing a static 3-D image on a screen or using an application such as an interactive computer game or pilot's flight simulator that must render many

Figure 7. Considering Evaluation Paradigms for Geospatial Data in GIS

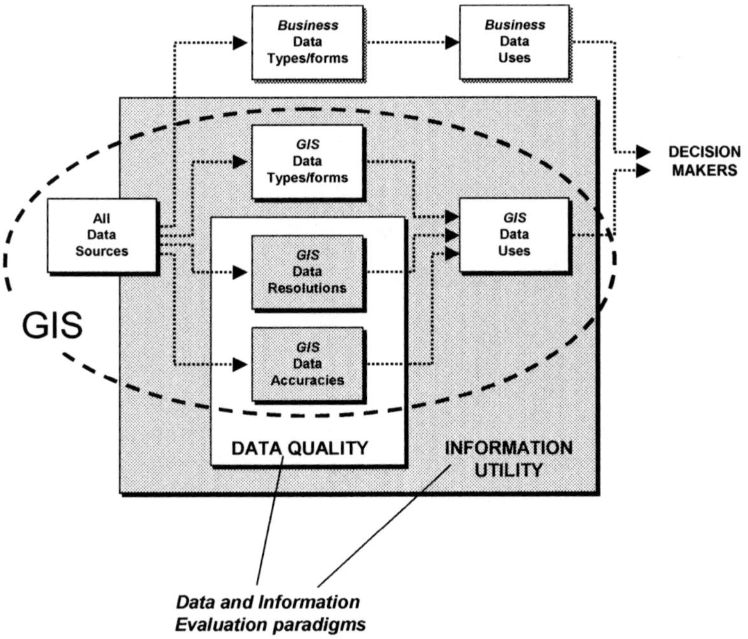

rapidly changing versions of the view per minute. The essential concept is the integration of raster, vector, and elevation data for complex presentation to the viewer. Examples of textual data that may be used within GISs and benefit business applications include street addresses of a firm's customer, vendors, other suppliers, or competitors. Examples of numerical data ingested into GISs are values for things that have no intrinsic shape, such as elevation, rainfall, temperature, slope, wind or current speed values, which relate to earth "things," or business data, such as sales volumes, which can be tied to business operations in a geographical place or region.

Earlier we discussed errors in analyses, with respect to organizational systems. The use of GIS and business data in decision-making necessitates a concern for data quality and errors induced into analyses from the source data. Evaluation paradigms become vital. *Figure 7* considers two different evaluation paradigms for considering the data used in GIS analyses: data quality, which focuses on data accuracy and resolution, and information utility, which extends data quality issues within a user's relevance context that also considers the source and form of the data as well as its intended use. We make a distinct difference between performing accuracy assessments or developing error matrices, and calculating the utility of the information content to the user.

An important point not yet addressed are some of the domain uses of GIS analyses with their supporting geospatial data. Rare is the field that cannot or does not benefit from GIS; shown here is partial list of fields using GIS technologies and output innovatively:

- Land management
- Telecommunications
- Agriculture
- Military operations
- Intelligence
- Transportation
- Law enforcement
- Recreation
- Marketing
- Operations
- Sales & retail
- Logistics

The focus of *Figure 7* is the scope of the data quality paradigm and the broader scope of the information utility paradigm. See Congalton & Green (1999) and Congalton & Plourde (2002) for excellent treatment of accuracy issues and data quality. See Meeks & Dasgupta (2003) for a detailed treatment of geospatial information utility.

Evaluating the Value of Geospatial Information in GIS

"Information is the substance from which managerial decisions are made," (Forrester, 1961, p. 427). This drives the need for valuing information in organizations. Information value is comprised of quantitative and qualitative aspects. Quantifying the value of information allows comparisons between the outcomes of organizational systems' performance or of decisions made with or without the "nth" piece of information. Semantically describing information allows managers and researchers to understand the complex and integrated ways information contributes to competitive value or otherwise improves performance. Reichwald (1993), in Wigand et al. (1997), identifies three levels of information exchange: syntactic, semantic, and pragmatic. Information transmission occurs at the syntactic level. The semantic level adds meaning to the symbols transmitted. The pragmatic level adds sender's intention and receiver's use to the meaning and transmission of information. The ways in which businesses, organizations, and consumers interact and value information are being transformed. As business managers interact with GIS analysts and other geospatial professionals, the full range of meaning to be derived from data and information of all types must be considered.

Lawrence (1999) provides a useful treatment of quantitative issues surrounding information value today. As managers are often concerned with resource allocation and financial performance, a current measure of the value of information is to calculate the dollar cost or benefit of using information within a decision-making context. This is called the value of the informed decision. Decision trees and expected value calculations are the staple of the value of the informed decision. Lawrence offers several models, primarily focused on treating value as a utility function. These useful methods emphasize decision-making outcomes. However, two common aphorisms are relevant: "the whole is greater than the sum of the parts," and one familiar to engineers and social scientists, "tell me how you will measure me and I will tell you how I will perform." The information value literature mostly addresses the measurement of components of information. However, there is a dearth of literature addressing how to holistically value the contribution of information to the organization. Though Lawrence and others address utility, considering the "whole" speaks to integrating diverse and complex types of information.

Modern decision-making, that is, analyze, decide, act, and evaluate, occurs in very compressed cycle times. Perhaps a new information valuation schema should consider a jigsaw puzzle metaphor. The need for and the use of information is rarely so binary that one piece of information crosses the threshold for a managerial "eureka!" Effective decision-making relies on the collection and analysis of many disparate pieces of information. Some are easy to find and fit into context. This is analogous to a puzzle's edge and corner pieces. The place (or "role") of other puzzle pieces is not so readily apparent. Though the puzzle is not complete until all pieces are in place, by the time the puzzle is 80-90% complete, it becomes pretty clear what the final image will look like. Similarly, in organizations, not all "puzzle pieces" are equal. It is precisely the unequalness that makes valuing information important: to be able to compare acquisition costs (sometimes known and often times unknown) against the presumed a priori benefits of collecting and using the targeted pieces of information to make decisions.

As the acquisition and use of information are costly, the optimal use of information involves economic tradeoffs. Therefore, valuing information is attracting research and thought. However, until now, little attention has been paid to integrated information valuation within the geospatial information domain, which is increasingly coming to the attention of decision makers seeking to improve decision models by considering spatiotemporal factors. In earlier work (2003), we proposed a metric called Geospatial Information Utility (GeoIU) to allow decision makers to assess the degree of utility incurred for accessed geospatial data sets when making decisions that incorporate those geospatial data and information. The GeoIU metric uses multi-attribute utility theory to assess, score, and weight metadata queries run against geospatial data and information discovered in distributed sources. When using spatial and temporal information to improve decision-making, attention must be paid to uncertainty and sensitivity issues (Crosetto & Tarantola, 2001). Attention must also be paid to spatial and temporal scales relevant to the decision being supported (Pereira, 2002) and to the quality and utility of available data, with respect to the intended use(s) of the data (Obermeier, 2001). This last issue defines the core problem GeoIU addresses: that decision makers collect and use geospatial data of varying spatial and temporal scales in order to improve decision-making. More attention needs to be paid to finding appropriate methods for assessing the utility of the geospatial data being used (Bruin et al., 2001). *Figure 8* restates the general systems model in terms of GIS processes.

It is often useful to broadly classify GIS and geospatial data users into two broad categories: (1) public-sector users and (2) private-sector and business users. Public-sector users are primarily interested in public domain uses of geospatial and spatiotemporal data: for example, military planners may require highly accurate, very current digital data sets for planning flight routes for cruise missiles. Flight route planning requires digital elevation models to support terrain contour matching algorithms within the missiles' guidance modules. To optimally employ so-called "smart weapons" such as these, planners and targeteers must have access to current, high quality digital data sets with minimal horizontal and vertical accuracy errors. In order to reduce operational risk in the development of missile flight routes based on new digital geospatial data sets, the "pedigree" or quality of the supporting data must be assessed (Johnson et al., 2001).

Figure 8. A Simplified Model of GIS as a System Supporting Decision Making

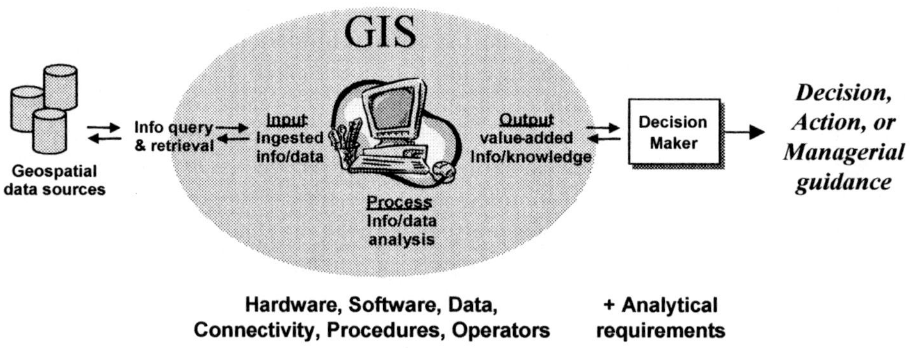

Many other public sector uses of GIS and geospatial data abound: fuel modeling to predict and prevent wildfires, cadastral records and land tax planning, information visualization of municipal government services via web-based GIS applications, and many more. Each of these uses of geospatial data has different requirements for data accuracy, currency, and form; some applications have stringent requirements while others less so.

Business users are primarily interested in commercial or business intelligence analyses of geospatial data and information. These users may also have varying needs for highly accurate and current data sets: for example, wireless telephone service planners may employ GIS technologies and data analyses in order to determine optimal site locations of a new array of digital wireless telephone signal towers. Their analyses might focus on making maximum use of both send and receive signal strengths vis-à-vis local terrain limitations (e.g., received signal strength is a function of transmitted power, number and locations of transmitter towers, radio frequency line-of-sight obstacles, etc.) in order to minimize the number of towers needed while providing a guaranteed quality of service for their wireless telephone subscribers. Using fewer, well placed towers may mean lower operating costs and higher operating margins. Similar to the missile flight route problem, a wireless telephone tower location analysis based on geospatial data of poor or uncertain quality is subject to errors, which may roll through the calculations, quite possibly resulting in improperly located towers, reduced systems performance, higher installation and operations and maintenance costs, and unhappy customers.

As successful and useful analyses depend on many factors, including data quality and data accuracy in many forms, GIS users must have adequate mechanisms to evaluate the relevance of their data to their analyses. As shown in *Figure 9*, GeoIU tools can be used as a "filter" between data sources and analytical engines and processes to give analysts and decision-makers insight into the uncertainty they face.

Aircraft- or satellite-based multi-spectral imaging (MSI) sensors have improved our understanding of the earth's surface and human activities on it (Lillesand & Kiefer, 2000). Depending on the resolution of the image, electro-optical (EO) (i.e., photographic) imagery of a stand of trees on a plot of land may or may not permit general classification of tree type. This question, and these sources of spatial information, may be pertinent to several types of decision makers. For example, local tax assessors may care about land use classification for tax purposes (Montgomery & Schuch, 1993). Forest rangers may care about tree, vegetation, and soil types and moisture contents to perform predictive fire-fuel modeling (Burrough & McDonnell, 1998). A paper company may care about assessing the density and maturity of certain tree types for determining the readiness of the harvest of a particular tract. The predominant tree types within a given image pixel would dictate how that pixel would be coded or classified. With the advent of early MSI sensors, such as NASA's LANDSAT-series, capable of imaging in seven spectral bands and displaying results on false-color images using three user-selected bands, greater understanding of the earth's surface became possible. From EO's one visible band to MSI's seven spectral bands, forest rangers using GIS and MSI data are able to interpret soil moisture or hardness, as well as more complete and accurate classification of the trees mentioned above. Newer hyper-spectral imagery (HSI) sensors sense the earth in 200+ bands, providing finer resolution represented by narrower "slices" of the electromag-

Figure 9. Assisting the GIS Process with Utility Assessments from GeoIU

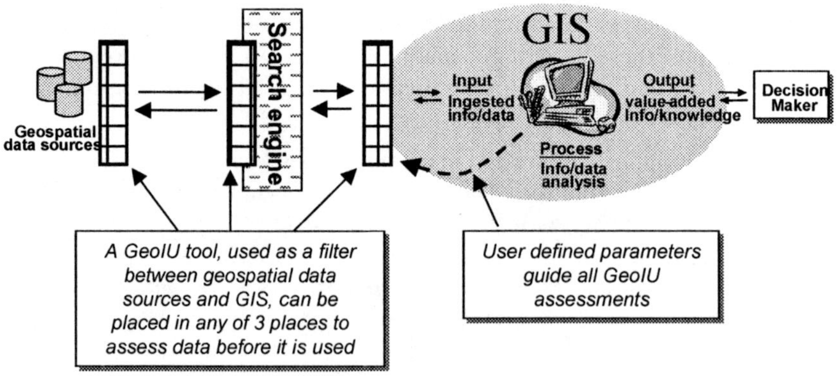

netic spectrum. This dramatic increase in spectral resolution is being accompanied with increases in spatial resolution (i.e., how clearly things can be seen) and accuracy (i.e., how correctly things can be located).

Looking into the Future

With sensing as a means of generating source data to feed business and organizational information processors, humans are thought to sense in five or more dimensions: sight, hearing, taste, smell, touch. In phenomenology-based sensing, we can think of sight as supported by the many forms of imagery: electro-optical (EO) visual images, radar images, motion video, moving target indicator (MTI), light detection and ranging (LIDAR/LADAR), etc. Similarly, hearing can be thought of as acoustic sensing and electronic or signals intercepts sensing; smell is represented as olfactory biological and chemical remote sensing; and touch is represented by seismic sensing and thermal sensing in many forms. The human sense taste seems to have no analog in the remote sensing world; however, there are other technical sensing phenomenologies that have no direct human analog either, such as magnetic measurement and signatures.

Not only are emerging and continually evolving technologies improved and exploited to expand data collection capabilities, but the order of magnitude increases in capabilities are being translated into the data processing realm as well. For example, a highly integrated multi-disciplinary approach (called Multi-Intelligence, or "Multi-INT") is being refined within a military/national security/national intelligence context. This Multi-INT approach represents a highly focused degree of integration of the different collection "senses" (remember this represents as many as 15 different forms of sensing versus the five that a human uses). The point is that order of magnitude increases in information richness (i.e., number, quality, and completeness of feature or entity attributes) and information accuracy (i.e., including spatial location accuracy and content accuracy), collected and analyzed dynamically over time are possible.

What was once the domain of GIS specialists is now falling into the realm of business managers who want to develop new decision-making constructs to improve their organization's strategic and tactical tempo of activities and performance within their respective industries. It is not necessarily the sources of spatiotemporal data (e.g., remotely sensed satellite imagery) that are critical — though they are becoming more and more useful in all sectors — but it is the uses of these data that are important and bear watching. This leads us to the conclusion that GIS is able to support business and improve managerial decision-making on several levels.

At the first level, GIS answers space and time questions. At the next level, GIS allows analysts, managers and decision makers to think differently about what constitutes useful data in evolving decision-making models. This includes admitting that heretofore-unused data sources (including remote sensing sources) may improve decision-making. At this same level of complexity, GIS allows analysts, managers, and decision makers to think differently in their current decision-making processes and to adjust these processes to accommodate the new reality of GIS. The evolving field of information visualization supports and is supported by advances in GIS. Incorporating GIS in decision-making forces managers and others to decide what they want to see and how they want to see it. Finally, at the most sophisticated level of GIS integration, using GIS permits managers to inculcate within their organizations a spatially- and temporally-oriented mindset. Analysts, managers, and decision makers who have developed a spatiotemporal mindset look at their problems, processes, input data and output needs completely differently. And this may be the greatest benefit GIS provides business managers: the help them to see their problems and solutions differently so that they may solve their problems more effectively.

Summary

This chapter encourages managers and decision makers in non-earth sciences organizations to consider using GIS to improve decision-making. We feel there are several innovative ways GIS can help make these improvements. As identified in Introduction with the "three themes to carry away," we believe:

- GIS can improve organizational decision-making through the awareness that all business decisions include space and time components. The benefit is that thinking spatiotemporally provides additional analytical approaches and methods.
- GIS use both business data and remotely sensed data. An awareness of the power of the different forms and sources of remotely sensed data and the ways their integration can transform organizational notions about how and where to collect business data helps improve both GIS-based and non-GIS based decision-making.
- Accessing many different data sources and types imply challenges with using these data; these challenges include determining the quality of the data ingested,

manipulated and outputted; and equally as importantly, determining the utility and relevance of the ingested and outputted data and information as they pertain to the result of the final decision or action.

References

Althausen, J.D. (2002). What remote sensing system should be used to collect the data? In J.D. Bossler (Ed.), *Manual of geospatial science and technology* (pp. 276-297). London: Taylor & Francis, London.

Bossler, J.D. (2002). An introduction to geospatial science and technology. In J.D. Bossler (Ed.), *Manual of geospatial science and technology* (pp. 3-7). London: Taylor and Francis.

Bruin, S.D., Bregt, A., & Ven, M.V.D. (2001, May). Assessing fitness for use: The expected value of spatial data sets. *International Journal of Geographical Information Science*, *15* (5), 457-471.

Burrough, P.A., & McDonnell, R.A. (1998). *Principles of geographical information systems.* Oxford: Oxford University Press.

Christakos, G., Bogaert, P., & Serre, M. (2002). *Temporal GIS: Advanced functions for field-based applications.* Berlin: Springer-Verlag.

Clemen, R.T., & Reilly, T. (2001). *Making hard decisions with decision tools.* Pacific Grove, CA: Thomson Learning.

Congalton, R.G., & Green, K. (1999). *Assessing the accuracy of remotely sensed data: Principles and practices.* Boca Raton: Lewis Publishers.

Congalton, R.G., & Plourde, L.C. (2002). Quality assurance and accuracy assessment of information derived from remotely sensed data. In J.D. Bossler (Ed.), *Manual of geospatial science and technology* London: Taylor and Francis.

Couclelis, H. (1992). People manipulate objects (but cultivate fields): Beyond the raster-vector debate in GIS. In A.U. Frank, I. Campari & U. Formentini (Eds.), *Theories and methods of spatio-temporal reasoning in geographic space* (pp. 65-77). Berlin: Springer Verlag.

Crosetto, M., & Tarantola, S. (2001, May). Uncertainty and sensitivity analysis: Tools for GIS-based model implementation. *International Journal of Geographical Information Science*, 15 (5), 415-437.

Decker, D. (2001). *GIS data sources.* New York: John Wiley.

DeMers, M.N. (2000). *Fundamentals of geographic information systems* (2nd ed.). New York: John Wiley & Sons.

Forrester, J.W. (1961). *Industrial dynamics.* Cambridge, MA: MIT Press.

Frank, A.U., Campari, I., & Formentini, U. (eds.). (1992). *Theories and methods of spatio-temporal reasoning in geographic space.* Berlin: Springer-Verlag.

Fujita, M., Krugman, P., & Venables, A. (1999). *The spatial economy: Cities, regions, and international trade*. Cambridge, MA: MIT Press.

Hoch, S.J., & Kunreuther, H.C. (2001). The complex web of decisions. In S.J. Hoch, H.C. Kunreuther & R.E. Gunther (Eds.), *Wharton on decision making* (pp. 1-14). New York: John Wiley & Sons.

Isaaks, E.H., & Srivastava, R.M. (1989). *An introduction to applied geostatistics*. New York: Oxford University Press.

Jankowski, P., & Nyerges, T. (2001). *Geographic information systems for group decision making: Towards a participatory, geographic information science*. London: Taylor & Francis.

Johnson, R.G., Watts, W., Meeks, W.L., & Fulton, T. (n.d.). *United States imagery and geospatial information service geospatial transition plan*. Bethesda, MD: National Imagery and Mapping Agency.

Lawrence, D.B. (1999). *The economic value of information*. New York: Springer.

Lillesand, T.M., & Kiefer, R.W. (2000). *Remote sensing and image interpretation* (4th ed.). New York: John Wiley & Sons.

Longley, P.A., Goodchild, M.F., Maguire, D.J., & Rhind, D.W. (2001). *Geographic information systems and science*. Chichester, UK: John Wiley & Sons.

Malczewski, J. (1999). *GIS and multicriteria decision analysis*. New York: John Wiley & Sons.

Meeks, W.L., & Dasgupta, S. (2003). Geospatial information utility: An estimation of the relevance of geospatial information to users. *Journal of Decision Support Systems, 17*, in press.

Montgomery, G.E., & Schuch, H.C. (1993). *GIS data conversion handbook*. Fort Collins, Colorado: GIS World, Inc.

Obermeier, J. (2001). Discussion of product adequacy and product evaluations at NIMA, and finding automated methods for determining the utility of geospatial information. Bethesda, MD: National Imagery and Mapping Agency.

Ormsby, T., Napoleon, E., Burke, R., Groessl, C., & Feaster, L. (2001). *Getting to know ArcGIS desktop: Basics of ArcView, ArcEditor, and ArcInfo*. Redlands, CA: ESRI Press.

Pereira, G.M. (2002). A typology of spatial and temporal scale relations. *Geographical Analysis, 34* (1), 21-33.

Raper, J. (2000). *Multidimensional geographic information science*. London: Taylor & Francis.

Reichwald, R. (1993). *Die wirtschftlichkeit im spannungsfeld von betriebswirtschaftlicher theorie und praxis*. Munich.

Wigand, R., Picot, A., & Weichwald, R. (1997). *Information, organization, and management: Expanding markets and corporate boundaries*. Chichester, UK: John Wiley & Sons.

Chapter IX

Strategic Positioning of Location Applications for Geo-Business

Gary Hackbarth, Iowa State University, USA

Brian Mennecke, Iowa State University, USA

Abstract

The chapter presents several conceptual models, each of which can be used to improve our understanding of whether spatially enabled virtual business is appropriate or not. The first model, the Net-Enablement Business Innovation Cycle (NEBIC), modified from Wheeler (2002), consists of the steps of identifying appropriate net technologies, matching them with economic opportunities, executing business innovations internally, and taking the innovation to the external market. The process consumes time and resources, and depends on organizational learning feedback. The second model, modified from Choi et al. (1997), classifies geo-business applications in three dimensions, consisting of virtual products, processes and agents. Each dimension has three categories: physical, digital, and virtual. The chapter discusses examples of spatially enabled applications that fall into certain cells of this model. The model is helpful in seeing both the potential and limitations for net-enabled applications. The final model classifies spatially enabled applications by operational, managerial, and individual levels. Examples are given that demonstrate spatial applications at each level.

Introduction

Using "location" information as part of a firm's competitive strategy to generate revenue, develop market share, extend services, and provide superior customer service is not a new idea. Decision makers have long used zip codes, census data, traffic flow patterns and the like to determine store locations, effect pricing, determine product mix, and the availability of services. Location or spatial components are critical factors that decision makers consider when approaching a problem or task. We should not be surprised that global positioning systems (GPSs) and digital mapping are being included as value-added information to a wide range of product and service offerings. Importantly, pervasive Information Technology (IT) and associated net-enabled architectures now provide the capability to share spatial information with business partners as well as offering customers similar access to location specific information. Firms must now strategize about new products and services inclusive of location information across the supply chain, not only to make better decisions, but also to provide location information as one more dimension of customer service.

Leveraging location to enhance firm profitability is a function of internal accelerators and external competitive pressures relative to a firm's position in a market (Schuette, 2000). Firms must have strategic leadership, employee technological expertise and sufficient IT resource availability to integrate the necessary geographic information systems within current business processes. Equally important are the environmental conditions essential to successful implementation of location applications. Firms must also have the right internal structure to integrate location technology with business partners, market share to justify resource allocations, the ability to overcome entry barriers, relationships with strong suppliers and the ability to affect customer preferences. Thus, this chapter will focus on developing a Geo-Business Application Model useful in positioning e-business firms in deploying location specific applications.

First, we introduce Wheeler's Net-Enabled Business Innovation Cycle (NEBIC), then develop, and present, the Geo-Business Application Model. We will conclude with a discussion of practitioner considerations of location technologies.

Net-Enablement Business Innovation Cycle (NEBIC)

Firms require a net-enabled strategy to assist them in leveraging information to support, enhance, differentiate, and substitute technology for physical processes (Straub et al., 2001). This would seem particularly true in dealing with digital location services that not only replicate physical location services but also enhance their capabilities by making them more efficient. Wheeler (2002) proposed the Net-Enabled Business Innovation Cycle (NEBIC) to measure, predict, and understand a firm's ability to create value using digital networks. As shown in *Figure 1*, firms follow an ongoing cycle that begins by exploiting a firm's dynamic capability to select and match IT to current economic opportunities, then reengineers relevant business processes to exploit IT to achieve some new business innovation, and then continuously assesses customer value (Wheeler, 2002). Importantly, a firm's unique capabilities allow them to make strategic changes in order to adapt to dynamic markets (Zahra et al., 2002a, 2002b). Clearly, the dynamic capabilities represented by digital location specific services fit the NEBIC well when firms have the time and value potential to leverage internal organizational capabilities into marketplace success. Firms must also consider the external market in that one may be dependent upon other suppliers to build Wi-Fi networks, position satellites, or the like, but also opportunities to build compatible software/hardware that complement or extend existing products or services.

Dynamic capabilities characterize change capabilities that help firms redeploy and reconfigure resources to meet customer demands and counter competitor strategies (Zahra et al., 2002b). Successful firms apply an in-house IT expertise concomitant with a cultural capacity for change that leverages available IT resources. The firm leadership is committed to dynamic change and a willingness to impact people by supporting paradigm shifts in thinking and allowing cross-functional decision-making. External observation of suppliers, customers, and the leveraging of market share to surmount barriers to market entry with both existing and potential products is just as important as having dynamic capabilities internal to the firm. To be useful, location information must enable other products and services that enhance economic opportunities. Firms must have the insight to seize these opportunities; however, exploitation of economic opportunities cannot occur without communication with the manufacturing, marketing, and IT groups for implementation. These groups must balance internal capabilities with a correct assessment of customer value. For instance, water-resistant handheld GPS receivers exist for anglers while hunters can use camouflaged units; yet, neither requires the text or spoken directions desired by motorists. As an organization learns, cooperation between the business units increases such that technological change is foreseen

Figure 1. Net-Enablement Business Innovation Cycle (NEBIC) (Wheeler, 2002)

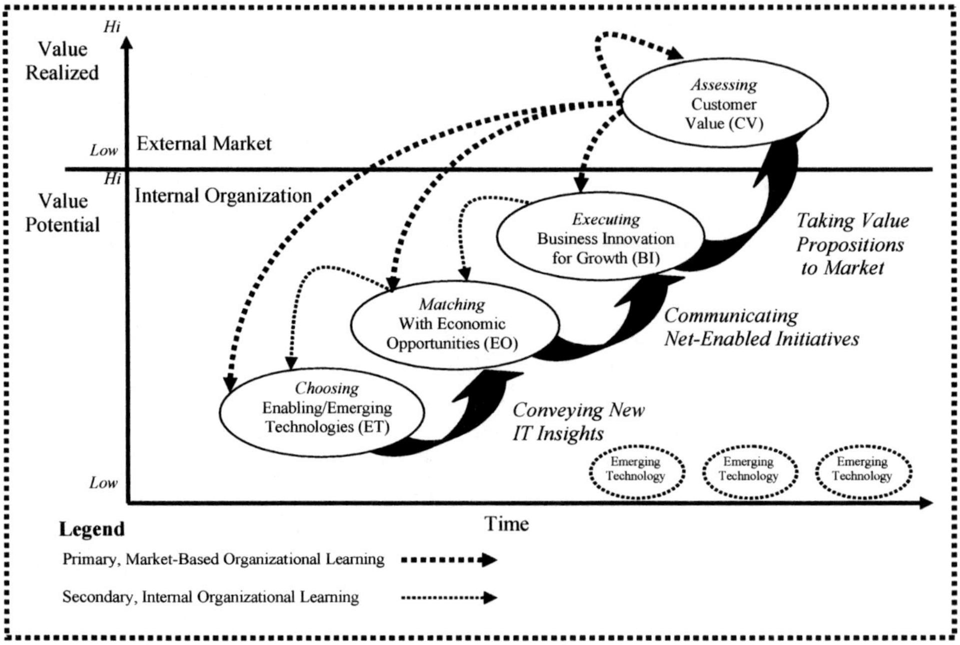

with strategies in place to maintain a trajectory of growth (Cockburn et al., 2000; Zahra et al., 2002b). Over time, better timing and more effective implementation of location-specific IT strategies would sustain a competitive advantage. At this point, IT Strategy leads Business Strategy in that access to technology and employee understanding of the value proposition of technology creates an escalating spiral of technology and knowledge acquisition that continuously creates newer and better applications that require new business strategies to further stimulate customer demand (Cockburn et al., 2000).

Geo-Business Application Model

The decade of the 1990s saw a tremendous growth in the number of geographic information systems (GISs) for stand-alone as well as client server applications. Supply chain management, marketing, shopping, business services, and information distribution all represent applications where geographic technologies are useful (Kalakota et al., 1996a; Kalakota et al., 1996b). Location services can support employees (e.g., a delivery driver locating a store in an unfamiliar territory) or a firm's customers (e.g., a customer seeking to locate the nearest store outlet). In addition, location applications represent functions that are not only operational, but also tactical or strategic.

Enhancing the operational capabilities of employees and building practical applications for customers is of economic value, but what is of more lasting value is finding a mechanism to contrast existing processes and products that leverage alternative technologies into additional economic opportunities. This is not too surprising given that the applications of GIS and spatial technologies to the Internet are a relatively recent phenomenon. When a new class of technology or application emerges within a particular context, early product introductions tend to be simpler and less sophisticated in their capabilities, such as an operational application that shows a GPS location on a digital map. Today, spatial applications access the Internet to focus on problem solving, data analysis, planning, and other tactical and strategic tasks. As organizations learn and technology matures, applications tend to become more sophisticated and specialized (i.e., many GPS mapping applications now include both voice and text directions for the fastest, most direct routing for product deliveries).

The NEBIC clarifies the cyclic development of internal resources and formulation of IT strategies essential to long-term competitive advantage, but it does little in providing guidance in choosing enabling or emerging technologies that match economic opportunities. One insight into this limitation is the Model of E-Commerce Market Areas (Choi et al., 1997), which provided a three-dimensional framework for understanding the relationship between resources, actors, and processes that exist in both electronic and physical markets. In general, this framework differentiates products based on whether they are "physical" or "digital" (i.e., electronic). For example, an electronic product is a mapping software program whereas a physical product would be a plastic foldout map. A second dimension differentiates digital from physical processes. A digital process is using the Internet to access a digital vacation road map while a physical process involves the physical act associated with reading a printed map and plotting a route of travel by hand. The third dimension is the nature of the agents involved in the transaction. A web-store would be digital while the corner travel store would be physical.

One might think that the Model of E-Commerce Market Areas identifies Geo-Business opportunities where there are digital agents, digital processes, and digital products. Unfortunately, this framework is only useful in identifying where Geo-Business opportunities exist in terms of automating existing products, processes, or agents (i.e., "Let's sell our maps online or download a digital map to a PDA for a small fee"). This is a normal process, in that firms typically first experiment with technology, then integrate technology into their business functions, before transforming themselves into firms that leverage information externally across the supply chain linking customers and suppliers with information to achieve a competitive advantage (Kettinger et al., 2000). The framework loses some of its predictive power with Geo-Business applications because it fails to show where knowledge acquired from emerging technologies conveys new IT insights that enable firms to take value propositions to the marketplace. Knowledge of geographic proximity, consumer or managerial behavior, and similar variables might provide insights that affect opportunities to leverage location for competitive advantage. For instance, location beacons exist for cars and trucks that pinpoint their location if stolen. This same technology could provide continuous location information for the real-time routing of shipping or rerouting of traffic in congested areas.

A few alterations to the Model of E-Commerce Market Areas may convey additional insights that lead to value added Geo-Business applications. For example, if the agent dimension broadens to encompass virtual actors that contrast differences between physical and virtual decision makers, automated ordering systems and not merely electronic storefronts, then this dimension would have greater utility. One might see a traditional buyer at one end of the dimension while a virtual ordering system that price matched and ordered on demand might lie at the other end. Similarly, the process dimension broadens to encompass virtual processes. The important distinction made here is whether the virtual interaction with Geo-Business technology is enabling existing electronic processes. For example, sailors navigate by marking positions on a chart. A GPS can provide real-time positioning displayed on a monitor. A recent innovation displays real-time weather information concurrent with GPS positioning information. This value proposition eliminated the redundancy of manually plotting GPS positions on a paper chart, along with real-time weather information, greatly enhancing the safety and effectiveness of local anglers and sailors in South Florida. Finally, the product dimension broadens to encompass virtual products and services. At one end of the spectrum, we would see a product such as a chart or map and at the other end a virtual product such as a tracking locator in a car or truck that keeps track of mileage and location to enable routing and maintenance decision-making while being displayed in an operations center.

The modified Geo-Business Application Model (Choi et al., 1997), shown in *Figure 2*, includes not just the automation of physical agents, processes and products but also value-added knowledge to existing electronic agents, processes and services. Looking at the front corner area of the model shown in *Figure 2*, which focuses in on physical products and resources, we may simply ask which physical processes or products do we automate? We may then include the third dimension and determine which agent would provide a more appropriate channel for distribution and sales. We could automate outward even more, looking at other digital and virtual processes and products. This procedure simplifies the process of examining various Geo-Business technologies in the context of needed internal capabilities and external competitive pressures. One could

Figure 2. Geo-Business Application Model

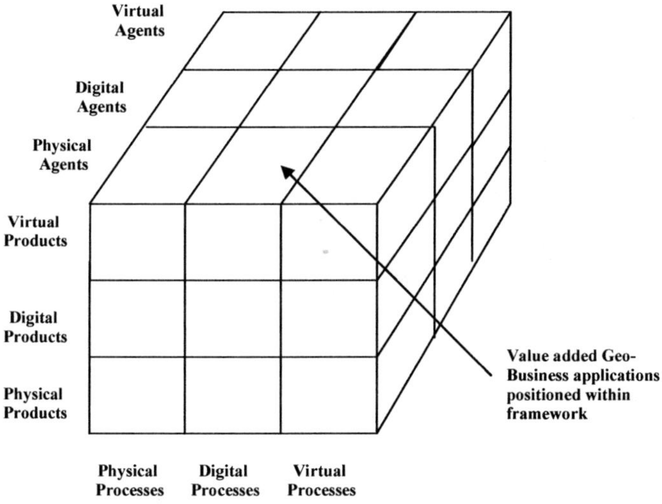

Modified substantially from Choi et al. (1997)

place an "order and transaction fulfillment" application in the digital product and physical process intersection. In this case, we have a firm with a website on which customers may order and pay for products online. Unfortunately, the sales department prints the order and hand carries the order fulfillment document to the warehouse for shipping. Clearly, further automation of this process is possible, as is the addition of location data. Knowing where sales originate could optimize the location of a warehouse to decrease delivery costs. It is now possible to decouple warehouse location from other firm functions. Customer location data provides additional justification and cost savings.

In fact, a logical next step is to "map" various business applications or products across these two dimensions. *Figure 3* represents the mapping of various processes for a fictitious firm. For example, fleet management pertains to managing physical resources (i.e., trucks) involved in physical processes (i.e., delivering goods and services) is positioned in the lower left corner of the diagram. Alternatively, information dissemination carried out via Geo-Business tools involving virtual information accessed using virtual processes is in the upper right corner of the diagram. Some applications such as customer relationship management involve a mix of digital and physical processes and products and therefore position themselves in the middle of the framework.[1]

A further examination of each process also highlights that applications in the upper-right quadrant represent what might be termed "higher-level" or strategic applications while those in the lower left are more "common" or operational applications. This terminology is reminiscent of the taxonomy that is often used to categorize information systems: operational, tactical, or strategic. Of course, how an actual firm chooses to position various processes may differ. This type of taxonomy is still valuable for considering and classifying the role of Geo-Business technologies. In fact, Information Systems (IS)

Figure 3. Dimensions of Geo-Business Processes and Products

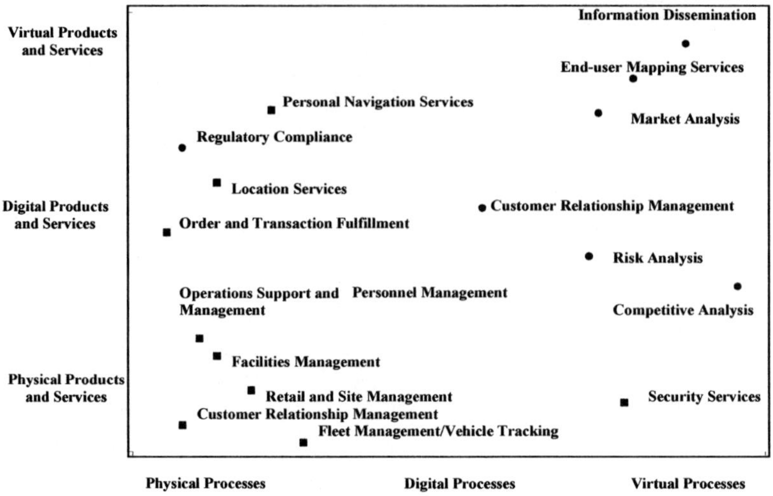

researchers have devoted considerable effort to examining and classifying various types of information and decision support systems (DSS) using this type of taxonomy. For example, Money, Tromp, & Wegner (1988) suggest that DSS usage divides into three groups: (1) those at the operational level, (2) those at the managerial level, and (3) those at the personal (individual user) level (Money et al., 1988). This type of framework could be useful if a third category were adapted to include market-focused benefits and applications: that is, services provided by organizations that are focused on providing customers and users with individual-level benefits. Thus, it is useful to expand the meaning of agents in the Geo-Business Application Model to include strategic decision makers, managers and customers.

The focus of the remainder of this chapter will be on discussing Geo-Business applications in light of this type of taxonomical framework. The next section discusses the role of geographic data in the context of e-business, followed by several representative examples of Geo-Business applications from each level.

Geo-Business Applications in Context

An important question to ask in the context of defining the benefits of Geo-Business technologies is "What is the role of location in managing individual and organizational activities?" Most people agree that location has historically been important and continues to have a critical role in most human endeavors. For example, a large proportion of corporate databases include some geographic (spatial) information. These spatial data may be as broad in geographic scope as the state in which a customer is located or as specific as the location of a customer's shipment at a particular time of day. Clearly, businesses value this information or they would not collect it.

While much of these data meet operational needs, the data do not entail producing a map or generating sophisticated geographic analyses. Rather, they encompass basic operational functions, such as shipping products to the customer that motivates organizations to collect location data. Interestingly, many organizations, in effect, have out-sourced their basic operational functions to the Post Office or Federal Express, separating themselves from the potential economic costs (as well as opportunities) that arise due to the complex interaction between geography and their business operations. Therefore, in many industries location information has a limited relevance to management decision-making activities.

This may help to explain the modest growth in desktop geographic technologies in a variety of private sector industries. Efforts in the last decade by GIS desktop software venders to encourage the adoption of their software largely was aimed at applications that use spatial data held in corporate operational systems; systems that are generally designed with the purpose of providing managers with a means of monitoring and controlling organizational activities. While a number of firms have wholeheartedly adopted and implemented the geographic technologies offered by these venders, it is often the case that mapping capabilities are critical to the success of these particular firms' operational systems. For example, public utilities, transportation companies, and firms managing natural resources have all adopted and used desktop GIS because of its ability to provide useful information about the status of geographically distributed resources (Lapalme et al., 1992; Mennecke et al., 1998). In many operational systems, however, the value added by geographic technologies does not outweigh the costs. Thus, the slow pace of desktop GIS adoption in some industry segments during the 1990s is likely due to this emphasis on positioning GIS technology for use in tasks where its functionality is not critical to achieving organizational success.

This begs an important question, "Where do geographic technologies, such as Geo-Business technologies, add significant value?" Traditionally, geographic technologies have been positioned as client and, in some cases, server-based tools used to support operational and managerial applications. In this context, successful geographic applications integrated with such functions such as facilities management, logistics, demographic analysis, and site location tasks. Today, organizations are increasingly recognizing that various facets of their business involve location-dependent problems and require mobile solutions. For example, end-user oriented mapping tools such as MapQuest have become pervasive among mobile devices representing extensions to traditional business desktop applications. The term mobile implies that location is an important business capability needed at the operational, managerial, and individual levels.

Operational Support Applications

Early Geo-Business technologies supported organizational functions at the operational level. For example, one of the first applications was in automated mapping (AM) and facilities management (FM) (Coppock et al., 1991). More recently, the widespread availability of tools like GPS and other triangulation technologies have enabled firms to apply mobile technologies to a number of operational applications. For example, transportation and logistic applications such as routing, asset management, and field-force management have become important, and in some cases mission critical, functions in a variety of firms.

AM is a broad term that generally refers to the use of GIS to capture spatial data using technologies that do not require the manual coding of geographic data. AM tools such as remote sensing and portable GPSs allow more accurate map reproduction by removing the paper map as a data source (Goodchild, 1992). FM is an application that supports the real-time monitoring of facilities, such as emergency management, security, and other applications requiring the management of resources that are geographically distributed. An extension of this technology in the utility industry is AM/FM. Pennsylvania Power and Light has located several million utility poles using geographic technologies. Furthermore, Boston Gas created an Automated Mains Management System project that integrates their distribution system with other types of information (e.g, soil conditions, leak histories, and construction projects). Similarly, Indianapolis Power and Light has developed an AM/FM system, a Work Management System (WMS), and a Customer Information System (CIS) to facilitate access to and the management of information about their physical infrastructure, their customers, their field force, and other resources (Davis, 2000).

Firms that manage or harvest natural resources also make significant use of GIS/AM technology for facilities management to manage well locations, lease information, groundwater mapping and seismic information. Companies like Shell Oil, Aramco, Texaco and Statoil (Norway) have adopted GIS and digital mapping for supporting their operational and exploratory activities. Similarly, IHS Energy Group, a firm that provides spatial information for the oil industry, has literally millions of well locations captured and stored in its spatial database. Champion International Corporation uses GIS in its Forest Products Division to manage forest stands and the lands associated with these resources (Gates, 1995). This task does not only involve harvesting resources, but also using GIS to manage the environment by, for example, buffering creeks, rivers, and lakes that are removed or restricted from production.

Of course, some firms must manage not only facilities that are outside, but also those that are inside of a building or in a local area. Some grocery firms, such as Kroger and Safeway, use GIS systems as space management systems for merchandise planning and the development of plan-o-grams (Garrison 1999). These systems allow retailers to manage shelf space and floor layout to better present products to customers and do so in a way that makes sense in light of the geo-demographic characteristics for individual markets. This type of technology, and the analyses that it enables, allows retailers to take control of their shelf space and actively customize the layout of the store for particular market segments. Similar systems in warehousing and distribution facilities manage space, product movement, and other operational activities. For example, automobile manufacturers use GIS-based technology to track products through the assembly process by using local position systems (LPS) to track containers and direct forklifts to the nearest container that needs to be repositioned (Kmitta, 1999). Similarly, the New York University Medical Center uses LPS to track wheel chairs, stretchers, and other resources. Distribution and Auto Service, Inc. (previously called Annacis Auto Terminals, Ltd.), which operates an automobile distribution terminal in Vancouver, BC, that services more than 20 international automobile firms (Docherty et al., 1996), uses a GIS to manage

13,000 parking spots on their facility as they receive and ship out imported automobiles. Similarly, United Parcel Service (UPS) also uses GIS to manage logistics and transportation problems associated with the loading and unloading of trucks, the movement of packages within distribution facilities, the identification of optimal locations for distribution facilities, and the routing of vehicles. These systems lower operating costs, improve customer service, and utilize firm resources more efficiently.

Many firms outsource their shipping to specialized firms like UPS, FedEx, and the Postal Service in order to leverage their specialized dynamic capabilities in GIS expertise. These firms do not have the desire or resources to develop the necessary internal capabilities to utilize geographic information. However, many firms cannot or will not outsource the tasks associated with managing their supply chain. For such firms, the task of routing vehicles and resources to distribute products and services can be as much as 16% of the cost of the product (Kearney, 1980). These firms must consider such factors as route characteristics, driver limitations, vehicle capacity and costs, supplier and customer schedules, and the nature of the merchandise, as well as the development or selection of an appropriate routing algorithm and the appropriate display or presentation of results (Lapalme et al., 1992). Importantly, the visualization capabilities of GIS make it a valuable tool for managing routing (Greenfield, 1996).

Managerial DSS Applications

Functions such as planning, decision-making, and tactical analysis are all areas where GIS functionality can provide the firm with a significant advantage. There are many reasons for this, but three stand out. First, because GIS is a tool for collecting and managing spatially defined data and linking this data with attribute data (i.e., data from a traditional database), it provides a unique platform for analyzing data based on geography. This is important because geography is often a natural schema for organizing data.

Second, GIS generally incorporates an array of tools that are used to display, analyze, or query the data based on spatial criteria, criteria derived from the attribute data, or upon some combination of these data. An important display capability of GIS is that each data set represents as a unique map layer. In this context, each layer is similar to an individual user view (or table) in a database. The important difference that makes GIS so powerful is that data sets can be overlaid one on top of another, thus creating one or more new layers (or user views) that contain images showing how data relate to one another. This capability is important because it allows a user to visualize the relationships among the data and thereby identify patterns or relationships that might not otherwise be obvious. These types of systems do more than answer "Where is this building or what is located at this address?" For example, GIS managerial tools allow users to examine data by pointing to an object, by defining a polygon, or by selecting records within a given distance (radius) of a location to determine whose property is in a flood plane or the space needed for roads in a new sub-division.

A third GIS capability is that of spatial analysis. GIS spatial analysis capabilities perform "what if" analyses. For example, GIS users may ask questions such as "What number of people will pass by our restaurant if we locate it at the corner of 5th Street and Evans

Street?" or "What will happen to the real estate market in our community if the textile plant moves to Mexico?" Most GISs incorporate a number of statistical tools and data manipulation functions used to test models and transform data. The important value-adding capability that GIS provides is for visualizing the components in the models.

Virtual Applications

Fleet management and other routing activities have supported supply chain applications for many years, however, these functions have evolved considerably in the last several years with the increasing integration of wireless location-based services and GIS technologies. Recently a great deal of convergence has occurred between wireless devices, location technologies, and spatial management and analysis tools with the result that many firms can now manage fleets in real time in a seamless manner. PepsiCo, for example, uses GPS and wireless modems to monitor individual bottle delivery trucks. Every 10 seconds, the system collects positional information and then feeds this information to the dispatcher every 15 minutes. The position/velocity/time (PVT) data monitor activities, streamline operations, and increase the amount of information available to both managers and customers. Proctor & Gamble uses the Analystics Center of Expertise (ACOE) to manage, coordinate and integrate geographic operations within the firm.

Burlington Northern and Union Pacific railroads have collaborated on the development of a system they call the Positive Train Separation (PTS) advanced train control project, which is designed to manage not only facilities, but also train locations, routing, and similar functions (Vantuono, 1995). Cox Communications uses GIS as a support technology in its scheduling and customer service system for its cable customers. Cox uses GIS in conjunction with its specialized fleet management system (FMS), Fleetcon, to provide customers with a two-hour window for customer service calls (Corbley, 1996). This service provided Cox with the ability to improve not only customer service, but also better utilize their service technician's time by optimizing schedules and routes.

Potential to expand government's use of GIS into the virtual world comes from Oregon's Department of Transportation hope to use GIS to fill dwindling highway maintenance coffers (Middaugh, 2003). They want to put a tracking device in every motor vehicle registered or transiting Oregon roads. Vehicles would be billed for each mile driven. Clearly, this technology could be adapted to prioritize road building and repair, track speeders, dispatch police to accidents, and locate missing vehicles to name a few possible innovations. Physical resources to monitor road use become unnecessary as road use decisions are made with more complete, timely and factual information.

Conclusions

GISs represent technological applications that may give firms a strategic competitive advantage. To do so, firms must have the dedicated internal capabilities and an appreciation of the external competitive pressures to react to the changing marketplace as suggested by the Net-Enhancement Business Innovation Cycle (Wheeler, 2002). The Geo-Business Application Model provides a framework to identify opportunities to leverage processes, products and actors to generate additional customer value. Even more importantly, GISs may interact with existing systems to augment and improve decision-making at the operational, managerial, and strategic levels.

The NEBIC suggests that firms incrementally and continuously evolve their IT. GISs are an IT that must evolve to complement existing processes and products in order to add customer value. While a firm may position processes, products and actors differently in the physical, digital, or virtual framework of the Geo-Business Application Model, the need to transform a firm's processes across the supply chain to leverage all the available information, including spatial information, is of paramount importance in the firm's seeking to break out from the old ways of thinking (Kettinger et al., 2000). Location data uniquely inter-relate the location of products, actors and processes, leveraging asymmetric information to provide a competitive advance. This makes sense in order to manage resources more efficiently and to provide greater value to the customer.

References

Choi, S.Y., Stahl, D.O., & Whinston, A.B. (1997). *The economics of electronic commerce.* Indianapolis, IN: Macmillan Technical Publishing, 626.

Cockburn, I., Henderson, R., & Stern, S. (2000). Untangling the origins of competitive advantage. *Strategic Management, 21* (10-11), 1123-1146.

Coppock, J.T., & Rhind, D.W. (1991). The history of GIS. In *Geographic information systems: principles and applications* (pp. 21-43). London: Longman Scientific & Technical.

Corbley, L. (1996, April). Cox cable tunes in customers with fleet management. *Business Geographics*, 42-44.

Davis, M. (2000, March). IPL's customer focus: It's all on the map. *Geo Info Systems*, 40-42.

Docherty, M., Scovell, L., T. Calvin, T., & Simons, H.A. (1996, April). Parking cars efficiently. *Business Geographics*, 28-30.

Garrison, S. (1999, January). Inside space: The new frontier. *Business Geographics*, 33-38.

Gates, K. (1995, February). Champion manages forests — enterprisewise. *Business Geographics*, 36-39.

Goodchild, M.F. (1992). Geographical information science. *International Journal of Geographical Information Systems, 6* (1), 31-45.

Greenfield, D. (1996). Point-and-click OR. *OR/MS Today*, 16-17.

Kalakota, R., & Whinston, A.B. (1996a). *Frontiers of electronic commerce.* Sydney: Addison-Wesley.

Kalakota, R., & Whinston, A.B. (1996b). *Electronic commerce: A manager's guide.* Sydney: Addison-Wesley.

Kearney, A. (1984). *Measuring and improving productivity in physical distribution.* Council of Logistics Management.

Kettinger, W.J., & Hackbarth, G. (2000). Reaching the next level in e-commerce. In D.A. Marchand, T.H. Davenport & T. Dickson (Eds.), *Mastering Information Management* (pp. 222-227). London: Financial Times/Prentice Hall.

Kmitta, J. (1999, October). Keep on tracking: When local assets move indoors, local positioning systems help a variety of industries keep tabs on them. *Business Geographics*, 36-38.

Lapalme, G., Rousseau, J.M., Chapleau, S., Cormier, M., Cossette, P., & Roy, S. (1992). GeoRoute: A geographic information system for transportation applications. *Communications of the ACM, 35* (1), 80-88.

Mennecke, B.E., Dangermond, J., Santoro, P.J., Darling, M., & Crossland, M.D. (1998). Responding to customer needs with geographic information systems. In S. Bradley & R. Nolan (Eds.), *Sense and Respond: Capturing Value in the Network Era.* Boston, MA: Harvard Business School Press.

Middaugh, K. (2003). Runnin' On Empty. In *Government Technology* (pp. 44-47).

Money, A., Tromp, D., & Wegner, T. (1988). The quantification of decision support benefits within the context of value analysis. *MIS Quarterly*, *12* (2), 223-236.

Schuette, D. (2000). Turning e-business barriers into strengths. *Information Systems Management*, *17* (4), 20.

Straub, D., & Watson, R. (2001). Transformational issues in researching IS and net-enabled organizations. *Information Systems Research*, *12* (4), 337-345.

Vantuono, W.C. (1995, March). Mapping new roles for GIS. *Railway Age*, 45-52.

Wheeler, B.C. (2002). NEBIC: A dynamic capabilities theory for assessing net-enablement. *Information Systems Research*, *13* (2), 125-146.

Zahra, S.A., & George, G. (2002a). Absorptive capacity: A review, reconceptualization, and extension. *Academy of Management Review*, *27* (2).

Zahra, S.A., & George, G. (2002b). The net-enabled business innovation cycle and the evolution of dynamic capabilities. *Information Systems Research*, *13* (2), 147-150.

Endnote

1 Customer relationship management is a difficult application to classify because customer relationships may involve virtual, electronic or physical channels for either physical, digital or virtual products or services.

Section III

Applications and the Future

Chapter X

Geographic Information Systems in Health Care Services

Brian N. Hilton, Claremont Graduate University, USA

Thomas A. Horan, Claremont Graduate University, USA

Bengisu Tulu, Claremont Graduate University, USA

Abstract

Geographic information systems (GIS) have numerous applications in human health. This chapter opens with a brief discussion of the three dimensions of decision-making in organizations — operational control, management control, and strategic planning. These dimensions are then discussed in terms of three case studies: a practice-improvement case study under operational control, a service-planning case study under management control, and a research case study under strategic planning. The discussion proceeds with an analysis of GIS contributions to three health care applications: medical/disability services (operational control/practice), emergency response (management control/planning), and infectious disease/SARS (strategic planning/research). The chapter concludes with a cross-case synthesis and discussion of how GIS could be integrated into health care management through Spatial Decision Support Systems and presents three keys issues to consider regarding the management of organizations: Data Integration for Operational Control, Planning Interorganizational Systems for Management Control, and Design Research for Strategic Planning.

Introduction

Geographic information systems (GIS) have numerous applications in human health. At the most basic level, entire research and practice domains within health care are strongly grounded in the spatial dimension (Meade & Earickson, 2000). Indeed, the pioneering work of Dr. John Snow in diagnosing the London Cholera Epidemic of 1854 not only launched the field of epidemiology, but did so in a manner closely linked with the visual display of spatial information (Tufte, 1997). The health care enterprise has become much more complex since the time of Dr. Snow and so have the technologies that are employed to conduct spatial analysis regarding heath care conditions and services (Dangermond, 2000).

This chapter opens with a brief discussion of the three dimensions of decision-making in organizations — operational control, management control, and strategic planning. These dimensions are then discussed in terms of the case study focus of the chapter, which includes a practice-improvement case study under operational control, a service-planning case study under management control, and a research case study under strategic planning. The chapter proceeds with the analysis of GIS contributions to three health care applications: medical/disability services (operational control/practice), emergency response (management control/planning), and infectious disease/SARS (strategic planning/research). The chapter concludes with a cross-case synthesis and discus-

Figure 1. John Snow's Map of the Broad Street Pump Outbreak, 1854[1]

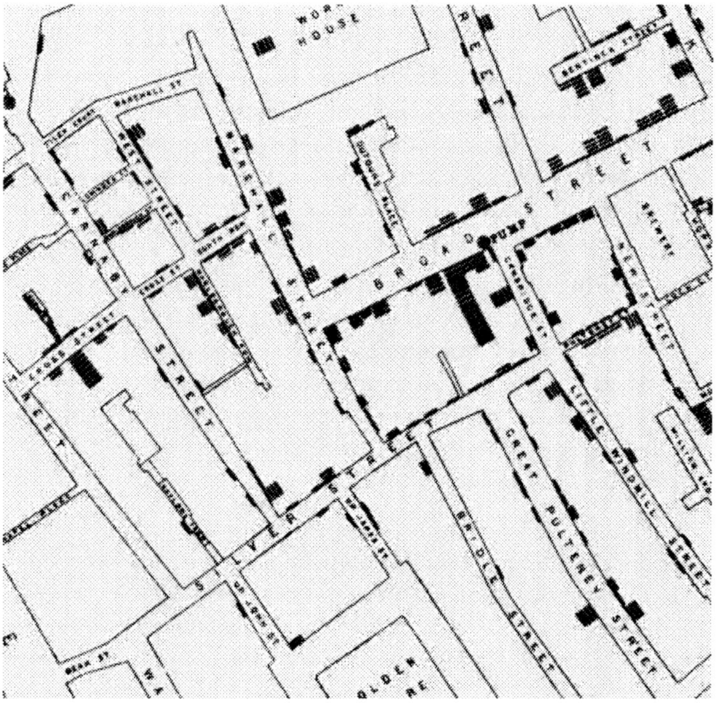

sion of how GIS could be integrated into heath care management through spatial decision support systems.

Background

One definition of a GIS is as "a group of procedures that provide data input, storage and retrieval, mapping and spatial analysis for both spatial and attribute data to support the decision-making activities of the organization" (Grimshaw, 2000, p. 33). One of the most well known models for thinking about the nature of these decision-making activities in the organization is Anthony's Model.

Anthony's Model implies a hierarchy of organizational decision-making. Here, a qualitative distinction is made between three types of decision-making: Operational Control, Management Control, and Strategic Planning (Ahituv, Neumann, & Riley, 1994). As GIS has developed, the range of applications for spatial data on human heath has grown dramatically (Cromley & McLafferty, 2002). In an effort to provide an in-depth understanding of these applications, this chapter considers three distinct application areas of GIS and Human Health: Practice, Planning, and Research. The combination of GIS and Human Health applications with the decision-making processes as defined in Anthony's Model is outlined below:

- **Operational Control** is the management of people, assets, and services using spatial information to ensure the delivery of the health care service while assuring that specific tasks are carried out effectively and efficiently. Our focus in this dimension is how spatial information can improve the *practic*e of health care.
- **Management Control** encompasses the management surrounding the health delivery system as a whole, and is specifically related to the needs and provisioning of health services, health promotion, disease prevention, and health inequalities while assuring that resources are obtained and used effectively and efficiently in the accomplishment of the organization's objectives. Our focus in this dimension is the use of spatial information to assist in the *planning* of health care services.
- **Strategic Planning** deals with the spatial distribution of diseases, their epidemiological patterns, and relation to environmental health risks and demographic characteristics while deciding on objectives of the organization, on changes in these objectives, on the resources used to attain these objectives, and on the policies that are to govern the acquisition, use, and disposition of these resources. Our focus in this dimension is how spatial-based *research* can affect the strategic design of health care delivery applications.

These combinations of GIS and Human Health applications and decision-making processes are used to present this particular series of case study summaries (*Figure 2*). The first case is an example of the *practice* of GIS regarding Disability Evaluation delivery at the Operational Control level. The second case is an example of *planning* regarding

Figure 2. GIS and Human Health Decision-Making and Applications

the use of a GIS for the delivery of Emergency Management Services at the Management Control level. The third case is an example of *research* regarding the conceptual design and development of a GIS as it relates to the National Electronic Disease Surveillance System at the Strategic Planning level.

Research Methodology

The research presented in this chapter draws on three case studies to illustrate various organizational scenarios in which a GIS was utilized, or could be utilized, to solve a particular Health Care Service problem.

Case Study

Case study methods can be used to explore the occurrence of a phenomenon with special attention to the context in which the case study is occurring. The most common working definition of this approach is offered by Robert Yin, who views case studies as "an empirical inquiry that investigates a contemporary phenomenon within its real-life context, especially when the boundaries between phenomenon and context are not clearly evident" (1994, p. 13). Yin's defining work and related treatments further note that there are several uses of case studies:

- **Exploratory Value** – To uncover the nature of the phenomenon of interest. This can often serve as a precursor to more quantitative analysis.
- **Explanatory Value** – To help explain a phenomenon, such as when a quantitative study has revealed statistical association between variables but a deeper understanding of why they are related is missing.
- **Causal Value** – To provide a rich explanation of phenomenon of interest, including "patterns" that are not easily discernible through more abstract and/or numerical analysis.

Case studies have been used throughout the social sciences, as well as in business studies. Single-case design often uses the *extreme* or *unique* case to illustrate those phenomenon that are acutely visible such that inferences can be easily drawn that can be generalized to less extreme cases (Yin, 1994). Pare (2001) recently summarized the widespread use of case studies to examine the influence of information technology and systems in a variety of fields. Moreover, in his work with Elam, they note the promise of building a theory of IT through multiple case studies (1997). In a similar manner, this chapter uses three case studies in an exploratory fashion to enhance the understanding of IT usage, specifically GIS usage, within the context of health care.

For each case relevant literature was reviewed. For the first and second cases, the authors obtained original empirical information as part of separate studies. In the third case, a new and timely application at the strategic level is proposed.

Case Studies

Operational Control Case Study

Background

One core practice area in medical services is the matching of patient/client services to providers (Cromley & McLafferty, 2002). The company in this case provides an array of disability evaluations, management, and information services nationwide. Of relevance to the subject of this chapter, this company (headquartered in Southern California) examined the use of a GIS to assist them in planning and marketing their disability evaluation services. With respect to Operational Control, this case study deals with appointment processing and the *practice* of ensuring the effective and efficient delivery of this service using spatial information.

Problem

The problem for this company was to provide an appointment for a claimant with a physician in a timely manner while meeting specific requirements and constraints. The existing workflow process was inadequate in meeting these requirements. *Figure 3* illustrates the workflow for this process, which begins when a case manager receives a request for an appointment. These requests are prioritized or "triaged" and the required exam sheets are generated using an expert knowledge base. Based on constraints such as physician specialty, availability, contract type, and location, an appointment is made for the claimant with the physician who is the "closest fit" (distance and travel-time) with minimal travel-time being the higher priority.

In this workflow process, the company was pleased with the efficiency and effectiveness of the first steps in the workflow process, which are computer-based. However, the last

Figure 3. Claim Process

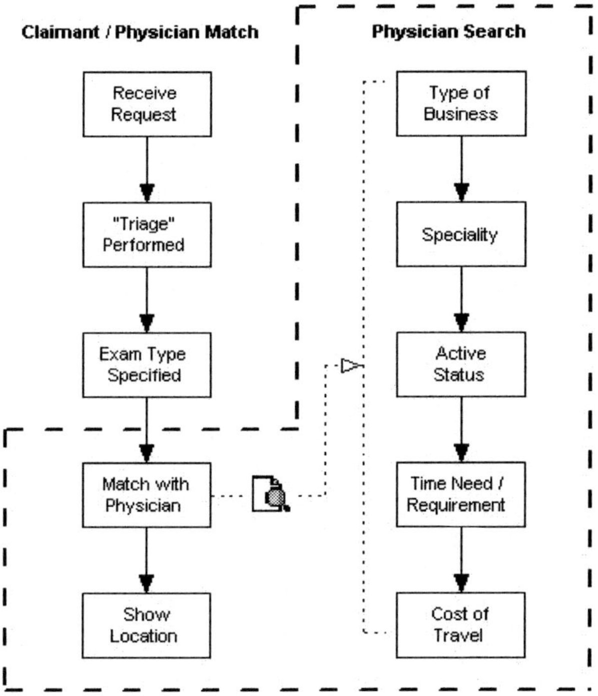

few steps were conducted using paper-based data sets and a number of Internet-based mapping websites (Yahoo Maps and MapQuest). As a result, these last few steps negatively impacted the amount of time a case manager interacted with a claimant, thereby increasing the company's costs in providing this service.

Solution

Using an Internet-based GIS application, a successful prototype was implemented for use within the company's Extranet. To develop this solution, a GIS planning process, also known as the GIS development cycle was employed (NYSARA & NCGIA, 1997). This process, illustrated in *Figure 4*, consists of a set of eleven steps starting with a needs assessment and ending with the on-going use and maintenance of the GIS system.

Outcome

Figures 5 through *8* provide an overview of the GIS based address-matching process that was developed. Now, when a case manager receives a request for an appointment they pinpoint the location of the claimant by "geocoding" the claimant's address (*Figure 5*). With the location of the claimant identified, the case manager then performs a physician attribute search, such as physician specialty, to identify only those physicians that meet

Figure 4. GIS Development Cycle

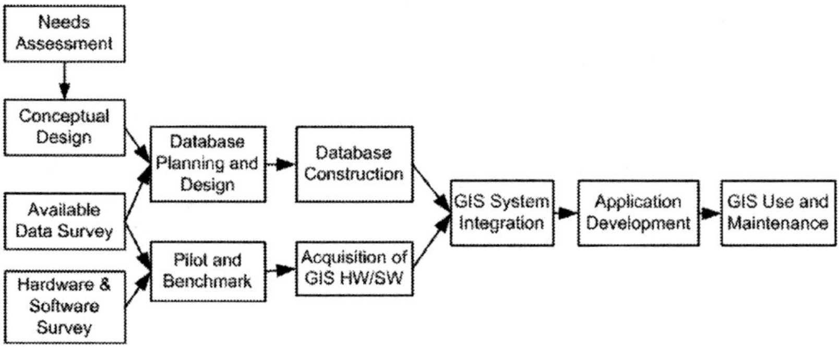

Figure 5. Step 1: Geocoding of Claimant

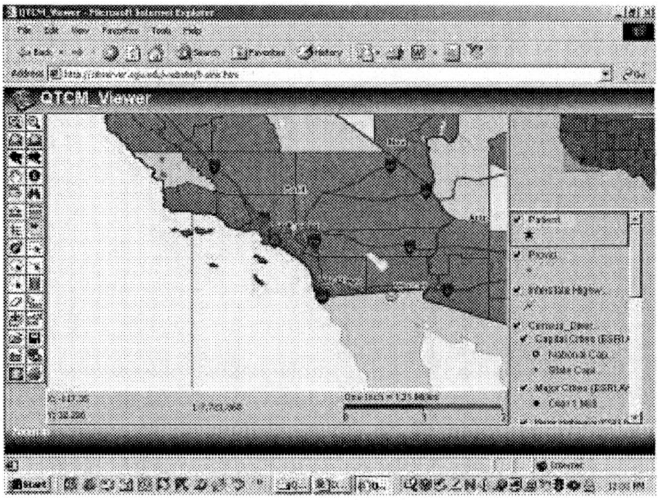

Figure 6. Step 2: Physician Attribute Search

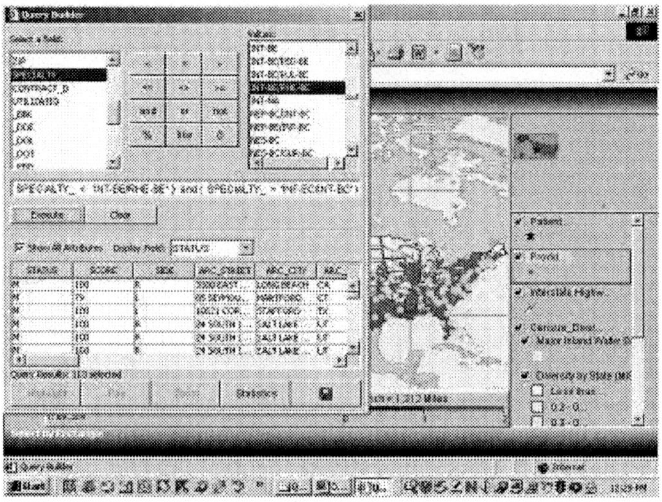

Figure 7. Step 3: Physician Spatial Search

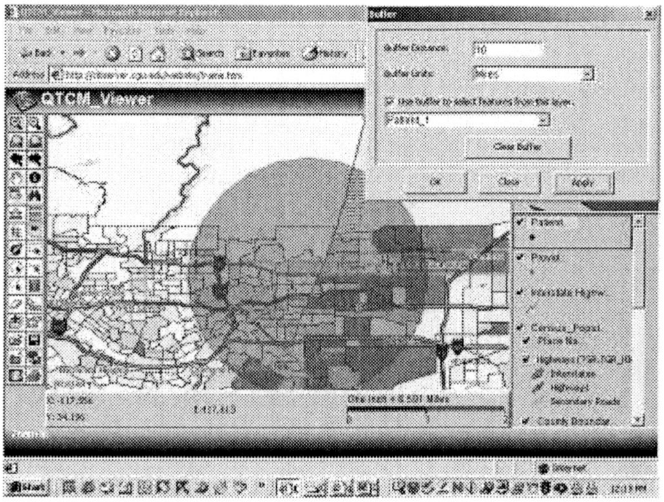

Figure 8. Step 4: Final Physician Selection

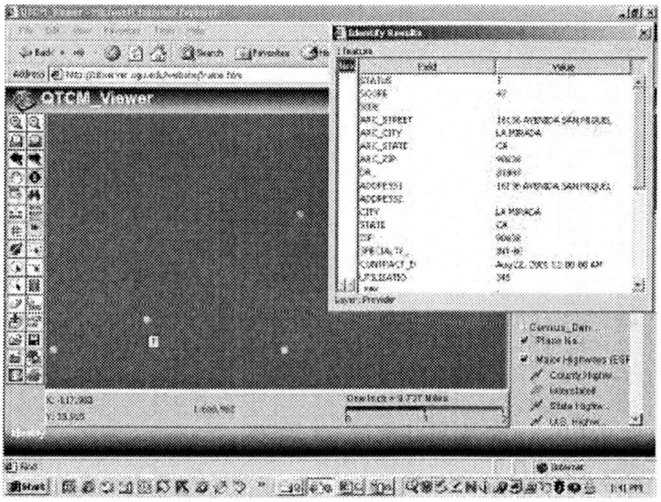

the claimants' requirements (*Figure 6*). A physician spatial search is then performed to narrow down this list even further by locating only those physicians that are within a specified proximity to the claimant (*Figure 7*). Finally, a physician is chosen from this group that most closely matches the claimants' requirements and an appointment is made (*Figure 8*). The most important outcome of this case was a reduction in time required to set an appointment location and date. An additional beneficial outcome of this case was the development of a "live" connection between this system and the company's Oracle database. This database, which is updated on a daily basis, enables the company to exchange information between geographically distributed offices in real time. Consequently, users using the new GIS are now able to view the latest data, geocoded on a map, from any company location. Another important outcome of this case is the fact that this shared information is in a visual format.

Management Control Case Study

Background

The United States Emergency 911 system, established more than 30 years ago, has become a cornerstone infrastructure for emergency management. However, the system is becoming increasingly stressed due to new wireless and digital communications technologies (Jackson, 2002; National Emergency Number Association, 2001). The original design could not anticipate the widespread use of mobile communications for emergency purposes seen today. Consequently, this growth in wireless telecommunications is forcing the Emergency 911 infrastructure to change (Folts, 2002; Jackson, 2002; National Emergency Number Association, 2001). This case considers the broad development of Emergency Medical Service (EMS) systems, within the specific context of rural Minnesota. With respect to Management Control, this case encompasses the spatial properties surrounding the health delivery system as a whole, and is specifically related to the *planning* of E-911 services with the goal of assuring that wireless telecommunications services resources are obtained and used to accomplish the objectives of the State of Minnesota.

Problem

The advent of competitive sector telecommunications services in the wireless arena has played a pivotal role in the fast growth and use of the safety information network. Wireless phones have rapidly become one of our most effective tools in improving emergency response time and saving lives. A wireless 911 phone call can shave valuable minutes from the time otherwise required for a caller to find a conventional phone to access emergency medical services (Tavana, Mahmassani, & Haas, 1999). In the past 10 years, wireless phone use has grown exponentially. There are more than 120 million wireless users making approximately 155,000 emergency calls a day across the United States. The steady increase in private sector wireless subscribership and resulting mobile EMS use has created a need to better understand the implications of this rapidly growing system.

One illustration of the spatial challenges confronting EMS providers is the lack of location information regarding E-911 accessibility. The U.S. Federal Communications Commissions (FCC) has enacted mandatory requirements for wireless communications carriers to provide automatic location identification of a wireless 911 (E-911) phone call to an appropriate Public Service Answering Point (PSAP) (Federal Communications Commission, 2001). Both private carriers and public agencies are working closely together to overcome this difficult requirement. Although the technical requirements for building these systems have been thoroughly outlined, the execution of the service has materialized slowly (Christie et al., 2002; Zhoa, 2002).

One possible reason for this is the difficulty involved in committing to one of several viable technology alternatives to provide E-911. For example, one E-911 technology choice is a satellite-based system, which places a GPS-enabled chip in mobile phones along with location readers at the receiving point. With the current rate of technological change, selecting the one best solution, or combination of solutions, for long term system planning and investment is a difficult and daunting task for system administrators and designers (Proietti, 2002). The consequence of this situation is that deploying end-to-end E-911 systems will require new spatial technologies and nontraditional partnerships, particularly among wireless carriers, emergency dispatch center administrators (e.g., PSAPs), law enforcement, fire and EMS officials, automotive companies, consumers, technology vendors, and state and local political leaders (Jackson, 2002; Lambert, 2000; Potts, 2000). From a spatial perspective, the result is that location-based (E-911) services will be differentially deployed across regions, leading to a need to understand which areas are well serviced and which may require additional policy attention.

From a *planning* perspective, the need for special attention to rural areas is evident from the following statistics. According to the U.S. Department of Transportation, more than 56% of fatal automobile crashes in 2001 occurred on rural roads (National Center for Statistics and Analysis, National Highway Transportation Safety Administration, & U.S. Department of Transportation, 2002). The Minnesota Department of Transportation (MnDOT) reports that only 30% of miles driven within the state are on rural roads, yet 70% of fatal crashes occur on them (Short Elliot Hendrickson Inc. & C.J. Olson Market Research, 2000). In addition, 50% of rural traffic deaths occur before arrival at a hospital. Appropriate medical care during the "golden hour" immediately after injuries is critical to reducing the odds of lethal or disability consequences. Crash victims are often disoriented or unconscious and cannot call for help or assist in their rescue and therefore rely heavily upon coordinated actions from medical, fire, state patrol, telecommunications and other entities (Lambert, 2000).

Solution

The solution to the rural EMS program entails a combination of responses. These responses were analyzed within the context of a specific case study; an analysis of Minnesota's E-911 system (Horan & Schooley, 2003). The first activity in this case was to analyze the entire system and to construct an overall architecture of the system. This architecture, presented in *Figure 9*, illustrates Minnesota's EMS system along several key strata, technology, organizations, and policy, and identifies possible critical links (shaded gray) in the overall system. A summary of each layer follows.

- **Technology** – The top layer of the architecture illustrates some of the essential networks and communications technologies used by Minnesota EMS organizations to carry out their individual and interorganizational functions. From a GIS perspective, the GPS-equipped wireless devices and infrastructure to determine spatial location are critical elements.
- **Organizations** – The middle layer illustrates some of the public and private organizations involved in the Minnesota EMS and the general interorganizational relationships between these organizations. There is a significant geographic dimension to the organizational layer: each of the major stakeholders has distinct service boundaries (for example, there are 109 PSAPs, yet nine rural transportation operation centers).

- **Policy** – For EMS interorganizational relationships (i.e., partnerships, joint ventures, etc.) to succeed, policies need to be developed that facilitate the interorganizational use of new and existing communications technologies. The overarching EMS technology-related policies, illustrated in the bottom layer, currently under development in the state are E-911 and 800 MHz radio. This includes the state-federal effort to develop standards and procedures for using location information received from mobile phones.

Outcomes

This case study raises several technological, organizational, and policy issues for planning EMS in rural Minnesota specifically, as well as for rural areas in general. The architecture highlights several critical areas that arose from this review and therefore have implications for future advancements. Areas, such as those denoted in gray in *Figure 9*, provide a focal point for discussing implications of this architecture for planning both EMS in general and GIS specifically.

GIS can assist greatly in understanding the extent of EMS coverage. Currently, this understanding is at a general level of detail, i.e., the level of compliance with new E-911 regulations. For example, *Figure 10* provides an example of the spatial dimension of E-911 deployment by a major provider. As displayed in this figure, the metropolitan region of Minneapolis has deployed location-identifying systems (e.g., E-911 Phase 2), while such systems have only been partially deployed in rural areas[2].

Figure 10 also provides an overview of the level of compliance with these new regulations, with lower compliance in rural areas. As shown in the figure, many regions

Figure 9. Interorganizational Architecture for Emergency Management Systems in Minnesota

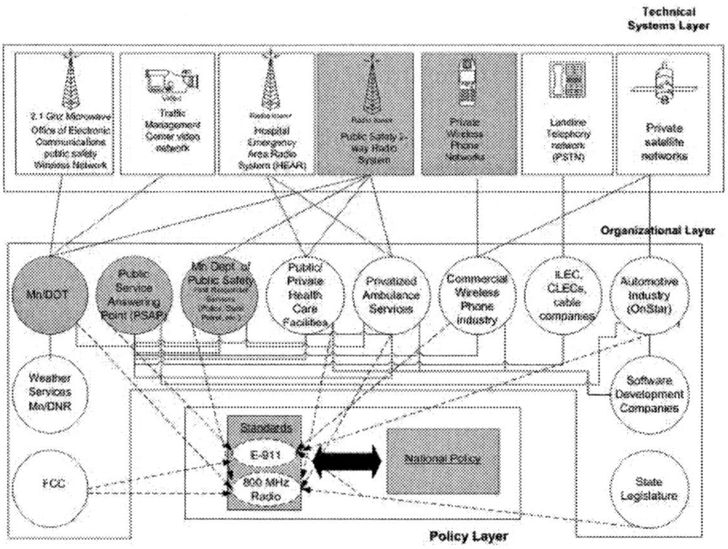

Figure 10. Spatial Distribution of E-911 Compliance Status (Qwest)

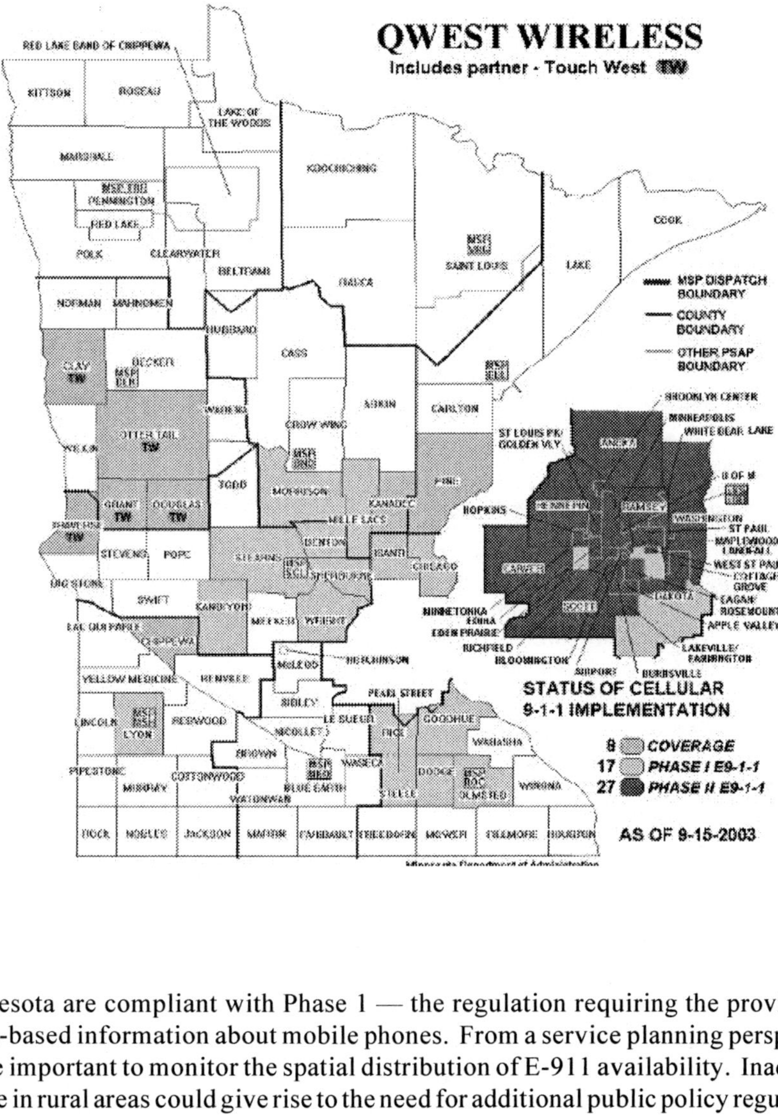

in Minnesota are compliant with Phase 1 — the regulation requiring the provision of location-based information about mobile phones. From a service planning perspective, it will be important to monitor the spatial distribution of E-911 availability. Inadequate coverage in rural areas could give rise to the need for additional public policy regulations.

The deployment of advanced 911 capabilities is however, only one aspect of integrated EMS services. Especially with recent concerns regarding homeland security, attention is now turning to how EMS is planned as part of an overall readiness strategy. In this context, GIS can play an important role in providing a spatial platform for EMS and related emergency services. One example of this is the GIS development work underway in Dakota County, Minnesota. This county, which includes significant rural as well as urban areas, has undertaken a comprehensive GIS-based approach to emergency preparedness with an Internet-based GIS platform that integrates both EMS related factors (e.g., Ambulance Service Areas and Cellular Towers) with other civic institutions

Figure 11. Internet-Based GIS Emergency Preparedness Application

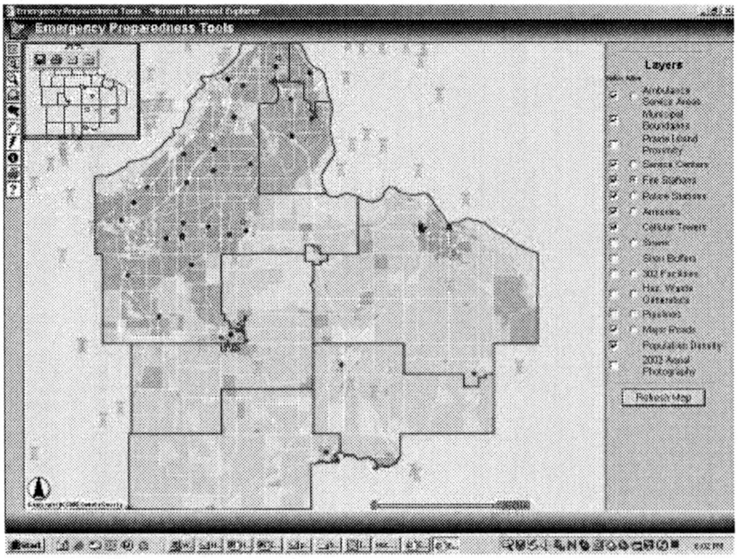

involved in emergency preparedness (e.g., Fire Stations, Police Stations, and Armories)[3]. As illustrated in *Figure 11*, the Municipal Boundaries (outlined in bold) within Dakota County and the locations of Fire Stations, Police Stations, and Armories (dots) are identified. The locations of Ambulance Service Centers (dots) and Cellular Towers (tower symbol) are identified as well.

Platforms such as this represent a critical new dimension of interactivity whereby emergency management systems can be accessed dynamically and across institutions as well as infrastructure systems. Returning to the original architecture outlined above, this platform can be used by institutions such as the departments of transportation, public safety, and emergency services to facilitate cross-agency partnerships in service and technology deployments.

In summary, this case demonstrates that the planning of emergency medical services is well suited to benefit from a dynamic GIS platform. Particularly in rural areas where resources are often scarce, such a platform can provide a common database to (1) monitor the spatial deployment of services, (2) facilitate resources sharing among institutions, and (3) provide a common understanding of "conditions" against which to plan for new technologies, systems, and policies.

Over time, it will be important to monitor the rate by which rural communities improve their "readiness." GIS can assist in this monitoring, including tracking of funding expenditures, regulatory compliance, etc. Finally, it is essential that methods of planning and analysis used to determine the form, level, and location of service and resource provision reflect the geographical components underpinning the health care system, i.e., the planning process should have an explicit geographical focus (Birkin, Clarke, Clarke, & Wilson, 1996).

Strategic Planning Case Study

Background

As noted in the introduction, spatial analysis and mapping in epidemiology have a long history (Frerichs, 2000), but until recently, their use in public health has been limited. However, recent advances in geographical information and mapping technologies and increased awareness have created new opportunities for public health administrators to enhance their planning, analysis and monitoring capabilities (World Health Organization, 1999b). Moreover, effective communicable disease control relies on effective disease surveillance where a functional national communicable diseases surveillance system is essential for action on priority communicable diseases (World Health Organization, 1999a). With respect to Strategic Planning, this case considers a *research* program regarding the spatial distribution of severe acute respiratory syndrome (SARS), its epidemiological patterns, and a theory to direct health organizations objectives and policies regarding the acquisition, use, and disposition of GIS.

The National Electronic Disease Surveillance System (NEDSS) program was initiated in the United States to provide an integrated, standards-based approach to public health surveillance and to connect surveillance systems to the burgeoning clinical information systems infrastructure (U.S. Department of Health and Human Services, 2002a). It is expected that the NEDSS will improve the nation's ability to identify and track emerging infectious diseases, monitor disease trends, respond to the threat of bio-terrorism, and other scenarios where the rapid identification of unusual clusters of acute illness in the general population is a fundamental challenge for public health surveillance (Lazarus et al.,2002).

To be effective with the rapid deployment of new health information systems, it is important to maintain effective mechanisms for rapid technology transfer to occur across government, academia, and industry (Laxminarayan & Stamm, 2003). The NEDSS program articulates an architecture that will enable public health information systems to communicate electronically, thereby decreasing the burden on respondents and promoting timeliness and accuracy (U. S. Department of Health and Human Services, 2002a). Stakeholders in the NEDSS include not only the Center for Disease Control (CDC) and other agencies within Department of Health and Human Services (DHHS), but state and local public health departments, healthcare providers, laboratories, health care standards organizations, health care product vendors, and healthcare professional services organizations.

The NEDSS Base System (NBS) is an instance of the NEDSS standards for use by all stakeholders that enable systems from different CDC program areas to be integrated. Disease-specific data and processes are incorporated and integrated within the NBS using Program Area Modules (PAM). Though this modular approach allows for the sharing of these data and processes, the NEDSS standards do not provide specific guidelines to support the development of these modules. Given that "GIS provides an excellent means of collecting and managing epidemiological surveillance and programmatic information, GIS represents an entry point for integrating disease-specific surveillance approaches" (World Health Organization, 2001, p. 7). An Information System Design Theory (ISDT) approach for the development of GIS-based PAMs would provide a useful guide for the various stakeholders in the NEDSS. This approach is detailed in the following section.

Problem

Severe acute respiratory syndrome (SARS) has emerged as a serious international occupational health disease. Since the disease was first reported it has infected numerous health care workers, some fatally (Centers for Disease Control and Prevention, 2003). Consequently, early identification of SARS cases is critical, as no specific treatment protocol exists. As a result, a GIS-based PAM for SARS is needed that could be used to manage the health care services required to combat this disease.

Unfortunately, current GIS-based attempts to mitigate the spread of this disease are lacking in respect to the NEDSS; that is, they lack an integrated, standards-based approach to public health surveillance and information dissemination[4]. The objective of this case study is to propose an ISDT such that organizations meeting the requirements of this theory will, by design, develop a GIS-based PAM compliant with NEDSS standards. Geographic information science has the potential to create rich information databases, linked to methods of spatial analysis, to determine relationships between geographical patterns of disease distribution and social and physical environmental conditions. As the core of a decision support system, geographic information science also has the potential to change the way that allocations of resources are made to facilitate preventive health services and to control the burden of disease (Rushton, Elmes, & McMaster, 2000).

Information System Design Theory Approach

While the overriding methodology in this chapter is the use of case studies, the case in this section is specifically concerned with the design of GIS as systems. As such, this analysis is informed by concepts deriving from a "design" approach to information systems. As noted in the ISDT approach introduced by Walls et al. (1992), the design process is analogous to the scientific method where hypotheses are to be tested by designing and building the artifact or product. They outlined several characteristics of design theories:

- Design theories are composite theories that encompass kernel theories from natural science, social science, and mathematics.
- While explanatory theories tell "what is," predictive theories tell "what will be," and normative theories tell "what should be," design theories tell "how to/because."
- Design theories show how explanatory, predictive, or normative theories can be put to practical use.

- Design theories are theories of procedural rationality. "The objective of the design theory is to prescribe both the properties an artifact should have if it is to achieve certain goals and the method(s) of artifact construction" (p. 41). Thus, the artifact must have all the characteristics identified in the design theory.

This case study explicitly draws upon their structure for creating a design product and process. Briefly, their first component of the design product involves a set of meta-requirements that describes the class of goals to which the theory applies. Their second component is a meta-design, which describes a class of artifacts hypothesized to meet the meta-requirements. Their third component is a set of kernel theories, theories from natural or social sciences governing design requirements. Their final component of the design product is a set of testable design product hypotheses that are used to test whether the meta-design satisfies the meta-requirements.

Beyond the design product, designers of GIS would necessarily need to be concerned with principles of the design process. According to the ISDT, the first component of the design process involves a design method which describes the procedures to be used in artifact construction (Walls, Widmeyer, & El Sawy, 1992). The second component is kernel theories, theories from natural or social sciences governing design process. The last component of the design process is the testable design process hypotheses, which are used to verify whether the design method results in an artifact that is consistent with the meta-design. While ISDT is explicitly applied in this third case study, the theme of effective design holds for all: that is, there is a design need to create GIS system designs in a manner that effectively contributes to the design goal of improving human heath.

Solution

The NEDSS has identified a number of enabling technologies for each element of the NEDSS technical architecture. One of these architectural elements is Analysis, Visualization, and Reporting (AVR). AVR capabilities support the epidemiological analysis of public health data and the communication of the analytical results of that analysis (U.S. Department of Health and Human Services, 2001). The specific requirements for this architectural element include tabular and graphical reporting, statistical analysis, and geographical information analysis and display. Also included are features such as the creation of pre-defined and ad-hoc reports, the ability to share results with colleagues and the public, and the extraction of data for use with standard analysis tools. The AVR Requirements are the foundation of the GIS Meta-Requirements as well as each of the specific GIS Meta-Design elements outlined in *Table 1*. The Meta-Requirements and Meta-Design are derived from the OpenGIS Service Architecture (Open GIS Consortium Inc., 2002). Taken as a whole, these design elements would constitute the foundation of an ISDT for GIS-based PAMs.

Table 1. ISDT for GIS-based PAMs

AVR Requirements	GIS Meta-Requirements	GIS Meta-Design
Geographical information analysis and display Tabular reporting Graphical reporting	Geographic human interaction services	Geographic spreadsheet viewer: Client service that allows a user to interact with multiple data objects and to request calculations similar to an arithmetic spreadsheet but extended to geographic data. Geographic viewer: Client service that allows a user to view one or more feature collections or coverages. This viewer allows a user to interact with map data, e.g., displaying, overlaying and querying.
Creation of pre-defined and ad-hoc reports	Geographic model/information management services	Product access service: Service that provides access to and management of a geographic product store. A product can be a predefined feature collection and metadata with known boundaries and content, corresponding to a paper map or report. A product can alternately be a previously defined set of coverages with associated metadata.
Extraction of data for use with standard analysis tools	Geographic processing services – spatial	Subsetting service: Service that extracts data from an input in a continuous spatial region either by geographic location or by grid coordinates. Sampling service: Service that extracts data from an input using a consistent sampling scheme either by geographic location or by grid coordinates.
Extraction of data for use with standard analysis tools	Geographic processing services – thematic	Subsetting service. Service that extracts data from an input based on parameter values. Geographic information extraction services: Services supporting the extraction of feature and terrain information from remotely sensed and scanned images. Image processing service: Service to change the values of thematic attributes of an image using a mathematical function. Example functions include: convolution, data compression, feature extraction, frequency filters, geometric operations, non-linear filters, and spatial filters.
Extraction of data for use with standard analysis tools	Geographic processing services – temporal	Subsetting service. Service that extracts data from an input in a continuous interval based on temporal position values. Sampling service. Service that extracts data from an input using a consistent sampling scheme based on temporal position values.
Statistical analysis	Geographic processing services – metadata	Statistical calculation service: Service to calculate the statistics of a data set, e.g., mean, median, mode, and standard deviation; histogram statistics and histogram calculation; minimum and maximum of an image; multi-band cross correlation matrix; spectral statistics; spatial statistics; other statistical calculations.
Sharing of results with colleagues and the public	Geographic communication services	Transfer service: Service that provides implementation of one or more transfer protocols, which allows data transfer between distributed information systems over off-line or online communication media.

Figure 12. Conceptual Picture of the NEDSS Base System and Program Area Modules

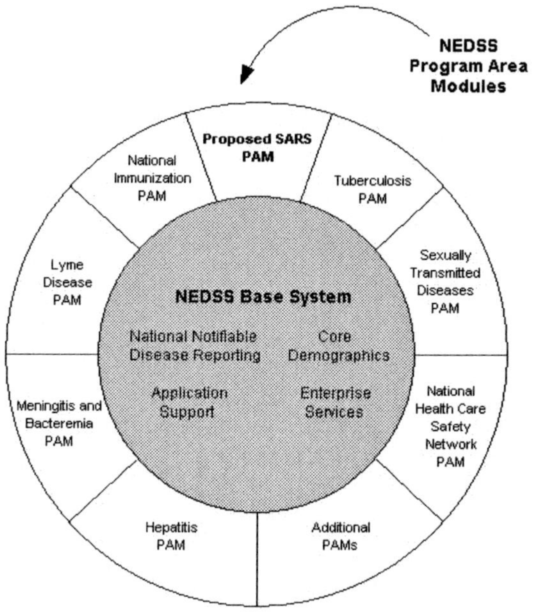

Adapted from U. S. Department of Health and Human Services (2002a)

Outcome

As seen in this case, "GIS offers new and expanding opportunities for epidemiology as they allow the informed user to choose among options when geographic distributions are part of the problem, and when used for analysis and decision-making, they become a tool with a rich potential for public health and epidemiology" (Clarke, McLafferty, & Tempalski, 1996, p. 1). GIS-based PAMs developed using the proposed ISDT would allow the sharing of common data and processes while incorporating disease-specific data and processes. As seen in *Figure 12*, a SARS-specific PAM developed in this manner would become one of many PAMs and part of the larger NBS.

Case Study Summary Analysis and Management Implications

A summary of the three case studies is presented in *Table 2*. Each of the three cases presented is based on the health care service provided, specific problem type, solution generated, and final outcome.

Table 2. Summary of Case Studies

Case Study	Health Care Service Provided	Problem	Solution	Outcome
Operational Control (Practice)	Disability Evaluation	Accuracy and time to schedule appointment location and date	Internet-based GIS application for use within company Extranet	Improved decision making and reduction in time required for scheduling physician appointments
Management Planning (Control)	Emergency Medical Service Delivery	Need for Ready EMS Delivery	Use GIS to provide efficient planning and management of E-911 health emergencies	Reduced EMS delay time and more rapid onset of trauma service delivery
Strategic Planning (Research)	Disease Surveillance System	Lack of Spatial Data Integration and Dissemination	Devise Inter-organizational Spatial Database for early disease detection	Program Area Module Design Theory for GIS

As discussed in the introduction, the visual display of spatial information in healthcare, starting with epidemiology, has a long history. The literature reveals, as do these case studies, that the role of the private sector becomes increasingly more important as you move from the strategic planning level down to the management and operational levels. For those charged with the management of organizations, the following issues should be considered:

- **Data Integration for Operational Control:** Geospatial location is considered a valuable organizing principle for architecting and constructing enterprise data stores where interoperable geospatial technologies play a foundational role in exploiting these data stores for enterprise missions (Open GIS Consortium Inc., 2003). As seen in the first case, this can occur at even the most basic level where demand from users at the operational level prompted the linkage of spatial and non-spatial data. As a consequence, the linkage of spatial and non-spatial databases will be among the challenges that organizations face as the spread of GIS applications and their extended functionalities drive organizations to integrate their existing structures with the increasingly important spatial dimension.
- **Planning Interorganizational Systems for Management Control:** Managing ad hoc inter-organizational networks are the nature of GIS data in health care environment. Emergency management applications are being developed in numerous areas around the country, ranging from instances where GIS-based emergency planning has been carried out by local governments, to cases where the Federal Emergency Management Agency has supported the creation and implementation of a GIS for this purpose (O'Looney, 2000). However, these diverse organizations collect spatial data for different purposes and goals. For successful planning, spatial data must be organized in such a manner as to appear as residing in one central database.

Thus, the issue becomes how to successfully manage the relationship between these loosely coupled organizations that have dependence on tightly coupled systems without compromising their organizational purposes and goals.

- **Design Research for Strategic Planning:** The global implication of geographic data, as seen recently with SARS, indicates that it is no longer confined to the local or national level. Unfortunately, global geographic information is currently little more than the sum of the highly varied national parts and is not readily available (Longley, Goodchild, Maguire, & Rhind, 2001). The Global Spatial Data Infrastructure Association was formed to address this issue and is dedicated to international cooperation and collaboration in support of local, national, and international spatial data infrastructure developments to allow nations to better address social, economic, and environmental issues of pressing importance (The Global Spatial Data Infrastructure Association, 2003). Their vision for a Global Spatial Data Infrastructure would support ready access to global geographic information. This vision would be achieved through the coordinated actions of nations and organizations through the implementation of complementary policies and common standards for the development and availability of interoperable digital geographic data and technologies to support decision-making at all scales for multiple purposes. The research challenge then is learning how to create a robust information system design process that considers spatial dimensions from global level international standard bodies to local level implementing organizations.

Discussion

IS, and spatial analysis more broadly, can inform a number of pressing health care service issues, both domestically and internationally. From a domestic perspective this includes issues such as treating highly distributed populations and efficiently targeting critical services. From an international perspective, this includes developing a rapid means to detect and treat human health outbreaks in developed and developing regions as well as the creation of spatial datasets to drive international health enterprises.

Toward a Spatial Decision Support System (SDSS) for Human Health

While the geographic dimension of human health has long been recognized, the integration of GIS into the health care industry has not received the level of attention as that of more cutting-edge medical technologies. Nonetheless, as revealed in these cases as well as others that are increasingly appearing in the literature (Lang, 2000), GIS can provide a useful tool at the operational, management, and strategic levels of health care resources. For this reason, researchers such as Cromley & McLafferty (2002) have begun advocating for Spatial Decision Support Systems (SDDS) in support of health care services. Furthermore, a fairly long history of decision support systems in health care has been recognized (Shortliffe & Perrault, 2001). The new view is to integrate the spatial element into these systems. It is hoped that this chapter provides yet another indicator of the value in using spatial analysis for health care delivery.

Acknowledgments

This chapter draws upon a series of research projects undertaken by the authors, as well as additional resources as appropriate. Findings related to the use of GIS and Disability Evaluations are based in part on the research conducted in collaboration with QTC, Inc. and have been reported in Hilton, Horan, & Tulu (2003). Findings related to the use of Emergency Management Systems and GIS are based in part on the research conducted in collaboration with the Humphrey Institute of Public Affairs, University of Minnesota and funded by the ITS Institute and have been reported in Horan, Kaplancali, & Schooley (2003) and Horan & Schooley (2003). The authors gratefully acknowledge the contributions of the following individuals to the chapter: Lee Munnich, Benjamin Schooley, and Dr. Subbu Murthy.

References

Ahituv, N., Neumann, S., & Riley, H. N. (1994). *Principles of information systems for management* (4th ed.). Business and Educational Technologies.

Birkin, M., Clarke, G., Clarke, M., & Wilson, A. (1996). *Intelligent GIS: Location decisions and strategic planning*. New York: John Wiley & Sons.

Centers for Disease Control and Prevention. (2003). *Public health GIS news and information* (No. 52). Washington, DC.

Christie, J., Fuller, R., Nichols, J., Chen, A., Hayward, R., Gromov, K., et al. (2002). Development and deployment of GPS wireless devices for E911 and location based services. Paper presented at the *IEEE Position Location and Navigation Symposium.*

Clarke, K., McLafferty, S., & Tempalski, B. (1996). On epidemiology and geographic information systems: A review and discussion of future directions. *Emerging Infectious Diseases, 2* (2).

Cromley, E., & McLafferty, S. (2002). *GIS and public health*. New York: Guildford Press.

Dangermond, J. (2000). Preface. *In L. Lang, GIS in health organizations*. Redlands, CA: ESRI Press.

Federal Communications Commission. (2001). *Fifth report and order; First report and order: Implementation of 911 Act* (Electronic version No. WT-Docket No. 00-110 and CC Docket No. 92-105). Federal Communications Commission.

Folts, H. (2002, July). Standards initiatives for emergency telecommunications service (ETS). *IEEE Communications, 40,* 102-107.

Frerichs, R. (2000). History, maps and the Internet: UCLA's John Snow site. *The Society of Cartographers Bulletin, 34* (2), 3-8.

Paré, G., & Elam, J.J. (1997). Using case study research to build theories of IT Implementation. In J. I. DeGross (Ed.), *Information systems and qualitative research* (pp. 542-568). Chapman & Hall.

Grimshaw, D. (2000). *Bringing geographical information systems into business.* New York: John Wiley & Sons.

Hilton, B., Horan, T. A., & Tulu, B. (2003). GIS-enabled health services call center system. Paper presented at the *14th Annual Information Resources Management Association International Conference,* Philadelphia, Pennsylvania, USA.

Horan, T. A., & Schooley, B. (2003). Stranded but connected: Case study of rural deployment of emergency management services. *Journal of the Transportation Research Board, forthcoming.*

Horan, T. A., Kaplancali, U., & Schooley, B. (2003). Devising a web-based ontology for emerging wireless systems. Paper presented at the *Americas Conference on Information Systems,* Tampa, Florida, USA.

Jackson, A. (2002). *Recommendations for ITS technology in emergency medical services* (No. PATH Record Number 26202). Washington DC: Medical Subcommittee of the ITS America Public Safety Advisory Group.

Lambert, T. (2000). The role of ITS in the emergency management and emergency services community. In U. of M. Center for Advanced Transportation Technology, ITE, USDOT, ITSA (ed.), *Intelligent Transportation Primer.* Washington, D.C.: Institute of Transportation Engineers.

Lang, L. (2000). *GIS in health organizations.* Redlands, CA: ESRI Press.

Laxminarayan, S., & Stamm, B. H. (2003, June 6-7). Biomedical engineering evolutions in combating bioterorism. Paper presented at the *5th International Workshop on Enterprise Networking and Computing in Healthcare Industry,* Santa Monica, California, USA.

Lazarus, R., Kleinman, K., Dashevsky, I., Adams, C., Kludt, P., DeMaria, A., et al. (2002). Use of automated ambulatory-care encounter records for detection of acute illness clusters, including potential bioterrorism events. *Emerging Infectious Diseases, 8* (8), 753-760.

Longley, P. A., Goodchild, M. F., Maguire, D. J., & Rhind, D. W. (2001). *Geographic information systems and science.* New York: John Wiley & Sons.

Meade, M. S., & Earickson, R. J. (2000). *Medical geography* (2nd ed.). New York: Guildford Press.

National Center for Statistics and Analysis, National Highway Transportation Safety Administration, & U.S. Department of Transportation. (2002). *National center for statistics and analysis fatality analysis reporting system (FARS) web-based encyclopedia.* Retrieved July 21, 2002 from: *http://www-fars.nhtsa.dot.gov.*

National Emergency Number Association. (2001). *Report card to the nation: The effectiveness, accessibility and future of America's 9-1-1 service.* Columbus, Ohio: The RCN Commission and NENA.

NYSARA & NCGIA. (1997). *Geographic information system development guides.* Retrieved from: *http://www.sara.nysed.gov/pubs/gis/gisindex.htm.*

O'Looney, J. (2000). *Beyond maps: GIS and decision making in local government.* Redlands, CA: ESRI Press.

Open GIS Consortium Inc. (2002). *The OpenGIS abstract specification, topic 12: OpenGIS service architecture, version 4.3.* Open GIS Consortium, Inc.

Open GIS Consortium Inc. (2003). *OpenGIS reference model.* Open GIS Consortium, Inc.

Pare, G. (2001). *Using a positivist case study methodology to build and test theories in information systems: Illustrations from four exemplary studies.* Working Paper.

Potts, J. (2000, Summer). Wireless phone calls to 911: Steps toward a more effective system. *Currents.*

Proietti, M. (2002). Carrier choices in location: The system integrator's view. *GPS World, 13* (3), 23-28.

Rushton, G., Elmes, G., & McMaster, R. (2000). Considerations for improving geographic information system research in public health. *URISA Journal, 12* (2), 31-49.

Short Elliot Hendrickson Inc., & C.J. Olson Market Research. (2000). *Advanced rural transportation information and coordination (ARTIC) operational test evaluation report*: Minnesota Department of Transportation (Mn/DOT).

Shortliffe, E. H., & Perrault, L. E. (2001). *Medical informatics: Computer applications in health care and biomedicine* (2nd ed.). New York: Springer-Verlag.

Tavana, H., Mahmassani, H., & Haas, C. (1999). Effectiveness of wireless phones in incident detection: Probabilistic analysis. *Transportation Research Record, 1683,* 31-37.

The Global Spatial Data Infrastructure Association. (2003). *Global spatial data infrastructure: Strategic development plan.* Retrieved from: *http://www.gsdi.org/.*

Tufte, E. R. (1997). *Visual explanations: Images and quantities, evidence and narrative.* Cheshire, CT: Graphics Press.

U. S. Department of Health and Human Services. (2001). *NEDSS systems architecture version 2.0.* Washington, D.-C.: Centers for Disease Control and Prevention.

U. S. Department of Health and Human Services. (2002). *NEDSS and NEDSS PAMS business discovery statement version 1.2.* Washington, DC.: Centers for Disease Control and Prevention.

Walls, J. G., Widemeyer, G. R., & El Sawy, O. A. (1992). Building an information system design theory for vigilant EIS. *Information Systems Research, 3* (1), 36-58.

World Health Organization. (1999a). *WHO Recommended surveillance standards* (2nd ed.). Department of Communicable Disease Surveillance and Response.

World Health Organization. (1999b). *WHO Recommended surveillance standards - annex 3: Role and use of geographic information systems (GIS) and mapping for epidemiological surveillance*. Department of Communicable Disease Surveillance and Response.

World Health Organization. (2001). *Protocol for the assessment of national communicable disease surveillance and response systems*. Department of Communicable Disease Surveillance and Response.

Yin, R. K. (1994). *Case study research, design and methods* (2nd ed.). Newbury Park: Sage Publications.

Zhoa, Y. (2002, July). Standardization of mobile phone positioning for 3G systems. *IEEE Communications, 40,* 108-116.

Endnotes

1 "Instead of plotting a time-series, which would simply report each day's bad news, Snow constructed a graphical display that provided direct and powerful testimony about a possible cause-effect relationship. Recasting the original data from their one-dimensional temporal ordering into a two-dimensional spatial comparison, Snow marked deaths from cholera on this map, along with locations of the area's 13 community water pump-wells. The notorious well is located amid an intense cluster of deaths, near the D in BROAD STREET. This map reveals a strong association between cholera and proximity to the Broad Street pump, in a context of simultaneous comparison with other local water sources and the surrounding neighborhoods without cholera" (Tufte, 1997. p. 30).

2 See *http://www.911.state.mn.us/911_enhanced.html.*

3 See *http://www.co.dakota.mn.us/gis/.*

4 SARS related GIS websites: *http://www.esrichina-hk.com/SARS/Eng/sars_eng_main.htm, http://www.sunday.com/Sunday/en/index.html, http://www.corda.com/examples/go/map/sars.cfm, http://www.cdc.gov/mmwr/preview/mmwrhtml/figures/m217a4f2.gif, http://www.mapasia.com/sars/default2.htm, http://www.info.gov.hk/dh/diseases/ap/eng/bldglist.htm, http://www.who.int/csr/sars/map2003_04_13.gif, http://spatialnews.geocomm.com/features/sars/, http://www.hku.hk/geog/hkgisa/sars.htm.*

Chapter XI

GIS in Marketing

Nanda K. Viswanathan, Delaware State University, USA

Abstract

This chapter examines the existing uses and potential uses of GIS in marketing. In examining the interaction of marketing and geography, the variables of demographics, space, and time are used as a framework. Specific applications of GIS in customer relationship management, market segmentation, and competitive analysis are illustrated with hypothetical and real world examples. Additional areas of GIS application include product strategy, price strategy, promotion strategy, and place or distribution strategy. The author hopes that understanding the existing and potential uses of GIS in marketing will spur the interest of marketing practitioners to integrate GIS into marketing strategy to create competitive advantage. Furthermore, it is hoped that this chapter will serve as an outline for the broader consideration of the applications of GIS in marketing.

Introduction

Geographic information systems (GIS) may be defined as a set of automatic tools and information systems, typically involving the use of computers, that are used to collect, organize, analyze, and use as an aid to decision-making all data that can be related to a specific geographical location on earth. While geographic data have always existed historically, the arrival of the personal computer and the development of mapping software that could process and present geographic information visually has enhanced the ability of the decision maker to leverage geographic data.

To understand the true power of GIS, consider the ease of decision-making using GIS in the following hypothetical situations.

A: Imagine a marketing manager of a restaurant chain attempting to perform competitive analysis of competing chains in a geographic market, incorporating demographic trends in the market, distances between customers and restaurants, the location of competing restaurants, freeways, and driving patterns.

B: Imagine a marketing manager of an automobile firm attempting to forecast demand for cars based on the demographic composition of a market, new neighborhoods that are being built in the area, and incorporating zoning demarcations, including information on residential and commercial establishments.

C: Imagine a marketer of detergents in a developing economy that has millions of small retailers and a salesperson that needs to plan sales routes for the coming month taking into account the layout of retail locations in a geographic area, the size of the retail outlet, the time of the day when retail outlets are open, and retailer inventory policies.

Prior to the arrival of GIS, in all the scenarios cited above, marketing managers had to rely on data formats, involving hundreds of rows and columns of information that could not be visually represented. This meant that decisions were difficult and made in the abstract. Visualizing zone demarcations, distances between outlets, and freeway and transportation patterns without a map is almost impossible.

The arrival of GIS gave decision makers the power to visually represent and combine data that were geocoded in any form in an interactive manner. For example, data pertaining to zip codes that were geographically contiguous could now be seen on a map next to each other rather than represented as abstract bits of information. The effect of different decisions, such as demand forecasting, retail location, route planning, or competitive analysis could be observed interactively on a map enhancing the decision maker's understanding of the spatial context of the market. GIS enables the decision maker therefore to convert data into information and knowledge that could aid in decision-making.

The importance of geography to marketing has been noted in prior research (Huff & Batsell, 1977). According to Huff & Batsell:

"Knowledge of the geographic location and aerial extent of a market is crucial in planning and evaluating marketing strategy. Examples of how such knowledge can be used include analyzing variations in sales penetration, determining sales territories, evaluating differences in promotional response, assessing the location of new facilities, pinpointing promotional efforts, forecasting sales, and analyzing market potentials."

Jones & Pearce (1999) suggest that geography is important to marketing since supply and demand vary with space, points of supply and demand are spatially separate, and space costs money to business. Though prior research points to the utility of geographic marketing knowledge, the use of geographic knowledge in marketing in the past has been limited by the lack of data as well as the tools to analyze such data. More recently, and in the last decade in particular, there has been an explosion in data. With the arrival of GIS, the ability to analyze data as an aid to decision-making has also expanded.

The initial impetus to the use of GIS came primarily from the public sector. Subsequently, the use of GIS in business and in marketing has grown in importance and will continue to do so in the future (Boyles, 2002). The primary reasons for the growth are not only due to the availability of data but also due to the interaction between a number of related human information processing factors and environmental factors. These factors include the availability of geocoded data, the growth of information systems in general, globalization, the integration of the Internet into GIS, the enhancement of both communication and the understanding of such communication when data are presented visually, rather than when presented in the form of text or tables. Consumer awareness and use of GIS has also increased (Witthaus, 2002).

In this chapter, we identify and expand on the primary areas in marketing where GIS could be used to enhance marketing efficiency and effectiveness. We expand on three variables that form the cornerstone of the use of GIS in marketing, namely: demographics, space, and time. In addition, the interaction of these variables and their utility to decision-making in the domains of marketing strategies and tactics related to product, pricing, promotion, place, segmentation, competitive analysis, customer analysis, and customer relationship management are analyzed. While some of the GIS application areas proposed in this chapter can be seen in practice today, other applications of GIS proposed in this chapter have seen relatively low levels of application.

Relationship between GIS and Demographics

The starting point for any marketing strategy is the customer. In business-to-consumer marketing and in business-to-business marketing, understanding the demographics of the consumer is critical to understanding the customer itself. Demographic variables such as age, income, gender, race, and household size are strongly linked to the demand for products and services in business-to-consumer marketing. In business-to-business marketing, the demographic variables of firm size, firm type, and firm income are linked to the demand for products and services that the firm would need to market its own products and services.

Today, a tremendous amount of geocoded demographic data for both consumers and businesses is available from governmental sources, such as the census, and private data providers, such as CACI. The quality and quantity of geocoded data available has increased. With the arrival of new technologies, such as the personal computer, Global Positioning Systems, and the Internet, data is now available at the individual level. This data has been useful in enabling marketers to understand and meet the needs of individual customers and to tailor personalized marketing strategies to segments of "one." This ability to target customers at lower levels of market aggregation is a source of enhanced customer satisfaction and enhanced competitive advantage. Firms such as Amazon and Reflect.com exemplify the success of such a personalization strategy. The interfaces with which to access data from public sources such as the census are also becoming more user friendly, as, for example, Maptitude Table Chooser Software (Comenetz & Thrall, 2003).

The availability of demographic data, along with the usefulness of such data in a variety of marketing applications, forms one of the primary forces propelling the use of GIS in marketing. In addition to demographics, two other variables that are critical to the use of GIS are space and time, and they are briefly discussed in the following paragraphs as they relate to marketing.

Spatial Relationships in Marketing

Space is an important variable in the area of marketing since the primary purpose of marketing is to facilitate exchange and space is a critical element in the facilitation of exchange (Jones & Pearce, 1999). A large number of services that today dominate the American economy and are increasingly likely to dominate the global economy are intricately tied to space. Many services are intricately tied to space in that the customer and provider of the service have to necessarily meet at a point in space in order for the service exchange to occur. Whether a customer is obtaining a haircut, staying at a hotel, eating at a restaurant, or holidaying in a theme park, the context in which the service is performed is a spatial one or, more precisely, the two dimensional version of space, i.e., location.

Beyond the notion that services are provided in a spatial context, the importance of location to the facilitation of exchange can be further elucidated based on the concept of utility. Different kinds of utilities that buyers and sellers receive in the process of exchange, such as place utility, time utility, information utility, and image utility, are linked to the dimension of space. A GIS system increases utility by providing increased meaning and understanding to the spatial context in which the exchange is facilitated. For example, imagine a customer in a car with access to a GPS system and a database of restaurant-related information accessible via the Internet making a decision on a restaurant to visit. For such a customer, the GPS system in combination with the Internet enables the customer to locate a restaurant nearby to his/her location, enhancing the customer's place utility, and enables the customer to get to the restaurant faster by providing directions, enhancing the his/her time utility. In addition, the Internet could provide added information about the restaurant, such as size and menu type, enhancing the customer's information utility. All the information provided could enhance favorably the customer's predisposition to try the restaurant, resulting in an increase in image utility.

In addition to the importance of space in facilitating exchange, the general drive towards increased efficiencies, increased levels of trade, and increasing expectations of customer service worldwide has made time as well critical to the exchange process. In other words, it is not only important to provide the necessary service or product at a specific location but also make sure that it is available at the appropriate time. This enhanced relationship between space and time further adds to the value of GIS in the marketing process.

GIS and the Dimension of Time in Marketing

The importance of time to the exchange process is hinted at by the large number of industries — the fast food industry, overnight mail delivery, online education, one hour photo processing, telecommunications, and retail banking — that offer timeliness as their most important benefit to the customer. In addition to industries organized around the notion of time, supply chain processes in all industries have timeliness as one of their central objectives. As we have mentioned earlier space is critical to exchange. However, the interpretation of the relationship between space and exchange is more meaningful when we add the element of time. An example of a non-marketing application of the space-time interaction is the analysis of individual accessibility to employment centers (Weber, 2003). Examples of specific marketing elements where time and space are critical would be competitive analysis and diffusion of innovation.

In order for a firm to carry out competitive analysis, the performance of competitors over time and space would provide an understanding of the dynamics of the marketplace and of historical trends, an understanding that would be critical to the evolution of marketing strategy. Monitoring trends in the social or economic environment would be critical to understanding the process of diffusion and the consequent nature of demand for a product or service. Further, the understanding of the nature of demand would have implications for demand forecasting and market planning. Assuming that both space and time are critical to the exchange process and consequently to marketing, the question arises as to how the criticality of space and time are related to GIS.

Traditionally, definitions of GIS have incorporated geography as the main element in the information system. Thus, when we have a database that contains geocoded information that can be visually represented, it is considered a geographic information system. However, as technology advances we need to move beyond these traditional definitions. Advances in Global Positioning Systems (GPS), telecommunications, and the Internet have now made it more feasible and important for GIS to go beyond a static incorporation of geographic data into an information system and integrate dynamism into the data, enabling the system to consider space and time. The logic of the marketplace will lead to the increased integration of traditional GIS with GPS, telecommunications, and the Internet, leading to substantial new uses in the area of business and specifically in the area of marketing. We next consider the applications of GIS in specific areas of marketing.

GIS and the Marketing Mix

The application of GIS to all marketing areas including the marketing mix can be examined from the viewpoint of the interaction of demographics, space, and time. In the following paragraphs we discuss the specific application of GIS in the four parts of the marketing mix — product, price, promotion, and place.

GIS and Product Strategies

Within product-related decisions, the specific areas influenced by space, time, and demographics, and consequently those areas where the application of GIS would be particularly useful, are product diffusion, product life cycle, and demand forecasting. The diffusion of new products has an extensive tradition of research through the use of quantitative models beginning with the Bass model. (See Mahajan, Muller, & Bass, 1990, for an extensive review of quantitative diffusion models in marketing.) Traditional diffusion models such as the Bass (1969) model are focused on identifying the rate of diffusion and specify equations that attempt to forecast the rate of diffusion based on a number of variables. In addition to quantitative models developed to understand and forecast the diffusion process, there have also been attempts to qualitatively describe the kind of customers who would adopt a new product at different points in time after the introduction of the product. These qualitative models (Rogers, 1962) divide customers in to five categories — innovators, early adopters, early majority, late majority, and laggards.

Studies on the socio-economic characteristics of the adoption process suggest that the similarity of adopter characteristics on demographic, economic, and social dimensions has an impact on the diffusion process (Morris, 1993; Valente & Rogers, 1995; Burt & Talmund, 1993). Given the geographic, time dependent, and demographic nature of the diffusion process it would seem that this would be a ripe area for the application of GIS.

Since the diffusion of innovation is a spatial process (Hagerstrand, 1967), GIS could be used to spatially map the diffusion of new products over time and provide the marketer with insights as to the pattern of adoption of new products, and the ability to predict the pattern in the future based on past data. Anecdotal and broad-based empirical evidence suggests that consumers within neighborhoods influence the pattern of consumption of each other, resulting in similarities in the pattern of consumption within neighborhoods. The tendency of people to congregate in neighborhoods consisting of people like them has also been commented on in the popular press (Brooks, 2003). The fact that consumers within neighborhoods influence each other is evidenced by marketing practices that divide America into geo-demographic clusters that consist of homogeneous clusters of customers with similar characteristics. Geography is also an important factor in customer's own classification of their neighborhoods (Coulton, Korbin, Chan, & Su, 2001). When people of similar characteristics live within a neighborhood the homophily principle suggests that they would influence each other socially and economically (McPherson, Smith-Lovin, & Cook, 2001).

Given patterns of consumption similarities based on geography, it would indeed be fruitful to spatially map such patterns. In addition, the pattern of such consumption over time for a new product essentially forms the diffusion process. Beyond a basic broad based understanding of the diffusion process, GIS could be used to add to the existing usefulness of qualitative and quantitative models of diffusion. In the case of quantitative models, GIS could be used to examine the impact of variables at a smaller level of aggregation than is typical. For example, quantitative models have attempted to capture the rate of diffusion of a new product as a function of the level of advertising expenditure in a specific country. These models then attempt to explain variation in inter-country diffusion rates partly as a function of variation in advertising expenditures. The availability of geocoded data at the block level from public data sources makes it feasible to examine the impact of advertising expenditures on diffusion at a level of aggregation much lower than the national level. Since advertising expenditures are incurred at local, regional, and national levels, and product adoption in effect occurs at the level of the household or the individual, GIS based adoption models can be used to examine the pattern of diffusion within a neighborhood as a function of the advertising expenditures within that neighborhood.

Qualitative characterization of consumers within a diffusion process can also be provided a richer context with the use of GIS. Classifying a customer as an innovator is a useful beginning to understanding diffusion. However, if we not only know that a certain percentage of consumers will be innovators, which is what current research helps us identify, but also know through GIS where the innovators live, who they influence, and how they influence other consumers, the marketer would be in a much better position to influence the diffusion process. Providing a spatial context through the use of GIS to the use of specific variables in diffusion, such as advertising expenditures and consumer characterization, is only illustrative of the power of GIS in the area of diffusion (see the present book chapter by Allaway, Murphy, & Berkowitz). One could easily extend the use of GIS to provide a spatial context to other variables that impact diffusion, including national economic characteristics, and other supply side factors such as competition (Robertson & Gatignon, 1986). In addition, GIS could be used for test marketing new products as part of the new product development process. For example, GIS could be used to compare different geographic areas on new product adopter characteristics. If test results indicate that a hypothetical geographic area G1 has a greater rate of innovators that would adopt the new product than a hypothetical geographic area G2, other geographic areas with demographic characteristics similar to G1 could be identified. Marketing efforts towards new product launch would then focus on geographic areas similar to that of G1 to ensure faster new product diffusion.

In addition to diffusion, another important area where GIS can play a potentially useful role in the product aspect of the marketing mix at the strategic level is the product life cycle. The Product life cycle concept suggests that products go through different stages from introduction to growth to maturity to decline. Typically these four stages have been conceptualized as a 'S' shaped curve depicting the pattern of sales over time. However, more than the mere identification of the stages that a product is presumed to pass through from introduction to decline, the usefulness of the product life cycle concept lies in the delineation of the marketing strategies that are appropriate at different stages of the life cycle (Robinson & Fornell, 1985; Porter, 1980; Enis, Lagarce, & Prell, 1977; Bayus, 1994).

A typical product life cycle identifies the intensity of competition as varying depending on the stage of the life cycle. While the intensity of competition is relatively low at the introductory stage of the life cycle with few or no competitors, a large number of new competitors are presumed to enter at the growth stage of the product life cycle, with a shake-out narrowing the number the number of competitors at the mature stage of the life cycle. The difference in the intensity of competition at the various stages of the life cycle consequently leads to the adoption of different marketing strategies at the different life cycle stages. An example of such differences in strategy would the type and amount of money spent on advertising at different stages of the life cycle. In the introductory stage, the primary focus of advertising would be on educating the customer and generating primary demand for the product category, while in the maturity stage the primary focus would be on generating secondary demand for the brand and on sales promotion rather than advertising. While these strategies are clearly generalizations that may vary in their degree of applicability to a specific situation depending on the other variables at play, they serve nevertheless as useful guideposts for a marketing decision maker. Typically, national and international markets have been considered the appropriate level of aggregation at which to view the product life cycle. However, this presumes that markets are in effect uniform within national boundaries, or international boundaries — a highly questionable assumption. The reality of the marketplace is that even in developed economies such as the U.S., there are likely to be significant differences in the product life cycle stage of a particular market. This would mean that a marketer would be well served by data that identifies such regional differences, an ideal situation for the use of a GIS. By identifying the product life cycle stage at which a product is in a particular market, GIS would help the marketer identify the appropriate marketing strategy to be adopted in that market, and thus fine tune a strategy at much lower levels of geographic aggregation, rather than assuming a market within an entire country is uniform.

At one level, it may be argued that regional salespeople who know their markets in effect ensure that marketers do not assume uniformity across national markets, but this presupposes that all products have regional sales forces sufficiently close to the market to be in a position to influence marketing strategy decisions at the local level. In today's global marketplace, such an assumption may be unrealistic for most products and services, so that a GIS would be a wonderful tool with which to integrate product life cycles into marketing strategy. In addition, even local sales forces many a time are surprised by the level of GIS data in a market that they thought they knew well, and a GIS could be a useful tool for local sales force to conceptualize their own input to marketing strategy.

Implementing GIS to create a product life cycle map involves both longitudinal and geographic data. A longitudinal comparison of the demand and competition in a particular geographic market may indicate a high level of growth in demand and competition, by which we could infer that the market is in the growth stage of the product life cycle. Mapping of demand and competitive profiles in such a manner for a large market such as the U.S. would be the basis for creating a product life cycle profile.

The mapping of such data is easily amenable to different levels of aggregation. If for example the data is available at the level of a census block group the data may be viewed at the level of the census block group, at the level of a census Metropolitan Statistical

Area (MSA), or any other level of aggregation based on the choice of the decision maker. Such flexibility may be important in identifying regional variations, which are revealed only at certain levels of aggregation. For example, data aggregated for the entire U.S. by choice will not reveal geographic variations in product life cycle stage. However, an analysis prepared state-by-state may reveal such variations enabling the marketer to vary the marketing strategy state-by-state. If a marketer of cell phones such as Nokia or Motorola finds that Northern and Southern California are in the mature stage of the product life cycle, while Central California is in the growth stage, the marketing strategy can be adapted.

GIS and Pricing Strategies

Pricing strategies are strongly linked to geography, demographics and time, since demand and supply that determine prices for most goods and services vary from region to region and from time to time. The dependence of price on supply and demand and, consequently, on the three variables that are a critical component of GIS — geography, demography, and time — make pricing strategies a potentially strong marketing area for the application of GIS. In addition, pricing strategies are inherently complex both due to the unpredictable nature of competitive actions and inadequate understanding of customer response (Hennig, 1994). While GIS may not be particularly helpful in improving the predictability of competitive actions on pricing, it can improve the marketer's understanding of customer response to price and lead to improved pricing strategies. Once again, as in the case of diffusion and product life cycle, GIS can be particularly helpful by reducing the level of aggregation at which data is examined and consequently lead to enhanced understanding of the consumer.

Instead of examining consumer response to pricing strategies at national or regional levels a firm may be able to vary its pricing strategies from census block group to census block group based on an understanding of local market conditions. A gasoline retailer for example could examine variations in demand as a function of variation in competitor price and location. The same set of competitor prices may produce very different demand patterns in different locations due to differences in brand perceptions and customer loyalties. Let us consider a hypothetical example with firm X, two competitors C1 and C2, and two locations L1 and L2. At location L1, variations in the price of competitor C1 may impact the price that firm X can charge more than the variations in price of competitor C2. However, in location L2 variations in the price of competitor C2 may impact the price that firm X can charge more than the variations in price of competitor C1. Such variations in cross-price elasticity may lead to focus on the pricing strategies of competitor C1 in location L1 and a focus on the pricing strategies of competitor C2 in location L2.

It could be argued that such variations in pricing strategies can be examined without the GIS. However, the use of GIS is very beneficial in two ways. First, the visualization of location specific price and competitive profiles makes it easier to interpret what is happening in the marketplace. Imagine a marketing manager viewing a table with hundreds or thousands of rows with each row representing the pricing and competitive profiles of a census block group, and the same information presented in the form of a map with census block groups possessing similar profiles identified on the map with the use of similar colors. Given the notion of bounded rationality (Simon, 1979) and the idea of a manager as a cognitive miser, a map of pricing and competitive profiles rather than rows of data is the obvious choice.

Second, the availability of large quantities of data at lower levels of geographic aggregation makes it possible to cross reference price and competitive characteristics with other characteristics of the geographic region. Consider our earlier example with the firm X, the two competitors C1 and C2, and the two locations L1 and L2. A cross referencing of the cross-price elasticity of the two locations with other location specific characteristics may show hypothetically, for example, that customers in location L1 are primarily loyal to full service gas stations such as C1 and customers in L2 are primarily loyal to price, which may account for the difference in the nature of cross-price elasticity. Such localized information would lead to a pricing strategy more attuned to the local marketplace and potentially improve competitive position and profitability. The use of localized information to fine tune pricing strategies would be relevant, not only for consumer products, but also in business-to-business markets — for example in the supply of meat or produce to restaurants, office supplies to small business, or timber to customers in real estate home and office construction.

When prices and pricing strategies are subject to frequent revision based on conditions of supply and demand, GIS would be a tool that reduces uncertainty by improving the precision of information and the format by which information is presented. In addition, examining price competitive maps on a global basis may provide more insights for global pricing strategies.

GIS and Promotional Strategies

Promotional strategy essentially involves all tools, methods, and processes by which a firm communicates about the product or service to the customer. The most common methods that form part of a promotional strategy include personal selling, advertising, sales promotion, direct marketing, public relations, and Internet marketing. More than almost any other area of marketing, demographics are central to any aspect of promotional strategy. Whether the promotional strategy involves the salesperson meeting the customer, advertising targeted at a specific market in a specific location, a coupon sent to specific types of customers, a political fundraiser using direct mail, or Internet marketing, understanding the demographics of the customer is an important determinant of the success of the promotional strategy. In the following paragraphs we outline the usefulness of GIS in various aspects of promotional strategy.

Personal Selling

The use of GIS in planning sales routes is an ongoing business use of GIS. In addition to sales route planning, GIS would be particularly useful to a salesperson for market analyses that can be incorporated into sales presentations improving the effectiveness of sales calls. For example, the salesperson of a lawn maintenance firm could easily identify potential prospects by combining geocoded data from existing customers, information on new housing developments, weather conditions, and other relevant information. When making the presentation to a prospective customer the salesperson could identify easily other customers in the neighborhood using the service in order to help the prospective customer make a decision. The salesperson could also perform market analysis to identify new prospects. An example of a company using GIS to perform such market analysis is Dr. Green Lawncare in Ontario, Canada (Boyles, 2002). Dr. Green uses GIS to visually map the market and divide the market into primary, secondary, and tertiary target markets that are represented by different color codes on a map. Based on the map, Dr. Green found that the primary market area was concentrated in and around a particular suburb of Ontario, the secondary market area was concentrated in a different suburb of Ontario, and tertiary market areas were more dispersed geographically. The geographic information was invaluable in helping Dr. Green in deciding on where to focus marketing and advertising efforts.

A number of studies in marketing show the importance of social comparison information in influencing decisions, and since so much of social comparison is neighborhood-based, GIS would be an important part of any such influence in the process of exchange between buyer and seller. In addition to local or regional businesses such as lawn care maintenance, industries that have a larger geographic scope such as travel and tourism would also benefit substantially from the use of GIS in sales presentations. Beyond the issue of sales presentations, GIS can also be used an aid to allocate sales territories. Instead of using traditional methods to allocate territories, such as salesperson expertise or need for accounts on the part of a salesperson, GIS can be used to allocate them based on efficiency of coverage and to maximize time spent with clients and minimize time spent on travel, besides ensuring balanced allocation of accounts among salespersons.

Advertising

Advertising is an aspect of promotional strategy wherein the importance of demographics is supplemented by the importance of space and time. In planning any advertising campaign a marketer needs to know whom to target (demographics), where they are located (space), when to target (time), and how frequently to target. The combination of demographics, space, and time makes the importance of GIS to advertising self-evident. All advertising campaigns are conducted through one or more of different mediums including radio, television, newspapers, magazines, billboards, or other non-traditional media. In deciding on which media to use in advertising, the marketer has to take into consideration the extent of overlap between the marketers' target market and the target market of the medium through which the marketer plans to advertise.

The use of GIS will enable the marketer to analyze the level of such target market overlaps at very low levels of aggregation such as census blocks or even individual households in the marketplace. Combining target-market overlap information with data on the level of advertising exposure that consumers have at different locations will result in decisions that can ensure a more efficient use of the advertising dollar. Research on advertising wear-in and wear-out suggest that the effectiveness of an advertising campaign grows as the proportion of the population exposed to the advertising increases. However, after the entire target market is exposed to the advertising, i.e., when the process of wear-in is complete, the process of wear-out begins and the advertising loses its effectiveness (Blair, 2000; Masterson, 1999). Repetition of an advertisement does not enhance customer purchase intentions. GIS can be used as a tool to map the wear-in process of an advertising campaign by combining geographical information on the target market exposed to the media with information on product sales in that market. When GIS shows that wear-in is complete, the marketer can decide to change the execution of an advertising campaign in order to maintain effectiveness and prevent or minimize wear-out.

Similar to pricing and competitive profiles, advertising profiles of different geographic regions can be prepared and used as inputs to an advertising campaign. Such an advertising profile would include data on the marketers' advertising spend in a region, competitive advertising spends in the same region, and customer brand awareness in the region. Advertising dollars can then be allocated on a region-by-region basis with the purpose of achieving advertising wear-in as a function of the marketer's history of advertising expenditure, competitive spending on advertising, customer's exposure to the marketer's brand, and other relevant customer characteristics for that region.

Direct Marketing

Direct marketing is a growing form of communication and includes direct mail, telemarketing, catalog marketing, home shopping networks, and the Internet. Direct mail is its most popular form. The success of any direct marketing campaign is primarily dependent on the quality of the mailing list and the quality of the offer made to the prospect. GIS can play a useful role in improving the quality of mailing lists. The process of developing a successful direct marketing campaign combines trial and error and rigorous market research. Before launching a national direct marketing campaign, many firms carry out trial runs with different versions of the direct marketing campaign targeted at small groups of customers. This process helps the firm to test different versions of an offer and test different types of mailing lists. The mailing list and the offer with the highest response rates are than chosen for replication on a larger scale. This process of market research to test the potential success of a direct marketing campaign would be helped by a geographic information system that could be used to spatially identify the most likely prospects or the early adopters, the most loyal customers, and the geographic areas within which neighborhood effects are likely to occur. Direct mail campaigns having relatively low success rates of 5% to 10% are considered good. The ability to increase the success rate even by a small percentage through better targeting could mean a big difference in profitability. The Credit Union of Texas, which had traditionally obtained a response rate of 1% to 2% for marketing mail sent to its 145,000 members, increased the response rate to between 8% to 9% by narrowing the mailing list to 10,000 using GIS and demographic data (Boyles, 2002). The credit union narrowed the mailing list by identifying specific block groups for target marketing based on the level of market penetration. A map of the Irving Independent School District, with the level of market penetration for

Figure 1. Level of Market Penetration of the Credit Union of Texas in a School District

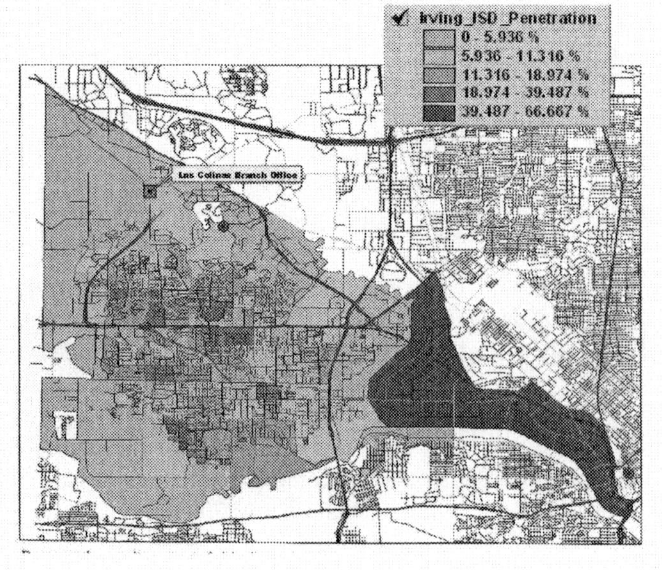

Source: Boyles (2002) with permission from ESRI Inc.

the credit union (Boyles, 2002) is indicated in *Figure 1*. The map of the school district was used since the membership of the credit union is based on school district boundaries.

This use of GIS can be done not only for direct mail campaigns but also for other forms of direct marketing including telemarketing, direct marketing through cable television such as the Home Shopping Network, and Internet-based campaigns. GIS can also help in synergistically combining multiple methods of direct marketing. For example, the customers who respond to a specific direct marketing campaign on television could be spatially identified using a GIS system. The neighborhoods of these customers could than be identified using GIS and subsequently targeted with a second direct mail campaign.

Sales Promotion

Sales promotion consumes over half the promotional dollars spent in the American economy. Sales promotion consists of trade and consumer promotion and usually involves the use of some form of incentives, usually monetary, to facilitate the sale of the product. Similar to direct marketing, the success of a sales promotions, especially consumer ones, are dependent on the ability of the marketer to target the right customer with the right incentive. The inappropriate targeting of sales promotional dollars many a time provides incentives to customer who would have purchased the product even without the incentive, thus increasing the cost of achieving additional sales. Using GIS to identify the sensitivity of geographic markets to promotion will help the marketer focus on those markets that are sensitive, reducing the probability of offering promotion where customers would have bought the product even without the promotion. The ability to target promotion-sensitive customers is particularly relevant when the focus of the marketer is on increasing quantities purchased, rather than on rewarding loyal customers.

In the case of sales promotional campaigns involving couponing, redemption rates tend to be in the reduced range of 1% to 2%. Using GIS to increase redemption rates would improve the effectiveness and profitability of the sales promotion campaign. In general, irrespective of the type of sales promotion, using GIS to improve customer targeting will result in better answers to questions, such as the store locations where promotional campaigns should be run, the size of the incentive to offer location-specific customers, and matching consumer characteristics to the nature of the incentive. For example, an office supplier that ran a sales promotion campaign sent free prepaid phone cards to secretaries and others identified as playing a role in the purchase of office supplies for the firm. With the use of GIS and a geocoded phone card, the company could track the customers who had used the card. Thus the salesperson could more easily get a foot in the door by targeting those customers who had used the prepaid phone card and speed up the adoption process of the promotion. The use of GIS to examine the spatial adoption of sales promotion programs is not new. Models of spatial diffusion of sales promotion programs have been applied to estimate the impact of distance, existing adopters, and the firms' marketing efforts on the diffusion process (Allaway, Berkowitz, & D'Souza, 2003).

Internet Marketing

Internet marketing and Internet-enabled GIS are evolving areas of marketing. The use of GIS in Internet marketing has only just begun. There are firms such as Geobytes.com that are now offering location specific information on visitors to a firm's website. The information on the visitor includes the IP address and the latitude and longitude of the location from which the visitor is accessing the website. While it may be possible for the visitor to a website to disguise his/her actual location, presumably most consumers do not bother to do so, thus providing rich information to marketers interested in using the information. Many firms would be keenly interested in knowing the geographic location of the customers to their website for the purpose of customizing their future market strategies. For example, an admissions office in a university will get valuable information about prospective students and their geographic location by examining geographic origin data of visitors to the website. This data can be subsequently used to plan direct mail campaigns.

GIS and Distribution Strategies

Space and time are central to any distribution strategy whether it is related to concepts of efficiency and the supply chain, to concepts of effectiveness and distribution structure, or to issues of retailing. The efficient management of a supply chain is concerned with the flow of material, information, and money and can save time and money (Davis 1993). The three flows are interrelated in that the efficient flow of material is dependent on and impacts the efficient flow of information and cash flow within the supply chain. GIS can be used to enhance the efficiency of the three flows by helping the decision maker monitor the supply chain and use the feedback from the monitoring process to initiate improvements.

For example, imagine a scenario wherein a transporter managing a national fleet of trucks uses GPS technology to locate the trucks and monitor speed of movement. The driver in the truck connects to a web-enabled GIS that provides real-time information on the next destination, the details of the customer, and other information related to transportation. The movement of the entire fleet is monitored from a single location; resources are allocated to ensure close match between supply and demand and minimize idle time. Efficient use of drivers, trucks, and inventory, and accurate monitoring of demand and inventory at different geographic locations are used to manage cross docking operations (cross docking involves the trans-shipment of goods from larger vehicles to smaller vehicles used for local deliveries without the use of an intermediate storage point). Such an imagined scenario is not far from reality.

Many transportation companies use one or more elements of this scenario. Sears for example has successfully used GIS for route planning and benefited from cost savings in the millions of dollars in the process (ESRI, 2003). GIS based monitoring has also been used in locating lost cargo (King, 2002).

Strategically GIS may also be used to plan the distribution structure of a firm. Traditionally distribution structures have been selective, intensive, or moderate. The decision as to which structure to adopt is based on a number of factors including the nature of the product, the nature of the market, competitive distribution structures and infrastructure in a particular market. Irrespective of the type of structure that is chosen, GIS can be used to overlay the structure on the demand profile of the market, and examine if the targeted market coverage is being achieved. For example, when Starbucks, which has adopted an intensive distribution structure, locates new outlets, the site-location decision would be better served by an understanding of the geographic characteristics of the market, the demographic characteristics of the market, the traffic patterns in a particular location, and future population developments in the area. Since all of these data can be tied to a specific geographic location, GIS can be used to combine data in a manner desired by Starbucks and serve as a significant input to the location decision.

Substantial research into building models of retail location already exists (Rust & Donthu, 1995; Donthu & Rust, 1989; Craig, Ghosh, & McLafferty, 1984; Huff & Batsell, 1977; Huff, 1964). Research in models of retail location suggests that the variables that impact store choice may differ depending on geography. Even when the variables are the same, their level of impact may vary depending on geographic location (Rust & Donthu, 1995). Attempts have been made to capture such geographic variations through estimating geographically-localized misspecification errors. One of the reasons why variables are left out in models of store choice is because they are difficult to measure. With the use of GIS and the easy and extensive availability of geographic information, the inclusion of geographic or location-related variables in retail store choice models is likely to improve predictive accuracy in store choice.

In addition to the use of GIS in store choice, GIS could also serve as a tool to measure and monitor performance and control resource allocation decisions in a distribution network. For example, a franchise operation such as McDonald's could monitor the performance of individual franchisees in comparison to their market potential for the geographic area that the franchisee is likely to draw from and reward franchisees for good performance. Additionally, GIS based data could be shared with franchisees to help improve their performance. McDonald's could monitor demographic trends, traffic patterns, and consumer behavior in specific geographic locations, and share this information with franchises. Such data could than be used by the franchisees to identify the most effective billboard locations, plan sales promotional campaigns, and perform competitive analysis.

While the quantity of geographic data available has been increasing, there is a caveat. There are practical problems in minimizing the errors connected with combining data, since different types of data may be available at different levels of aggregation and the data are also generated by different sources. Data on traffic patterns may be generated by the federal transportation department, demographic data by the census, population development projections data by local governmental authorities, and data on consumption patterns from private firms such as CACI and Claritas. In spite of these difficulties, the current trend has been to facilitate the integration of data and increase the availability of small-area data that reduces possibilities of error.

GIS is also seeing widespread use in almost all areas of retailing from analysis of trade areas (Thrall, 2003), site location, and inventory management. One innovative retail grocer that the author is familiar with uses GIS to identify geographic areas from which to draw potential employees.

Applications of GIS to the Understanding of Specific Marketing Elements Beyond the Marketing Mix

In addition to the four areas of marketing discussed so far; product, price, promotion, and place, GIS is also a beneficial tool in three other areas of marketing: market segmentation, relationship marketing, and competitive analysis. In the following paragraphs we discuss the relevance of GIS to these three areas.

GIS and Market Segmentation

Market segmentation is a process by which the market is divided into segments based on the homogeneity of response of the customer to marketing mix strategies. The variables that are used to identify homogeneity of response may include demographic variables, psychographic variables, uniformity of benefits desired, uniformity of usage situation and many other variables. For GIS, demographic and psychographic variables are particularly relevant.

One of the more popular GIS based segmentation systems in use by marketers, the ACORN segmentation system, combines geography and demographics to create a neighborhood classification system. GIS is used to create clusters of neighborhoods that are similar to each other on demographic characteristics and aspects of consumer behavior. The neighborhoods are analyzed at the block level, which consists of an average of 400 households. The classification of neighborhoods in the U.S. into over 40 different clusters and analysis of the buying behavior of these clusters helps marketers identify the segments that are most appropriately targeted for the products and services marketed by them. For example, each cluster's propensity to consume wine is available as part of the database — the kind of information that would be of use to a wine marketer identifying a target market. In addition to demographic characteristics, ACORN's lifestyle segmentation combines psychographics and geography. Prior to GIS, segmentation systems such as VALS divided an entire market into psychographic segments, but the knowledge of where consumers live who belong to a specific psychographic segment makes the segmentation system much more powerful and enhances the marketers' ability to target these segments.

It is not inconceivable that in the future it may be possible to link other market segmentation variables such as benefits and usage to geography, at least in the case of some products and services. Examples of such products or services would include real estate, travel and tourism, house care services such as lawn care, the hospitality industry, banking, and maintenance services. The benefits that a customer living in a small rural town desires from real estate would be different from the benefits that a customer living in an urban downtown would want. Such differences in geography in combination with benefits desired could then serve as basis to segment the market.

The ability to combine geography with other segmentation variables such as demographics, psychographics, and benefits improves the marketer's understanding of the marketplace. It enables the marketer to target the market, to develop an appropriate promotional strategy linked to the target, and to make the product or service available at the appropriate location. In effect, the value of GIS cuts across all aspects of the marketing mix. The application of GIS in market segmentation is likely to grow in the future.

GIS and Relationship Marketing

Customer satisfaction has always been an important part of marketing due to strong relationships between customer satisfaction, repeat purchase, and word of mouth. This has been further enhanced by the increased availability of data and decreasing costs of data processing technology. GIS can play a useful role in further cementing the relationship between the firm and the customer. Since the lifestyles of most customers are related to location, a GIS system that can monitor changes in lifestyle based on changes in location can help the marketer cater to customer needs as the needs change.

For example, let us consider the situation of a hypothetical student "Jane Doe". Information about Jane Doe is first entered into a GIS when she is a student. After completion of her studies, Jane Doe takes up her first job and moves into a new apartment. Subsequently, she gets married and moves into a new house. As Jane Doe progresses through different life stages, her income, family status and composition, and other aspects of her life change. In many cases these changes may be inferred from the location where she lives. By tracking such changes in location, a firm would be in a position to tailor appropriate marketing messages and offerings as she progresses through different life stages.

An automobile manufacturer or a bank may be particularly interested in Jane Doe as a prospective customer when she graduates, while a mortgage company or a home appliance retailer may be particularly interested in her as a prospective customer when she buys her first house. A baby foods manufacturer or a toy firm may be especially interested in Jane Doe when she has her first baby. Since many neighborhoods have substantial commonality in the life stages of the residents of that neighborhood, location would be an important marketing variable for firms that offer products that meet the needs of customers in different life stages. Consequently, such firms could maintain long-term relationships with customers through the use of a GIS. Some firms have begun to use localized birth and death rates based on GIS data to plan their marketing strategies (Hayward, 2001).

Going back to the example of the automobile manufacturer, most of them offer a range of vehicles that cater to the needs of customers in different life stages. A young student graduating out of college would be inclined to purchase a Chevy Cavalier, while an older retiree would prefer to buy a Cadillac. By tracking the progress of a customer through the different life stages with the help of a GIS, a firm could promote the Cavalier to the student and Cadillac to the same student, as the student grows older. Similarly, a bank may offer basic services to a student, and change the nature of product offerings as the student progresses through different life stages.

GIS can also be used to improve the efficiency of customer service that in turn would have a beneficial impact on long-term customer relationships. Many field representatives of service providers, such as telecommunication companies and utilities, use hand held devices that provide information on work orders and network facilities, improving the ability of the field representative to have the most up-to-date information and reducing delays in the process of carrying out repairs (Geospatial Solutions, 2002). Since the hand-held device can be connected to a database on customer service requests, the device can automatically access any update on the database, and record changes in customer appointments and schedules. This should help the service person better plan his or her service route for the day.

GIS and Competitive Analysis

The field of competitive analysis is an appropriate and promising area for the application of GIS due to the location specific and dynamic nature of competition in a free market economy. Location specificity refers to the notion that the nature of competition varies depending on the geographic market being considered. This is the reason why location decisions for many services such as banks are impacted by the competitive environment

(Miliotis, Dimopolou, & Gianikos, 2002). The primary competitors that Proctor & Gamble faces in the U.S. would be different from the primary competitors the firm faces in Asia. For a specific product such as Tide, the primary competitors in the West Coast of the U.S. may be different from the primary competitors in the East Coast. Geographic differences in the nature of competition are inevitable in a market economy due to the existence of disequilibria and "creative destruction" (Schumpeter, 1934). In addition, such geographic differences in competition are appropriately analyzed through GIS.

As part of an exercise in competitive analysis, a firm would benefit from the preparation of a competitive profile map that identifies the primary competitors in different locations. On this map could be superimposed other variables such as the competitive advertising and sales promotional spend, competitor's distribution intensity, competitor's sales force strength, and competitor's customer satisfaction in different regions. An information system that provides competitive information on a regional basis would be more useful to a firm than a system that relies entirely on information aggregated across the entire marketplace. Factoring in regional variations will enable a firm to fine-tune its competitive strategy to the nature of the competition in a particular location (Grether, 1983). For instance, after performing a regional competitive analysis, a firm may realize that it would be worthwhile to enhance advertising spend against competitor A in region 1 while enhancing distribution intensity against competitor B in region 2. Such location based competitive decisions will be substantially facilitated by the use of a GIS.

GIS and the Dynamic Nature of the Market Economy

This chapter earlier touched upon GPS and the need to integrate GPS into traditional notions of GIS. The importance of this integration is more easily evident if we examine the true nature of marketing in the economic marketplace. Marketing involves the facilitation of exchange between two parties. In a market economy, the exchange process is truly dynamic, with variations in the market economy across regions and time. The implication for information systems in such a market is that the information system also needs to be dynamic and possess the ability to incorporate such market variations. A traditional GIS system captures variations across regions by virtue of the nature of the data that is geocoded. However, when a traditional GIS is combined with GPS and the Web, the GIS captures variations across space and time and truly becomes dynamic. Users of GIS have suggested that mobile GIS uses, such as field mapping services, and distributed GIS implementation on the Internet are likely to be the growth areas of the future (Barnes, 1999, 2003). This user prediction is now borne out by newer technologies such as Simple Object Access Protocol (SOAP) that make it increasingly possible for small firms to provide web-based GIS services (Gonzales, 2003) and other technological advancements that simplify access to corporate data (Fjell & Gausland, 1999).

In many areas of marketing, the movement of materials and customers will have an impact on marketing decisions and marketing practice. These areas include supply chain management, market research, retail location, customer service, and marketing promotion. One of the major objectives of supply chain management is the efficient and effective flow of materials. By enabling the firm to track movement of materials, the firm can more efficiently manage inventory and reduce logistics costs in the process.

In the area of market research, a GPS system could potentially help the marketer better understand customers' lifestyles by tracking their movements. This may be carried out at the level of the individual consumer or at a higher level of aggregation such as a neighborhood. The GIS/GPS system may for example suggest that in neighborhood A, a large proportion of consumers spend most of their time in the vicinity of the neighborhood, while in neighborhood B, a large proportion of consumers spend most of their day far away from the neighborhood. Such differences of behavior may be found to vary depending on the day of the week, holidays, and time of the year. The ability of a GIS/GPS system to capture consumer behavior data on a continuous basis is an immense advantage over static information systems like the census.

The customer service function is also appropriately served by dynamic information systems. One of the first firms to provide customer service through the innovative use of GPS was Federal Express. Through the combination of GPS and the Internet, Federal Express customers can track products that have been mailed. However, many firms still do not have the ability to provide this kind of service. This author has personally experienced poor customer service from a home products retailer that was unable to provide information on the expected availability of a product overdue by over thirty days.

The use of GPS also holds much promise for direct marketing. Tailoring messages on billboards depending on the nature of traffic at a point in time, and identifying a dynamic retail product mix based on the profile of customers shopping at a time point are potential applications in the area of promotion.

Overall, combining the Internet that makes a GIS accessible from anywhere in the world, and a GPS that transforms a GIS into a real-time system, enhances in many ways the utility of GIS to marketing and to business in general.

GIS and Ethical Challenges

New technologies and innovations like the Internet raise new ethical issues and GIS is no exception. The arrival of the Internet led the American Marketing Association (AMA) to develop a separate code of ethics for marketing on the Internet. The code of ethics related to the Internet specifically concerns issues of privacy, information ownership, and information access. The use of GIS and GPS-related information raise similar concerns with regard to privacy and information access. The specific AMA code of ethics with regard to information privacy and information access is as follows:

"Privacy: Information collected from customers should be confidential and used only for expressed purposes. All data, especially confidential customer data, should be safeguarded against unauthorized access. The expressed wishes of others should be respected with regard to the receipt of unsolicited e-mail messages.

Access: Marketers should treat access to accounts, passwords, and other information as confidential, and only examine or disclose content when authorized by a responsible party. The integrity of others' information systems should be respected with regard to placement of information, advertising or messages."

The AMA code of ethics for Internet marketing could be applied to the use of GIS as well.

Conclusions

This chapter summarizes the major areas of GIS applications in marketing. From the four P's of marketing: product, price, promotion, and place to market segmentation, relationship marketing, and competitive analysis, the use and potential use of GIS has been outlined.

While the use of GIS in marketing has so far received little research attention from marketing academics, applications in business have been growing at a quicker pace, though not as fast as the applications in government, whether at the local or the federal level. In business, the use of GIS has still not fulfilled its potential (Borroff, 2002). One of the factors that would potentially influence the diffusion of GIS in marketing is the knowledge of GIS that college graduates and other professionals bring to the practice of marketing. It is important for marketing academics to examine the use of GIS in marketing from a theoretical and applied perspective. Some preliminary directions of research in this area suggest themselves. These include, at a general level, the examination of the role of geography in product diffusion, relationship marketing, customer service, pricing, competitive analysis, distribution, and promotional strategy. More specific issues that need to be examined include the impact of a particular aspect of geography on diffusion. Are diffusion rates higher in geographic areas with a large number of apartment complexes as opposed to those with numerous individual homes? The answers to both general and specific questions require more research attention. Grether in 1983 suggested that spatial analysis in marketing had been given relatively short shrift at that time. Hopefully with the explosion of data and the availability of technologies such as GIS and GPS, interest will revive on the role of geography in marketing.

Acknowledgments

Special thanks to the reviewers for their many comments and suggestions.

References

Allaway, A. W., Berkowitz D., & D'Souza, G. (2003). Spatial diffusion of a new loyalty program through a retail market. *Journal of Retailing*, *79*, 137-151.

Allaway, A. W., Murphy, L., & Berkowitz, D. (2004). The geographical edge: Spatial analysis of retail loyalty program adoption. In J. B. Pick (Ed.), *Geographical information systems in business*. Hershey, PA: Idea Group Publishing.

Barnes, S. (1999, November). Round Table: GIS in the next millennium. *Geo Info Systems*, *9*(11), 58.

Barnes, S. (2003, January). Market Map 2003: Standards, open systems drive spatial marketplace. *GeoSpatial Solutions*, 28-32.

Bass, F. M. (1969). A new product growth model for consumer durables. *Management Science*, *18*, 215-227.

Bayus B. L. (1994, September). Are product life cycles really getting shorter? *Journal of Product Innovation Management*, *11*, 300-308.

Borroff, R. (2002, June 21). Waiting to shine. *Precision Marketing*, 28-32.

Boyles, D. (2002). *GIS means business*. Redlands, CA: ESRI Press.

Brooks, D. (2003, September). People like us. *The Atlantic Monthly*, 1-6.

Burt, R., & Talmund, I. (1993). Market niche. *Sociological Networks*, *15*, 133-149.

Comenetz, J., & Thrall, G.I. (2003). Maptitude table chooser: Census data fast and easy. *Geospatial Solutions*, *13* (6), 48-51.

Coulton, C. J., Korbin, J., Chan, T., & Su, M. (2001). Mapping resident's perceptions of neighborhood boundaries: A methodological note. *American Journal of Community Psychology*, *29* (2), 371-384.

Craig, S. C., Ghosh, A., & McLafferty, S. (1984, Spring). Models of the retail location process: A review. *Journal of Retailing*, *60* (1), 5-36.

Davis, T. (1993). Effective supply chain management. *Sloan Management Review*, *34* (4), 35-45.

Enis, B.M., Lagarce, R., & Prell, A.E. (1977). Extending the product life cycle. *Business Horizons*, *20*, 46-60.

Fjell, K., & Gausland, I. (1999, August 30). Combining GIS and web simplifies access to corporate data. *Oil & Gas Journal*, 58-62.

Gas Utilities' Top 10 GIS technologies, applications. (2003, May). *American Gas*, 8-9.

Gonzales, M. (2003, February 1). Landscape. *Intelligent Enterprise*, 21-24.

Grether, E.T. (1983, Fall). Regional-spatial analysis in marketing. *Journal of Marketing*, *47*, 36-43.

Grimshaw, D. J. (2000). *Bringing geographical information systems into business* (2nd ed.). New York: John Wiley & Sons.

Hagerstrand, T. (1967). *Innovation diffusion as a spatial process*. Chicago, IL: University Press.

Harder, C. (1997). *GIS means business*. Redlands, CA: ESRI Press.

Hayward, C. (2001, October). The child-catchers. *Marketing Week*, 45-46.

Hennig, H. C. (1994, September). Price, quality, and consumers' behavior. *Journal of Consumer Policy, 17*, 335-348.

Huff, D. L. (1964). Defining and estimating a trade area. *Journal of Marketing, 28*, 34-38.

Huff, D. L., & Batsell, R.R. (1977, November). Delimiting the areal extent of a market area. *Journal of Marketing Research, 14*, 581-585.

Japanese Telcos deploy mobile GIS. (2002, August). *Geospatial Solutions*, 26.

Jones, K., & Pearce, M. (1999, March/April). The geography of markets: Spatial analysis for retailers. *Ivey Business Journal*, 66-70.

King, M. (2002, August). Finding lost cargo containers. *Geospatial Solutions*, 28.

Mahajan, V., Muller, E., and Bass, F. (1990, January). New product diffusion models in marketing: A review and directions for research. *Journal of Market Research*, 54(1), 1-26.

McPherson, M., Smith-Lovin, L., & Cook, J. (2001). Birds of a feather: Homophily in social networks. *Annual Review of Sociology, 27*, 415-444.

Miliotis, P., Dimopolou, M., & Giannikos, I. (2002, September). A hierarchical location model for locating branches in a competitive environment. *International Transactions in Operational Research, 9* (5), 549-566.

Mitchell, A. (1997*). Geographic information systems at work in the community: Zeroing in*. Redlands, CA: ESRI Press.

Morris, M. (1993). Epidemiology and social networks: Modeling structured diffusion. *Sociology Methods Research, 22*, 99-126.

Porter, M. E. (1982). *Competitive strategy*. New York: Free Press.

Robertson, T. S., & Gatignon, H. (1986) Competitive effects on technology diffusion. *Journal of Marketing, 50* (3), 1-12.

Robinson, W.T., & Fornell, C. (1985). Sources of market pioneer advantage in consumer goods industries. *Journal of Marketing Research, 22* (3), 87-95.

Rogers, E.M. (1962). *Diffusion of innovations*. New York: Free Press.

Rust, R. T., & Donthu, N. (1995, February). Capturing geographically localized misspecification error in retail store choice models. *Journal of Marketing Research, 32*, 103-110.

Schumpeter, J. A. (1950). *Capitalism, socialism, and democracy*. New York: Harper & Row.

Simon, H. A. (1979, September). Rational decision making in business organizations. *American Economic Review, 69*, 493-512.

Thrall, G. I., & Fandre, M. (2003, April). Trade areas and LSPs: A map for business growth. *Geospatial Solutions*, 48-51.

Valenete T.W., & Rogers E.M. (1995). *Network models of the diffusion of innovations.* Cresskill, NJ: Hampton.

Weber, J. (2003). Individual accessibility and distance from major employment centers: An examination using space-time measures. *Journal of Geographical Systems, 5* (1), 51-71.

Witthaus, M. (2002, September 6). Leaders of the pack. *Precision Marketing*, 24-25.

Chapter XII

The Geographical Edge: Spatial Analysis of Retail Loyalty Program Adoption

Arthur W. Allaway, The University of Alabama, USA

Lisa D. Murphy, The University of Alabama, USA

David K. Berkowitz, The University of Alabama, USA

Abstract

This chapter demonstrates important insights gained by adding spatial capabilities to marketing analyses. Four steps are described to produce a geographically enabled data set of the first year's daily use for a major retailer's loyalty card program at one store in a mid-western U.S. city. Traditional analysis is contrasted with results from a geographic information system (GIS). Probabilities of adoption were clearly tied to the geographic variables generated by the GIS; for example, over the whole year, the likelihood of someone adopting on a given day decreased 13.4% for each mile they resided away from the store, while each Innovator (adopted in the first two days) located within .6 mile of a prospective adopter increased adoption likelihood by 13.2%. Further, three very distinct spatial diffusion stages are visible showing adoption as a function of distance to the store itself, to the billboards, and to the earliest adopters.

Introduction

Today's retail marketing managers have access to better information than ever before. In particular, the spread of point-of-sale automation in retail stores has turned what used to be a trickle of data into a flood. For many retailers, this technology has become the basis for the development of innovative, customer-centered loyalty card programs. A battalion of intercept interviewers in a store for weeks or a buyer's panel operating for months can capture only a small portion of the data gathered by a point-of-sale loyalty card program every day. To make sense of this data deluge, marketers are having to rely on a battery of both familiar statistical techniques such as regression analysis and newer ones such as chaid and diffusion modeling.

Much of the value of the data generated by a POS-based loyalty card program is its ability to capture the speed and duration of market reaction to new store openings, product launches, advertising campaigns, promotions, and so on. As such, loyalty programs often lend themselves to a *diffusion of innovations* analysis approach. Yet, even though retailing (except web-retailing) is necessarily a geographically anchored activity, diffusion research has typically ignored geographic factors. The reasons for this neglect have for the most part been practical. Prior to the advent of geographic information systems (GIS), spatial data was difficult to use, expensive to collect, and often of uncertain quality. The most common tools for analyzing spatial data were paper maps and overlays — both cumbersome to use and difficult to update and refine. As a result, even marketers who clearly recognized the importance of geography in both their and their customers' decision-making seldom received the tools or training that would make geography worth addressing at the individual consumer level (Murphy, 1996).

The application of GIS to retail point-of-sale data holds great promise in allowing retailers to gain greater insights into consumer spatial behavior. With that in mind, this chapter attempts to add to the body of retail theory and practice by demonstrating how a GIS-centered spatial approach can expand researcher understanding of the diffusion of a new loyalty card program. Household-level data from the entire first year of a new loyalty program launched by a very large retailer in a major U.S. city is combined with GIS-generated measures to explore the effect of distance, marketing efforts, and other adopters on the diffusion process of consumer adoptions.

This chapter will demonstrate how adding spatial analysis to traditional market innovation approaches can help make sense of a huge volume of data, provide insights into the patterns of adoption and the influences on adopters, and ultimately help improve decision-making. Our goal is not to demonstrate the absolute superiority of spatial techniques over other approaches nor to develop new theory about spatial influences on diffusion, but to illuminate for both practitioners and researchers some areas where new insights may await both discovery and application. We consider this to be a particularly relevant goal with the new opportunities presented by having significant actual purchasing data and geographic analysis tools.

The chapter proceeds with an overview of background material followed by a description of this study and its data sources. Keeping with the objectives of this book, the steps required to utilize GIS with this particular data set are described step-by-step. The results are presented in tables and figures to facilitate the comparison between the insights that would be gained with and without GIS. The chapter concludes with a few implications for researchers and practitioners in this area.

Background

Geographic considerations have been central to the study of retailing. All retail marketing decisions have to take into account their probable impact on the size, shape, depth, and/or dynamics of the market area of the firm. From the consumer side, such personal decisions as willingness to travel, impediments to arrival, relative visibility of location, reaction traffic patterns, and the influence of competitive locations are also geographic in nature.

Three specific research streams have concentrated specifically on the geography of retailing. One research stream has concentrated on delimiting trade area boundaries so that business decisions that affect the sizes and shapes of those market areas can be evaluated more precisely (see, for example, Huff & Batsell, 1977; Donthu & Rust, 1989). Another stream of geographical research in retailing has involved the modeling of consumer choices in spatially defined markets (see, for example, Huff, 1962, 1964; Ben Akiva & Lerman, 1985). A third, although still emerging, stream is concerned with the spatial diffusion of consumer response to marketing efforts. Although a significant body of spatial diffusion theory does exist in geography and sociology (beginning with Hagerstrand, 1967), little of it has focused on retailing (Allaway, Berkowitz, & D'Souza, 2003).

Diffusion of innovations has proved a useful and durable explanation of how communication affects human behavior. This theory, pioneered by Everett Rogers (1962, 1983, 1995), is based on the notion that a new innovation is first adopted by a few innovators, who, in turn, influence others to adopt it, typically via word of mouth. Continuing influence of adopters on potential adopters explains the shape of the sales trajectory curve over time (Rogers, 1995). Spatial diffusion research adds the geographical element to this research, explaining the patterns of adopter interaction with potential adopters spatially as well as temporally.

Spatial diffusion research in marketing has been hampered in the past by large-scale requirements for spatially coded data, which has been traditionally difficult and time-consuming to acquire and use (Murphy, 1996). However, two new technologies have emerged that make the potential for doing spatial research faster, easier, and more accurate. This paper demonstrates the payoff that bringing these two technologies together can have for both retail practitioners and academics. One of these technologies — point-of-sale-based customer loyalty programs — delivers vastly improved customer-specific behavior data. The other technology — geographic information systems — improves the capability for analyzing and interpreting these data via their inherent spatial characteristics.

Point-of-Sale Data Capture and Customer Loyalty Programs

The widespread adoption of point-of-sale (POS) automation technology has given retailers the opportunity to improve nearly every aspect of their businesses. With electronic POS systems, detailed time, product, and price data are captured for every transaction, which has made planning, inventory management, buying, theft prevention, in-store promotion, and so on much more reliable. In addition, point-of-sale automation has opened the door to the development of individual consumer-based loyalty programs. Modeled after frequent flyer programs offered by airlines, retailer loyalty programs confer such benefits as immediate cost savings, members-only deals, rebates at some threshold level of spending, redeemable points, and/or eligibility for drawings and contests, all to "reward" shoppers for giving up alternative shopping opportunities. Schneiderman (1998) reports that nearly half of the U.S. population belongs to at least one loyalty program and that such programs are growing at a rate of approximately 11% a year.

More importantly for researchers, most retail loyalty programs involve the use of specially coded credit/debit cards or other special scanner-readable cards, which contain consumer-specific identification information. When these cards are scanned at the point of purchase, data is captured which links the consumer to the time, day, products bought, prices, and so on. Analysis of these data over time can yield invaluable insights into consumer shopping processes, reactions to marketing efforts, and long-term patterns of behaviors at the individual as well as at the aggregate level. A variety of techniques being applied to this data include various forms of regression, factor analysis, cluster analysis, time series analysis, and chaid analysis. In addition, the fact that loyalty programs typically require members to provide name, address, and other relevant information about themselves gives researchers the opportunity to use an arsenal of geographic analysis tools to better understand the shopping and buying behaviors of current customers and to target new ones.

Geographic Information Systems

Geographic information systems (GIS) are the second technology bringing radical change to retail-oriented research. Prior to GIS, spatial data was difficult to obtain in a form necessary for meaningful marketing research (e.g., addresses) and expensive and subject to error when collected in a more analytically usable form (e.g., accurate relative distances). Awkward to handle and time-consuming to create, the number of paper maps needed to cover the market areas of a major U.S. retailer could reach the thousands. In addition, a map-centered approach did not lend itself to easy physical reproduction, and the analyses were more difficult to replicate or extend than non-spatial analysis results (e.g., a trade area map is specific to a particular store location; a regression model of consumers based on census data is not).

While clearly based on cartography (the science of mapmaking), GIS technology goes much farther than a computerized map. The capability of a GIS is significantly expanded by combining the association of non-spatial descriptive data (attribute data) with spatial features in a visually interactive mode that supports changes in scale (e.g., zoom in/zoom out), the overlaying of different types of spatially-encoded information (i.e., like paper map overlays), and the creation and display of new geographic information (e.g., identifying the set of elements with certain attributes and within a specified distance of a geographic feature). Traditional (non-spatial) querying and analysis tools can be combined with spatial information once a geographic coordinate is associated with the item of interest (Murphy, 1995). For U.S. consumer data, the technique of geo-coding uses a pre-defined list of addresses and their spatial components along with a searching/matching algorithm (Densham, 1991; Keenan, 1995).

Retailing was an early adopter of GIS, primarily for store location decisions (e.g., Baker & Baker, 1993; Daniel, 1994; Foust & Botts, 1995). Once the store was located, however, the role of GIS often gave way to traditional analysis approaches (e.g., media-revenue recovery models) in which geography was a constant (e.g., the location of the store) or only slowly varying (e.g., a store's trade area). With the increasing amount of customer-specific data being collected at the store level, however, the analysis of customers can increasingly exploit the spatial aspects of consumer behavior.

This Study

This chapter demonstrates some of the additional insights that a GIS-based analysis approach can offer in the study of the spatial diffusion in the context of a new loyalty card program. We show that the study of spatially-oriented consumer behaviors and the business strategies that result from analysis of these behaviors both benefit greatly from the application of GIS technology. The situation involves the launch and testing of a new loyalty card program by a very large U.S. retailer within a major metropolitan area. This loyalty card program constituted a major effort on the part of the retailer to build store traffic, increase basket size, and increase shopping frequency while creating deeper relationship ties with its customer base. The large-scale launch effort for the loyalty card program included city-wide radio, a number of billboards, and professional in-store solicitation. Data capture was via checkout scanner, and every transaction in which the consumer "swiped" the card was recorded. According to company records, the launch of the program was highly successful, with an increase of nearly 30% in sales during the first few weeks of the program compared to the prior year.

Customer Loyalty POS Data

Detailed information on the full first year of the program was provided to the researchers, including the launch campaign, a name and address database of every cardholder, a purchasing occasion database, and a stock-keeping-unit (SKU) level sales database (both identified to the cardholder level) for three separate stores. When combined, the resulting data set covered well over one million distinct shopping trips and several million

SKU-level product purchases. After narrowing the focus to consumers of a single representative store, and combining and organizing information, a data set was created which included a cardholder identifier, detailed street address, date of first card usage, date of last card usage (both recoded from day one through day 365), number of purchasing occasions during the year, total dollars spent in the store over the year (using the card), average amount spent per shopping visit, highest and lowest dollar amount spent on a shopping trip, duration of shopping activity (last day of card use minus first day) and shopping interval (average time between purchasing occasions).

Traditional Non-Spatial Analysis

While most analyses of customer loyalty programs do not take advantage of GIS technology, loyalty program data are valuable to retail decision-making. Sales data at the aggregate and at the individual levels can be tracked over time and patterns noted and modeled. Time series analysis, cluster analysis, logistic regression, and other tools can be used to visualize the timing of shopping, to distinguish between loyalty groups, and to estimate the impact of different marketing efforts on the loyalty base and the subgroups within it.

We first demonstrate a traditional diffusion-of-innovation approach to search for insights about the growth of and the prospects for the loyalty card population. Using Rogers' (1962, 1983, 1995) and Mahajan, Muller, & Srivastava's (1990) frameworks, each of the nearly 18,000 adopters was classified into an innovator, early adopter, early majority, late majority, or laggard group. Because these are assigned categories based on timing of the adoption relative to the pattern of overall adoptions, classification of particular individuals into these categories is accomplished by examining the temporal distribution of adoptions and looking for transition points following the percentage distribution guidelines of Rogers (1995).

Compared to some other innovations (telephone, automobiles, air conditioning), a loyalty card program has a short adoption cycle (less than 180 days versus decades), which is a factor in making the decisions about the cut-off between adopter groups. The classification that captured the dynamics of this data most accurately was to set the innovator cut-off after two days, which yielded 1,073 persons, or 6.0% of all eventual adopters. The early adopter stage of the process began on day three of the program and ran through day seven, when 18.5% of all eventual adopters had made their first purchase. The cutoff for the early majority was made after the day 31 of the program, when 50.5% of all eventual adopters had made their first purchase. The late majority category cutoff was made after the 120th day, with 84.7% of the total, while the last 15.3% of adopters were relegated to laggard status. The comparison of the percentages by adopter group for this study vs. Rogers (1995) is shown in *Table 1*.

As shown in *Table 2*, there are significant differences among the adoption groups in nearly every category of basic descriptor. This, in and of itself, is interesting and can lead to additional insights relevant to retailers. Such phenomena as cross-shopping, the number of new adopters as well as the number of deserters each week, the increase in new adoptions following a radio blitz or new round of promotion, increases or decreases in

Table 1. Comparison of Adopter Group Classification

ADOPTER GROUP	Rogers (1995) Percent in Category	Rogers (1995) Cumulative Percent	This Study Percent in Category	This Study Cumulative Percent
Innovators	2.5	2.5	6.0	6.0
Early Adopters	13.5	16	13.5	18.5
Early Majority	34	50	32.0	50.5
Late Majority	34	84	34.2	84.7
Laggards	16	100	15.3	100

Table 2. Statistical Profile of the Three Diffusion Stages and Five Adopter Groups: Pre-GIS Insights

DIFFUSION STAGE	Stage One		Stage Two	Stage Three		Total
ADOPTER GROUP	Innovators	Early Adopter	Early Majority	Late Majority	Laggards	Total
NUMBER IN GROUP	1,070	2,216	5,646	6,045	2,698	17,675
			Pre - GIS Insights			
Profile Characteristics	Mean	Mean	Mean	Mean	Mean	Mean
Day of Adoption	1.47	4.85	18.40	68.62	167.29	55.58
Length of Loyalty Card Use (Days)	238.57	226.99	200.68	154.72	97.00	174.73
Interval Between Card Uses (Days)	23.23	25.17	28.05	29.26	22.35	26.94
Total Dollars on Card	$772.90	$614.61	$426.85	$319.34	$240.22	$406.08
Number of Purchase Occasions	27.67	18.50	11.39	7.53	5.65	11.07
Dollars Spent per Trip	$36.95	$38.99	$40.84	$47.39	$47.43	$43.62

overall card use, SKU's bought, and customer loyalty can all be tracked without the use of geographical data.

Application of GIS

However, just the fact that address-specific information exists in the loyalty card database enables retailers to expand the value of this data many-fold. The insights available by the application of GIS technology to these data open the door to a level of analysis far beyond those of traditional retail researchers. To take advantage of the potential inherent in the spatial data captured by the customer loyalty/POS program, it was necessary to begin by preparing the customer data set for loading into a GIS program. This involved the creation of a single data set from the three separate databases kept by the retailer — a customer ID-coded name and address database, a customer ID-coded purchase event database, and a customer ID-coded products purchased database.

Together, these databases held over ten million lines of customer ID-coded information. After combining by customer ID and isolating a single store's activity, a data set of approximately 23,000 cardholding customers was produced.

Step 1: Geo-Coding & Creation of Distance Variables

The first task was to check and correct the coding of address characteristics and zip codes so geo-coding could proceed. The database was then loaded into a popular PC-based GIS. Using a built-in search algorithm, which matches addresses ranges with those in the national streets database, these loyalty program customers were geocoded to yield detailed eight-digit latitude and longitude figures (called lat-long) on each cardholder. Approximately 15% percent of the addresses could not be matched, either because of address spelling errors, new construction (new streets not yet in the national street database), double-named streets, or colloquially named streets. These addresses were either hand-located or discarded, resulting in a final geo-coded database of 17,675 cardholders. Using similar address data, lat-longs were generated for the store, its competitors, and each billboard that advertised the loyalty program. Features of the GIS application were used to compute and add to the database additional spatial variables for each of the cardholder records including Euclidean distance from residence to the store, to the nearest billboard, and to each competitor, and the number of billboards and the number of competitors within 2.5 miles of the customer's residence.

Finally, a "Neighborhood Interaction Field" (NIF) was created around each of the adopters of the loyalty card. After testing dozens of distance measures, a figure of .1 kilometers, or .06 miles was selected as the appropriate NIF radius around each adopter. This distance covered approximately five to seven houses in all directions, more in tightly compressed housing configurations and fewer in areas with more distance between neighboring houses. Note that the spatial dispersion characteristics of other environments (e.g., more urban or dense, more rural or distributed) can and should affect the radius chosen; the goal was to identify a practical measure to capture the likely residence-based communication influences on adoption effectively for the mid-western U.S. suburban setting of the data set. All economic activity (previous adoptions, loyalty card-specific shopping behavior, spending, and so on by any of the other adopters) was captured for each cardholder and added to the data set as additional variables. None of these measures could have been generated without a GIS.

Step 2: Adding All Households in Market Area

A second data set of every household within a 35-mile radius of the store was created using geo-coded and mapped data from a direct mail list vendor. To truly understand the adoption process we need to study the innovation's effect on not only the nearly 18,000 adopters but also on the approximately 300,000 households in the greater market area who did not become adopters of the loyalty program. The same distance-based measures were computed for non-adopters and added to the data set.

Step 3: Adding Block Group Variables

Finally, a third data set was created using data provided with the GIS itself. Over 1,200 Census Block Groups exist in the metropolitan area. Adding Block Group data from the most recent Census of Population brought several hundred additional U.S. government and vendor-generated variables to the analysis, including population, income, education, housing, and commuting characteristics. The Block Group is the smallest census-generated geographic unit for which significant population-related information is available, and it is configured so as to capture relatively homogenous population clusters. Block-group-level demographic data were added to the adopter data set and non-adopter population data set to help profile adopter and non-adopter characteristics. Each adopter and non-adopter was classified by his or her lat-long coordinates into an appropriate Census Block Group.

Step 4: Convergence

The data from the geo-coded summary of the loyalty program behavior was combined with the market area household data and the Block Group data via the coordinates of the customers. By this point, the original loyalty card data set had been expanded to include (for each cardholder):

- His or her distance from the location of the store and each competing store;
- His or her distance from the location of nearest billboard;
- His or her distances to the nearest other cardholder, the nearest very early adopter, the nearest "very loyal" cardholder (based on purchasing characteristics);
- The density of cardholders in the immediate area of his or her residence;
- Latitude and longitude locations of all households within 35 miles of the store (whether or not they were loyalty program card holders);
- A variety of U.S. Census data (at the block group level) on income, ethnicity, occupation, education, housing, and commuting patterns.

Results of Analysis

We first discuss the descriptive results for the data set without the use of geographic data or spatial analyses and then the patterns visible with "the geographic edge" of spatial data and GIS. These initial insights are then extended by modeling the adoption behavior also with and without spatial capabilities.

Table 3. Statistical Profile of the Three Diffusion Stages and Five Adopter Groups: Additional Information from GIS

SPATIAL DIFFUSION STAGE	Stage One		Stage Two	Stage Three		Total
ADOPTER GROUP	Innovators	Early Adopter	Early Majority	Late Majority	Laggards	Total
NUMBER IN GROUP	1,070	2,216	5,646	6,045	2,698	17,675
	Additional Information Available from GIS					
Profile Characteristics	Mean	Mean	Mean	Mean	Mean	Mean
Distance from Store (Miles)	4.57	4.95	5.30	5.98	6.72	5.66
Median Household Income	$44,599	$43,221	$43,520	$42,970	$40,078	$42,834
Percent of Households headed by Executive or Professional	7.89%	7.67%	7.49%	7.16%	6.55%	7.28%
Percent Households headed by Minority Race or Latino	40.68%	41.87%	42.70%	43.68%	46.45%	43.38%

Descriptive Results Without Geographic Data

Table 2 shows a subset of the basic characteristics of the adopter groups for the new loyalty card program. As shown, there was significant consumer response to launch campaign, with nearly 3,300 adoptions during the first week of the program. By the end of the first month, approximately 9,000 adopters had already made their first purchase using the loyalty card. This type of analysis can yield insights about the size and speed of the adoption process, typical behaviors, differences in behaviors by adopter group, and more.

Geographical Edge: Descriptive Results

The use of spatial data and GIS allowed us to investigate the spread of adoptions over space and time, the influence of billboards, and the influence of prior adopters.

Outward Spread of Adoptions over Time

The ability of the geo-coded data to yield deeper marketing insights than the non-geographic data is evident. *Table 3* shows the same diffusion stages and the additional measurements and descriptors possible via the utilization of GIS technology, while *Figure 1* shows the spatial configuration of adopters at various distances and directions from the launching store. Those adopting during the first day of the program are labeled as innovators and coded with black stars. Early adopters, those making their first purchase during the remainder of the first week of the program, are coded with a darker gray bounded square. All other adopters are coded with the lighter gray squares.

Figure 1. Analysis of Early Adoption Period (with five-mile ring superimposed)

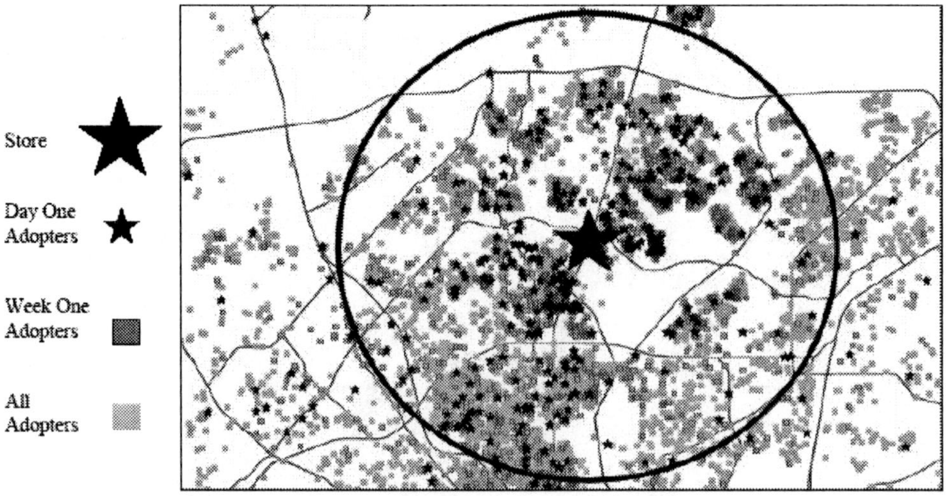

For example, over the duration of the program, there was a steady spread of the adoption center outward. As shown in *Table 3*, the mean distance from the store of new adopters increased from 4.8 miles during Stage One to 5.3 miles during Stage Two to 6.2 miles during Stage Three. For the 297,000 nonadopters within the 35-mile radius of the store, the mean distance from the store was 9.9 miles. There is clearly a proximity effect within the loyalty program, and managerial strategies based on radio, newspaper, or other broad-scope coverage are not likely to be as effective as those focusing in on consumers within the store's primary trade area.

To demonstrate further, the area surrounding the store was divided into concentric rings[1] – 0 to 3 miles from the store (encompassing 29,000 total households), 3.01 to 6 miles from the store (76,000 households), and 6.01 to 35 miles from the store (209,000 households). As *Table 4* indicates, 43% of all Stage One adopters were residents within three miles of the store. During Stage Two, 35.3% of the adopter group was drawn from the three mile ring nearest the store, 34.5% from the 3-6 mile ring, and 29.8% from the 6 to 35 mile ring. The largest group of Stage Three adopters (39.3%) was drawn from the 6 to 35 mile ring, while an additional 35.1% reside within the 3 to 6 mile ring. Interestingly, by Stage Three, 25.6% of the last group of adopters still came from the three-mile ring nearest the store. Overall, 7.9% of nonadopters lived in the 0 to 3 mile ring, 23.6% lived in the 3 to 6 mile ring, and 68.5% lived in the 6 to 35 mile ring.

Billboard Influence

In addition, adopter distances from the nearest of the program-specific billboards increased steadily from 2.6 miles to 2.8 miles to 2.9 miles over the three stages of the diffusion process, and the number of billboards within 2.5 miles of each adopter declined as well. For nonadopters, the mean distance from the nearest billboard was 5.1 miles. While the slow spread out from the billboards is partly accounted for by their location

Table 4. Characteristics of Diffusion Stages by Distance From the Store

Basic Adoption Data by Stage		0-3 mile Ring	3-6 mile Ring	6-35 mile Ring	1-35 mile Market
STAGE ONE ONLY (DAY 1-7)	NUMBER	**1,412**	**1,012**	**862**	**3,286**
Mean distance to store		1.810	4.290	10.400	4.825
Mean distance to nearest billboard		1.120	1.880	5.780	2.575
Number of billboards in 2.5 miles		2.320	1.150	0.000	1.350
Distance to the nearest innovator (e.a. only)		0.036	0.113	0.471	0.175
Number of innovators in .06 miles (e.a. only)		4.670	1.110	0.490	2.480
Percent of Stage One Adoptions in ring		43.0%	30.8%	26.2%	100%
STAGE TWO ONLY (DAY 8-31)	NUMBER	**1,991**	**1,974**	**1,681**	**5,646**
Mean distance to store		1.900	4.320	10.480	5.301
Mean distance to nearest billboard		1.120	1.900	5.680	2.751
Number of billboards in 2.5 miles		2.030	1.110	0.000	1.100
Distance to the nearest innovator (e.a. only)		0.055	0.166	0.706	0.288
Number of innovators in .06 miles		4.200	0.760	0.170	1.800
Number of stage one adopters in .06 miles		5.430	0.650	0.180	2.200
Percent of Stage Two Adoptions in ring		35.3%	34.5%	29.8%	100%
STAGE 3 ONLY (DAY 31-365)	NUMBER	**2,235**	**3,069**	**3,438**	**8,742**
Mean distance to store		1.930	4.428	10.570	6.205
Mean distance to nearest billboard		1.110	1.853	5.000	2.901
Number of billboards in 2.5 miles		2.220	1.100	0.000	0.950
Distance to the nearest innovator (e.a. only)		0.569	0.176	0.727	0.362
Number of innovators in .06 miles (e.a. only)		3.960	0.700	0.130	1.310
Mean distance to store		5.120	0.650	0.160	1.600
Percent of Stage Three Adoptions in ring		25.6%	35.1%	39.3%	100%
NON ADOPTERS	NUMBER	**23,508**	**70,066**	**203,187**	**296,761**
Mean distance to store		2.010	4.650	12.610	9.885
Mean distance to nearest billboard		1.170	1.930	5.960	5.112
Number of billboards in 2.5 miles		2.070	0.980	0.000	0.830
Distance to the nearest innovator (e.a. only)		0.073	0.234	1.340	2.006
Number of innovators in .06 miles (e.a. only)		1.860	0.210	0.030	1.020
Mean distance to store		3.940	0.480	0.070	2.120
Percent of Non-adopter population in ring		7.9%	23.6%	68.5%	100%
Loyalty Card Penetration Rate in ring		*24.0%*	*8.6%*	*2.9%*	*6.0%*
Total Number of Households		**29,146**	**76,121**	**209,168**	**314,435**

nearer the store than much of the market area (which makes their impact collinear with the store spread effect), there does appear to be a billboard effect. Even within the 0 to 3 mile ring nearest the store, nonadopters were significantly further from the nearest billboard than adopters.

This information, especially combined with maps and block group data, has the potential to provide marketers with exceptional insights for new billboard campaigns, direct mail efforts, neighborhood-wide as opposed to city-wide newspaper advertising, and other insights not possible without spatially captured data.

Influence of Prior Adopters

Importantly, however, the outward spread of adoptions was not absolute. There were significant percentages of adoptions still taking place near the store long after the launch (recall that more than 25% of Stage Three adopters lived less than three miles from the store). These continuing adoptions are evidence that consumers are very different in their adoption mentality, and that significantly more information and motivation are required to change some peoples' behavior than other peoples'.

A closer examination of the spatial configuration of adopters at a variety of distances from the store indicates that the store effect and radio/billboard effects alone cannot explain the distinctive spatial pattern of adoptions. As shown in *Figure 2*, even at 5.5 miles from the store there are clusters of significant adoption activity separated by areas with little or no adoptions. Using the unique capabilities of the GIS, a set of "rings" was generated around every adopter who did not have a prior adopter within his or her Neighborhood Interaction Field. These people were designated as potential Cell Drivers who might or might not influence the people living around them to adopt as well. The circled number one (1) represents those cardholders adopting during the first day of the loyalty program. Other adopters are coded with their diffusion stage (2 through 5).

In addition, many of the clusters of activity appear to be around the locations of the very earliest adopters of the loyalty program. In *The Anatomy of Buzz*, Rosen (2000) describes Whyte's (1954) classic study of the "neighborhood effect" at work in the spatial diffusion of air conditioners across Philadelphia neighborhoods. Aerial photographs were used to distinguish houses on each block with air conditioners from those without - "one block might have eighteen air conditioners, while the next block over might only have three." According to Whyte, these adoption clusters were the result of small word-of-mouth networks operating among neighbors. Where the early consumers in an area were vocal about their new purchases in a positive way, they influenced people around them to try the innovation as well. We observed the same phenomenon in this research. For every ring (distance groupings from the store), a much higher percentage of Stage Two and Stage Three adopters had Stage One adopters within their Neighborhood Interaction Fields than nonadopters (see *Figure 2*).

Modeling without Geographic Data

Although descriptive results are very useful managerially and for generating hypotheses, the typical treatment is to model the temporal curve of the diffusion process. A vast tradition of statistical modeling in marketing has been used to generate information about the influence of innovation and imitation within the adopter population (there are hundreds of studies; a useful starting place is Bass, 1969). Although beyond the scope

Figure 2. Diffusion Cells at 5.5 Miles from the Business

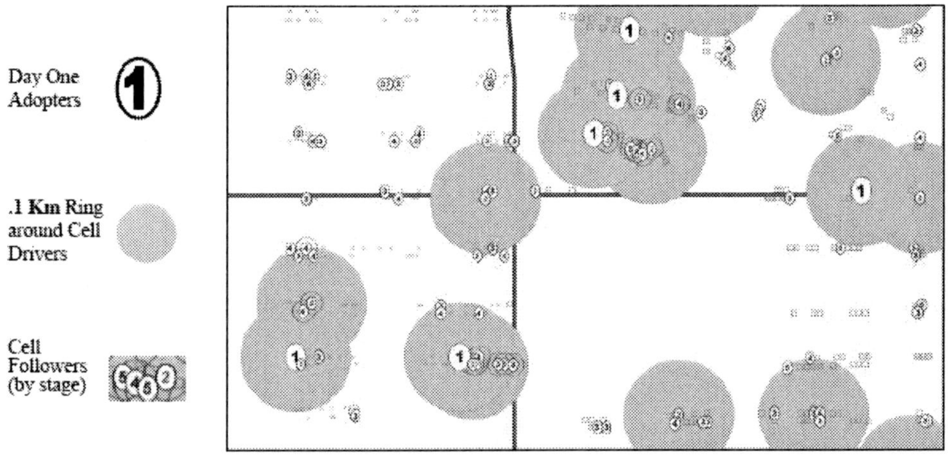

of this research, an entire book devoted to this topic is Mahajan, Muller, & Wind (2000). Most diffusion models can be categorized as event history models, which predict the time-occurrence of specific "events" which reflect our focus — the adoption of a new loyalty program (Kalbfleisch & Prentice, 1980). A number of recent papers (see, for example, Bass, Jain, & Krishnan, 2000; Roberts & Lattin, 2000) have treated diffusion as an event-history.

Geographical Edge: Modeling

The use of spatial data and spatial analysis allowed us to apply two different modeling approaches: an event-history approach to spatial diffusion, and a simulation of alternate billboard locations to test for the significance of spatial factors.

Modeling Spatial Diffusion

Modeling spatial influences on the diffusion process, even with ample high-quality data and sophisticated statistical software, is nearly impossible possible without the ability of the GIS to generate a wide variety of location-based and distance-based variables. Using the loyalty data via its conversion in a GIS, we set up a model to estimate and predict each potential adopter's likelihood of adopting the loyalty card on each day of the launch period as a function of (1) his or her distance from the loyalty card store, (2) distance from the nearest billboard advertising the program, and (3) the number of very early innovators living within their Neighborhood Interaction Field. Similar event history models of spatial processes are used in Allaway, Mason, & Black (1994), Pellegrini & Reader (1996), and Allaway, Berkowitz, & D'Souza (2003).

The results of the modeling effort were highly significant and very telling. Probabilities of adoption were clearly tied to those geographic variables generated by the GIS. When the entire year was considered, the likelihood that a person would adopt on a given day decreased by 13.4% for each mile further they lived from the store (the store-distance effect). For example, a person five miles from the store was 13.4% more likely to adopt than a person six miles from the store, all other things equal. The billboard effect was also important — each billboard located within 2.5 miles of his or her residence raised potential adopters' probability of adoption by 7.2%. In addition to these impersonal launch-based stimuli, the neighborhood-effect (the influence of previous adopters in stimulating people around them) was significant. Over the 52-week period, each Innovator located within the Neighborhood Interaction Field of a prospective adopter raised his or her likelihood of adopting by 13.2% (all estimates were significant at the .01 level).

Further Analysis of the Billboard Effect

To test whether the billboard effect was simply an artifact of distance from the store rather than the placement of the billboards, we created an experiment. The locations of each of the five billboards (the sixth was too close to the store to move) were shifted in the GIS to another location an equal distance from the store. The model was estimated with the new billboard locations, but the billboard effect for the artificial spatial configuration was not significant. This indicates that the specific locations were influential in adoption decisions, a result that could not have been determined without the ability of the GIS to generate new prospective sites, measure their distances from over 300,000 potential adopters, and prepare the data for modeling.

Implications for Researchers and Managers

Spatial diffusion processes should be an important subject of study for both academicians and managers. A wide range of retail activities generate spatial and temporal consequences and lend themselves to this research approach. Such studies could range from forecasting the development of a customer base around a new location under different launch strategies to tracking the negative mouth-to-mouth networking process resulting from a poor service encounter. Our modeling results indicate at least for this particular innovation that:

1. There are three very distinct spatial diffusion stages involved in the response of a market area to a new innovation;
2. The spatial pattern is established very early in the diffusion process;
3. Likelihood of adoption at any time is influenced by distance to the store itself, to the billboards, and to the earliest adopters of the program;

4. Members of this earliest adopting group, the Innovators, appear to be the key neighborhood effect "drivers" in influencing new adopters to follow them.

These results lead to important implications about more effective launch strategies for retail innovations. In this case, a city-wide, coordinated launch effort apparently attracted a large number of potential adopters who may have "been waiting for" just such an innovation. Well over 1,000 people reacted during the first two days (dozens of whom came from as far as 25 miles from the diffusion site) to an innovation that appeared to offer immediate and significant value. These are apparently the people who were influential in getting other people around them to adopt over the next several days or weeks. For managers, it appears clear that the value of "early advocates" of an innovation cannot be overstated, because they will be the drivers of that innovation through its market. They apparently deserve the extra benefits, attention, or whatever it is that keeps them as "salespeople" for the retailer.

Finally, the widespread availability of scanner data and the development of large numbers of loyalty programs and other customer relationship management programs tying scanner data to individual customer characteristics should make this type of research easier in the future. However, none of these analyses is possible without the ability of GIS technology to *capture* data spatially, *portray* incredibly complex multidimensional data easily and clearly for hypothesis generation, and *manipulate* data sets geographically. The combination of GIS software and the powerful data storage products and statistical modeling software available to managers and researchers can yield significant empirical results. In the future, retail managers may be able to choose from and develop strategies for a variety of spatial diffusion patterns for their particular marketing initiatives in the same way they currently select reach and frequency patterns for advertising strategies.

The contribution of this paper is to demonstrate that spatial analysis can be used to gain both practical and theoretical insights into retailing behavior. Our goal included removing some of the mystery around creating spatial data and in using it as a complement to traditional non-spatial analyses. We hope our readers can see that old questions about retailing behavior may be addressed in new ways with spatial analyses (especially combined with new data sources and the appropriate tools), and that these insights will themselves raise questions of both managerial and research significance.

References

Allaway, A. W., Berkowitz, D., & D'Souza, G. (2003). Spatial diffusion of a new loyalty program through a retail market. *Journal of Retailing*, *79*, 37-151.

Allaway, A.W., Black, W.C., Richard, M., & Mason, J.B. (1994). Evolution of a retail market area: An event history model of spatial diffusion. *Economic Geography*, 23-40.

Baker, S., & Baker, K. (1993). *Market mapping*. New York: McGraw-Hill.

Bass, F. M., Jain, D., & Krishnan, T. (2000). Modeling the marketing mix influence in new product diffusion. In V. Mahajan, E. Muller, & Y. Wind (Eds.), *New product diffusion models*. International Series in Quantitative Marketing. Norwell, MA: Kluwer.

Daniel, L. (1994, September/October). Enhanced modeling helps position business geographics in the retail industry. *Business Geographics*, 37-39.

Densham, P.J. (1991). Spatial decision support systems. In D.J. Maguire, M.F. Goodchild, & D.W. Rhind (Eds.), *Geographical Information Systems* (Vol. 1: Principles, pp. 403-412). London: Longman.

Donthu, N., & Rust, R. T. (1989). Estimating geographic customer densities using kernel density estimation. *Marketing Science, 8* (2), 191-203.

Foust, B., & Botts, H. (1996). Using market vectors to evaluate multiple sites in a single market. *Proceedings of Business Geographics for Educators and Researchers* (pp. 87-91). Chicago, IL.

Hagerstrand, T. (1967). *Innovation diffusion as a spatial process*. Chicago, IL: University of Chicago Press.

Huff, D. L., & Batsell, R. (1977). Delimiting the areal extent of a market area. *Journal of Marketing Research, 14* (4), 581-565.

Huff, D. L., & Rust, R. T. (1984). Measuring the congruence of market areas. *Journal of Marketing, 48* (1), 68+.

Huff, D. L., & Rust, R. T. (1986). Estimation and comparison of market area densities. *Journal of Retailing, 62* (4), 410-430.

Kalbfleisch, J.D., & Prentice, R. L. (1980). *The statistical analysis of failure time data*. New York: John Wiley & Sons.

Keenan, P. (1995). *Using a GIS as a DSS generator* (Working Paper MIS 95-9). Dublin, Ireland: Michael Smurfit Graduate School of Business, University College.

Mahajan, V., Muller, E., & Srivastava, R. (1990, February). Determination of adopter categories by using innovation diffusion models. *Journal of Marketing Research, 27*, 37-50.

Mahajan, V., Muller, E., & Wind, Y. (2000). *New product diffusion models*. International series in quantitative marketing. Norwell, MA: Kluwer.

Morrill, R., Gaile, G., & Thrall, I. (1988). *Spatial diffusion*. Beverly Hills, CA: Sage Publications.

Murphy, L. D. (1995). Geographic information systems: Are they decision support systems? *Proceedings of the Twenty-Eighth Annual Hawaii International Conference on System Sciences* (pp. 131-140). Maui, Hawaii.

Murphy, L. D. (1996). GIS in business schools: A status report. *Proceedings of Business Geographics for Educators and Researchers* (pp. 126-135). Chicago, IL.

Pellegrini, P.A., & Reader, S. (1996). Duration modeling of spatial point patterns. *Geographical Analysis, 28* (3), 219-243.

Reichheld, F. F. (2001). *Loyalty rules! How today's leaders build lasting relationships*. Cambridge, MA: Harvard Business School Press.

Rogers, E.M. (1962). *Diffusion of innovations*. New York: The Free Press.

Rogers, E. M. (1983). *Diffusion of innovations* (3rd ed.). New York: The Free Press.

Rogers, E. M. (1995). *Diffusion of innovations* (4th ed.). New York: The Free Press.

Rosen, E. (2000). *The anatomy of buzz*. New York: Doubleday.

Schneiderman, I. P. (1998, March 30). Frequent flyer reward programs starting to fly high. *Daily News Record*.

Whyte, Jr., W. H. (1954, November). The web of word of mouth. *Fortune*, 140.

Endnote

[1] Rings are one of several available methods for portraying a market area (e.g., polygons and probability surfaces are others). Rings are typically appropriate in the absence of competition and where traffic patterns or natural barriers do not influence or limit consumer access to the site. Here, the rings used are not proposed as representative of the firm's market area. We use the three, six, and 35-mile concentric rings only as a convenient classification tool to demonstrate spatial effects that would be difficult to expose without GIS.

Chapter XIII

Geospatial Analysis for Real Estate Valuation Models

Susan Wachter, Wharton School, USA

Michelle M. Thompson, Lincoln Institute of Land Policy, USA

Kevin C. Gillen, Wharton School, USA

Abstract

This chapter provides an overview of a major contemporary issue in real estate valuation — the use of geographical data to improve valuation outcomes. The spatial nature of real estate data allow the development of specialized models that increase the likelihood for better predictions. This chapter examines how using spatial data, with geographical information systems (GIS), can improve the accuracy of real estate valuation models. Contemporary theory in economics, planning, housing, and appraisal influences the model application that underlies the new field of GIScience and the use of Automated Valuation Models (AVMs) in practice. Exploratory methods of model development are also considered in the presentation of a case study along with a discussion of the changing history, development and future of AVMs and GIS.

Introduction

This chapter examines how spatial data and Geographical Information Systems (GIS) can be used to improve the accuracy of real estate valuation models. In recent years, there has been significant progress in the use of statistical models to value residential real estate. In particular, statistical models developed by academic researchers have been integrated into fast-developing Automated Valuation Model (AVM) technology. Historically, many municipal assessors have used a related technology for mass appraisals, Computer Assisted Mass Appraisal (CAMA). However, neither AVMs nor CAMAs fully exploit the potential of geographically related information to improve the accuracy of real estate valuation models.

AVMs and CAMAs attempt to model spatial and temporal variation in house prices. These models are used to mark residential property values to market, that is, to estimate the sales value of properties that have not been transacted recently. In particular AVMs are used by lenders to underwrite mortgage loans in lieu of full market real estate appraisals. The estimation process involves taking known sales prices and using this data to project the unknown. Academic researchers have developed statistical valuation models to do this. This methodology is being incorporated into AVMs and increasingly being used in the private sector.

There are two basic types of econometric valuation models used to estimate real estate market values. Hedonic models relate house prices to characteristics of the lot, the structure, and the neighborhood (Houthhaker, 1952; Rosen, 1974). Repeat-sales models produce an index through linking sale prices from the same properties over time (Bailey et al., 1963; Case & Shiller, 1987, 1989). Hybrid models combine hedonic and repeat-sales specifications to obtain more efficient parameter estimates (Case, Pollakowski & Wachter, 1991; Quigley, 1991, 1995; Hill et al., 1997; Case, Pollakowski & Wachter, 1997). However, most AVMs to date do not incorporate specific information on location (latitude and longitude). The key to an accurate valuation model is precise location data. Location is essential for valuation of all classes of property. Location can be used as an explicit and fundamental element within the modeling process by utilizing autocorrelation based statistical methods and GIS.

The introduction of GIS technology into statistical property valuation models has great potential. When applied to a geo-coded dataset of single-family properties, this technology allows the user to estimate and exploit the spatial relationships in property values to build improved automated appraisal models. The result is a more expansive class of models with significantly more predictive power.

Traditional automated appraisal models postulate that the value of a property is a function of its physical and neighborhood attributes. These models typically estimate the statistical relationship between transaction price and such variables as square footage, lot size, number of bathrooms, frontage, age of the property, area income, and other neighborhood indicators. Models might also include time series methodology; indexing a given property's value to a regional index of price change. While there is indeed a relationship between a home's value and these aforementioned variables, this specification is incomplete.

The application of GIS technology allows the user to explicitly account for locational[1] effects on a property's value. The computation of the spatial relationships in property values allows the user to expand model specification to include these spatial variables in the prediction algorithm. It is only very recently that such computations have been made possible by increased computer speed and capacity for data analysis and geospatial software such as ESRI's ArcView.

This chapter provides an empirical example of the power of integrating spatial information into traditional AVMs and CAMAS, using transaction and property characteristic data from San Bernardino, California. To demonstrate the tool's efficacy, we estimate a basic hedonic regression to characterize the relationship between the total value of a property and its individual attributes. We then add spatial variables to demonstrate the power of geospatial data and methods to improve the accuracy of valuation outcomes.

The chapter discusses integration of GIS and real estate models building on AVM research conducted at The Wharton School's GIS Lab. The Lab's research GIS-based AVM improves upon aspatial models, and also offers a potential solution to data inadequacies, which limit the accuracy of model-based appraisal estimates for markets where attribute data are limited. In particular, the GIS-AVM incorporates a spatial algorithm to exploit the latent information contained in the geographic proximity of properties in the same market. Via an interactive procedure, the AVM explicitly computes the spatial covariance structure of geographically proximate properties and incorporates this information into the model. The result is a significantly higher degree of predictive accuracy in estimates of house prices compared to models with limited spatial data. ArcView GIS v. 3.2 is used for geocoding, address matching and analysis.

The chapter is organized as follows: we first present background on CAMAs and AVMs. Next, we describe the basic hedonic model and its limitations, and then provide the theoretical explanation for why spatial solutions work to improve predictability of such models. Then we turn to the California empirical example. We augment the basic hedonic model by adding spatial components and we measure and compare the predictive accuracy of the models. We conclude with a conceptual discussion of the promise and challenge of the new technology.

Current State of CAMA and AVM Usage

The assessment community has historically relied on the statistical and appraisal method properties. More recently, the mass appraisal method of valuing properties, using large databases and statistical techniques, has provided the assessor with a means to expedite valuation. Internationally, municipalities that have begun exploring the adoption of CAMAs find that there are potentially significant cost savings in their use. The International Association of Assessing Officers (IAAO), an organization for the professional development of assessing officers, has taken the lead in providing guidelines and interpreting state mandates and standards for the creation of valuation models. The IAAO is currently involved in studying the expanded use of AVM to improve CAMAs.

The rise in the use of CAMA method of valuation by assessors has minimized error in subjective analysis of data typically found when using traditional appraisal methodology. In 2002, the Lincoln Institute of Land Policy and the Computer Assisted Appraisal Section (CAAS) of IAAO conducted a nationwide study of IAAO users to better understand the level of CAMA usage and the integration of GIS within the valuation process. The final report is pending but early indications are that CAMA is still not integrated in most assessing offices, with few integrating GIS in their practice.

While CAMAs were the first application of statistical modeling in appraisal, in the last decade lending institutions have made substantial advances in the use of AVMs.

At the heart of AVM and CAMA is the multi-linear regression (MLR) model. The establishment of the MLR is derived from value estimation theory using econometric models. Economists who recognized real estate (specifically housing) as a "bundle of goods" which have qualities that are significantly different than other "pleasure" goods contributed to the development of "hedonic models," which have been adopted for this sector. At its simplest, as discussed above, a hedonic equation is a regression of market values on housing characteristics (Malpezzi, 2002). The coefficients obtained by regressing house prices on the house characteristics are the hedonic prices and are interpreted as the households' implicit valuations of different housing attributes (Bourassa et al., 1999).

AVMs based on hedonic models have been developed by applied economists and have been implemented as an accepted method of valuation analysis by public and private interests. AVMs are now being used by government sponsored enterprises, Fannie Mae and Freddie Mac, and large banks for desk review of appraisals used in mortgage underwriting. The use of hedonic models for estimating values has been considered a significant advance in the appraisal industry. The issues of equitable and impartial value estimation have increased the professionalism and credibility of both assessors and appraisers. In particular, these models are useful in fraud detection. There are, however, questions about the relative accuracy of such models.

The success in estimating value generally is determined by evaluating relative accuracy based on the R-squared or taking an actual sample of estimated values and comparing them with existing sales. According to Case et al. (1997), "traditional hedonic pricing models... often exhibit prediction errors with a standard deviation in the range of 28% to 50%" while, "appraisers following ad hoc procedures often exhibit prediction errors with a standard deviation around 10%" (Pace et al., 1998). For the appraiser who faces constant public scrutiny, minimizing prediction error is critical. Thus, the utility of AVMs is called into question where the level of error has not been fully examined or explained.

Nonetheless, AVMs are typically far less expensive by an order of magnitude and are not vulnerable to subjective bias. Moreover, since they are statistical algorithms they avoid issues of subjective judgment. Some assessors and appraisers resist the use of AVMs and consider them a detriment to the appraisal industry. The Appraisal Standards Board states that, "the output of an AVM is not, by itself, an appraisal. An AVM's output may become the basis for appraisal review, or appraisal consulting opinions and conclusions if the appraiser believes the output to be credible and reliable for use in a specific assignment" (Advisory Opinion 18)...The IAAO recommends "...a third type of (appraisal) report...Appraiser-Assisted AVM (AAVM)...in which the report combines the most desirable parts of the AVM (unbiased market analysis and consistently applied model formulas) with the most desirable parts of the field appraiser (property inspection, local knowledge and experience)" (IAAO, 2003). Proponents and users of AVMs contend that the public substantially benefits from the lower cost of the AVM ($15) to that of a real estate appraisal ($250) (Geho, 2003). Detractors suggest that AVMs are not as accurate as field appraisal. Comparatively, many municipal assessors and real estate appraisers are concerned with the inability of the AVM to accurately estimate the market without appropriate "model calibration" (Gloudemans/IAAO, 1999), particularly for local markets. Knowledgeable valuation professionals who understand their local market feed data into the model, which reflect the current state of the market. Assessors cannot easily add adjustments for micro-areas and since CAMA generated price estimates are historical and not dynamic. Thus the models that exist today are reflective of these constraints.

The wide variety of multi-linear regression (MLR), mixed models and specialized models, such as feedback, allow the assessor-modeler to better define a model for their market (Kane et al., 2000). The main issue, however, is that there are a limited number of assessors who have been able to adopt the existing assessing models within CAMA. Based upon a recent IAAO-LILP survey, a vendor provided model is purchased by the assessor, where the model is developed based upon a stock model then "calibrated" by the assessor (Ireland et al., 2003).

The models in CAMA are being implemented in order to meet the increasing demands of the public to expedite and systematize the valuation process (Kane et al., 2000). Nonetheless, in the public sector, a universal and mandatory standard for assessing practice has yet to be enacted.

In the private sector, the real estate appraisal industry was turned on its head with the certification (through required state licensing) of appraisers and Uniform Standards of Professional Appraisal Practice (USPAP). States now require appraisers to meet minimum education and experience which, in the aftermath of the S&L crisis, increases the assurance of a reconciled value which weighs the "validity, accuracy and applicability" of the value in relation to the subject property (Mills, 1988). The AVM industry is in the process of developing such standards of accuracy.

Despite these advances, many AVMs and almost all CAMAs "nearly always totally ignore the number one criteria in the determination of real estate value — location. These computer programs could care less if your home is located in a much superior neighborhood — an inferior neighborhood is often separated by just one street from you. These computer programs don't care if you have an ocean front view or a 'crack house' view" (Appraiser Central, 2003).

AVM implementation arose from data availablity that was previously either too expensive or not accessible to the public. Data warehouses by public management or private valuation entities contribute to market data that can be obtained through a variety of electronic media. Today such data can be augmented by the addition of location information. Similarly, the advent of systems of spatial data management, retrieval and analysis in a single distribution center is the crux of geographic information systems (GIS).

The potential of GIS in supporting the appraisal industry is significant. The power of GIS lies in the ability to combine spatial and attribute data to account for location's impact on property values.

In the following, we discuss the traditional hedonic valuation methodology and its limitations, and provide a description of spatial methodologies and demonstrate their utility.

Hedonic Models and their Limitations

Traditional statistical models of property transaction prices postulate that the sale price of any given property (say, single-family homes) is a function of its hedonic characteristics. That is, for a given property-level dataset, a regression is estimated with the following econometric specification:

$$y_i = \beta_0 + \sum_{j=1}^{k} \beta_j H_{ji} + \varepsilon_i \qquad \varepsilon_i \sim iid(0,\sigma^2)\ \forall i = 1,2,\ldots,n$$

Where:

n = the total number of properties in the dataset;

y_i = transaction price of the *i*th property;

H_{ji} = the value of the *j*th hedonic characteristic for the *i*th property;

ε_i = the residual for each observation;

A hedonic characteristic is typically a physical feature of a property. Examples include characteristics such as square footage, number of bedrooms, or number of stories. Other examples include categorical indicator variables that can be created to capture the effects of hedonic characteristics that are non-numeric in nature: the type of exterior siding, or whether the property has swimming pool. Additionally, it is often desirable to incorporate characteristics of the surrounding neighborhood, such as Census tract median income, average SAT scores of the property's school district, and crime rates. While these variables are not typically classified as "hedonic" *per se*, they can nonetheless be thought of as attributes of the property that affect its value. The above econometric specification models the linear statistical relationship of a property's value as a function of its characteristics and attributes by computing the β's of the equations, given the sale price and a vector of hedonic values associated with each property.

Problems with the Traditional Hedonic Specification

In order to derive the most accurate predictions possible, the basic regression model makes use of some strong statistical assumptions to estimate the β coefficients. Namely, it assumes that the residuals of the regression, the ε's, are not autocorrelated across all *n* observations. Formally, this is written as:

$$\varepsilon_i \sim iid(0,\sigma^2) \qquad \forall i = 1,2,...,n$$

which defines the ε_i's to be distributed independently and identically ("i.i.d.") with a mean of zero and a finite variance for all *n* observations.

However, neighborhoods are typically characterized by local homogeneity. That is, the probability that a given home is very similar in both physical characteristics and value to its neighbor is very high. The implication of local homogeneity is that the value of a given property is not completely independent of the values of surrounding properties, where the influence of one property on the value of another declines with the distance between the two properties. Consequently, the measurement error, or ε_i, associated with the model's predicted price for a given property exhibits spatial dependence. Thus, the assumption that the ε_i's are distributed independently is violated, with negative consequences for the correct estimation of the β's.

The consequences of violating this assumption are real. By not accurately controlling for the underlying structure of spatial covariance, the model's estimation of the β coefficients is inefficient. So when the estimated equation is used to make out-of-sample predictions, the predicted house values may be inaccurate.

Another consequence is the failure to fully exploit all available information that is latent within the data. All applied researchers endeavor to make the maximum use of observable information to make current valuations. To ignore relevant information is equivalent to accepting a substandard model. Even if the estimated β coefficients are correct, the model is still under-specified. The result of this error of omission is to have greater variation (wider confidence intervals) of predicted values than would otherwise be the case.

A corollary to this problem of under-specification is the case of missing variables. As a practical matter, not all variables (e.g., proximity to a nuclear power plant) that influence a home's value are recorded by the local jurisdiction. For example, age of the property, the number of bathrooms or a description of its physical condition may not be observable to the researcher. However, the values of these variables are most certainly capitalized into a home's value. As long as a housing stock is locally homogenous (homes in the same neighborhood share similar attributes) then the inclusion of spatial terms indirectly corrects for the problem of omitted variables by capturing the capitalization effects. Since the goal is prediction rather than estimation of particular attributes, then spatial methods are sufficient to the task since they capture the *total* effect of omitted variables rather than their individual contributions (like the β's do). Again, the result is more accurate predictions.

Spatial Information for Improved Valuation Models

To correct for these problems, it is possible to incorporate spatial information and specifically to estimate the covariance in price between properties that are 'near' to each other. For our spatial algorithm, we compute the average price of surrounding properties for different categorical distances, and then enter these values into the model specification. Formally, we estimate the following equation:

$$y_i = \beta_0 + \sum_{j=1}^{k} \beta_j H_{ji} + \sum_{j=1}^{5} \lambda_j V_{ji} + \varepsilon_i \qquad \varepsilon_i \sim iid(0, \sigma^2)\ \forall i = 1, 2, ..., n$$

Where:

n = the total number of properties in the dataset;

y_i = transaction price of the ith property;

H_{ji} = the value of the jth hedonic characteristic for the ith property;

V_{1i} = the average value of all properties within 1/8 mile, for the ith property;

V_{2i} = the average value of all properties beyond 1/8 mile but within 1/4 mile, for the ith property;

V_{3i} = the average value of all properties beyond 1/4 mile but within 1/2 mile, for the ith property;

V_{4i} = the average value of all properties beyond 1/2 mile but within 1 mile, for the ith property;

V_{5i} = the average value of all properties beyond 1 mile but within D miles, for the ith property;

ε_i = the residual for each observation;

The inputs to the model are the y_i, H_{ji}, and V_{ji}, and the parameters of β_j and λ_j are estimated. The effect of introducing these spatial variables into the model's specification is to account for local spatial covariance in property values, and in the process, remove any spatial dependence in the residuals. The five categorical spatial variables are not arbitrary, since the influence of one property on another is declining with distance. Consequently, we estimate five different parameters for each of the five categorical distances:

$$\lambda_j \quad \text{for } j = 1, ..., 5 \qquad \text{with the expectation that: } \lambda_j > \lambda_k \text{ for } j < k$$

Stated informally, we would expect the λ coefficient on the average property value(s) for the shorter distances to have a higher value than the λ coefficient on average property values for longer distances. The net result of this specification may be a model with more predictive power that also simultaneously satisfies the underlying statistical assumptions that govern all regression models.

Identifying the Maximum Distance D

It remains then to identify the maximum distance D for which home values are autocorrelated. Concentric circles are then drawn inside this circle, centered on the subject property. The radius of each of these circles is guided by the shape of the semi-variogram; that is, the rate at which spatial dependence converges to zero as a function of distance. It is also important to distinguish what the shape of the semivariogram is — negative exponential, gaussian or spherical — in order to better determine the model structure and its constraints (Dubin et al., 1999).

Typically, we would expect the radius of each circle to increase with distance from the subject property. So, for example, the first circle would be 1/8 mile radius, the second ¼, then ½, then 1, then 2. For each donut created by these concentric circles, the algorithm computes the average (updated) sales price of all properties in that donut. For example, the average sales price of all homes beyond ¼ mile but within ½ mile. Finally, the algorithm estimates a regression that models each home's current value as a function of the average sales value of surrounding homes, weighted by distance. The coefficients from this regression typically should sum to one, or even slightly less than one. This algorithm is known as ordinary kriging, which "is a minimum mean squared error statistical procedure for spatial prediction"[2] (Dubin et al., 1999). Then, for each subject at time t, the model applies the coefficients to compute that property's value.

Choosing a Final Model

After computing the spatial averages for the five categorical distances, an exact set of hedonic spatial and regionalized variables must be determined prior to input into the final model. Regionalized variable are composed of "drift" (weighted average of points within a neighborhood) and "residual" (difference between the regionalized variables) (Dorsel et al., 2003). Typically, in any given property-level transactions dataset, there is a wide variety of variables to choose from. Many of these variables co-vary with each other and/or are redundant and must be transformed (e.g., instead of gross living area, use price per square foot), so it is undesirable to include every possible hedonic variable in a model. Moreover, it is common to take transformation and/or combinations of the given variables to compute new variables that are better suited to explaining variation in house price, when the relationship between variables is expected to be nonlinear. For example, computing the natural logarithm of building square foot, or taking the difference between average surrounding prices: $V_{ji} - V_{ki}$, where $j \neq k$.

To decide upon a final, parsimonious specification, we utilize a stepwise regression algorithm that estimates the model with different subsets of variables, adding or removing variables at each iteration for a given statistical significance criteria. Use of such an algorithm facilitates convergence to a particular subset of explanatory variables that have the highest predictive power and also comply with underlying statistical requirements. Then, a researcher may further develop this model in accordance with his/her own judgment and knowledge of the underlying theory and empirical evidence that govern this class of property valuation models.

After settling upon a preliminary model, the researcher then computes the predicted price and residual (actual price minus predicted price) for each property. If this particular preliminary model predicts sufficiently well, and there is no evidence of any spatial dependence in the residuals, then a final model has been reached. If not, the researcher may continue to experiment with alternative specifications until a sufficiently "good" model has been developed. The end product is an equation where the β_i's and λ_j's have been estimated, and are now actual numbers. This product can now be used to predict for an out-of-sample property, which has not transacted. This is accomplished by entering the values of the hedonic characteristics (the H_j's), the values of the spatial averages (the V_j's), and then computing the predicted price if this property were to sell today: the y.

With the given dataset of homes that contains each dwelling's sales price, sales date, and locations and attributes, standard statistical software packages can empirically compute the relationship of each attribute to total value. The groundbreaking econometric work in this area was done by Kain & Quigley (1975), who used home sales data from the St. Louis housing market in the 1960s to empirically estimate such a relationship. From their estimation, the regression coefficients then give an implicit price for each attribute.

Adding Spatial Components to the Model

The next step in this process is to characterize the spatial covariance structure of home values. This is done by estimating a semi-variogram, which models the spatial dependence in property values as a function of distance, and also computes a range parameter. The range parameter varies by market and determines at what distance in that market property values are correlated (e.g., two miles). For every sale in the market, the model draws a circle with a radius equal to the market's range parameter.

Finally, the algorithm estimates a regression that adds a vector of variables measuring the average values of nearby properties to the current specification. Such a specification is typically termed a "Spatial Autoregression" (SAR). For stationarity purposes, the coefficients from the spatial terms in this regression typically should sum to one, or even slightly less than one. If this condition holds, regression specification has the intuitive interpretation of simply modeling each home's current value as a weighted average of nearby homes' values, where the weights decline with distance to the subject property.

Testing the Model: Yucca Valley, San Bernadino, CA

The test case regression uses transactions of 3,585 single family home sales in the Yucca Valley area of San Bernardino County during 1999. *Map 1* shows an image of this area of the county, with the transactions symbol-shaded by price. Major roads and earthquake fault lines are also depicted.

The area of Yucca Valley lies along the southern border of the county, approximately halfway between Los Angeles County to the west and the Nevada border to the east. The two other significant urban centers in the country are Riverside MSA and the Hesperia-Apple Valley-Victorville MSA, both of which lie at the major confluences of roads in the western half of the county. Originally founded as a ranching center, the area is now home to the nation's largest Marine Corps base. In addition to becoming a relatively popular retirement area, the city's economy also benefits from tourists visiting nearby Joshua Tree Park and Palm Springs, as well as motorists passing through on their way to Las Vegas.

Yucca Valley was chosen as the subject area for two reasons. First, it is the smallest of the three urban centers, which makes the estimation of an SAR more tractable. Second, the other two MSAs lie within the sphere of influence of the greater Los Angeles metropolitan area. Since L.A. is a highly polycentric metropolitan region with many different employment centers, it is also likely to have a very complicated spatial covariance structure that would be empirically difficult to parameterize. Half of the area's housing stock was constructed prior to 1975.

Map 1. San Bernardino County, California

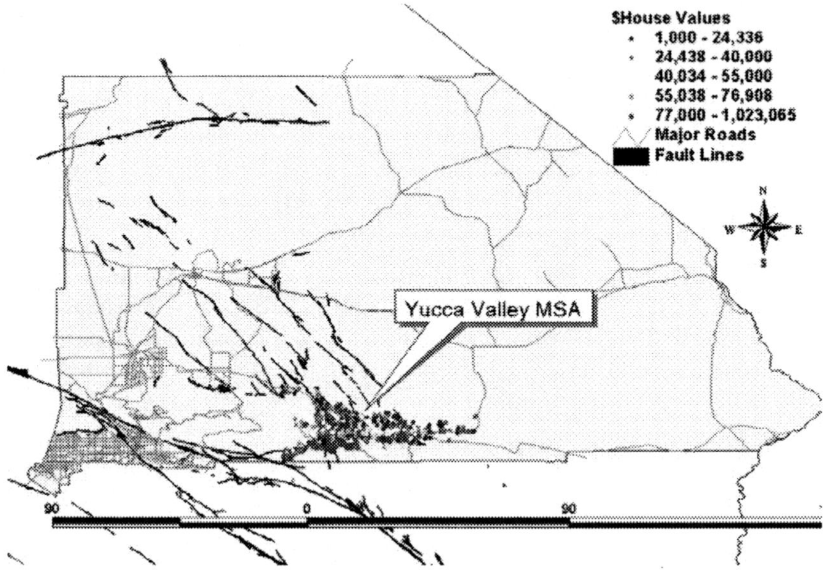

To measure the gains from including spatial terms in the regression specification, *Table 1* compares the results of three different regressions: hedonic variables only, spatial variables only, and both hedonic and spatial variables.

Adding spatial terms to the model clearly improves its predictive power. While the purely hedonic specification has an R^2 of 48.1%, the hybrid hedonic-SAR model achieves an R^2 of 71.1%. Consistent with urban economic theory's predictions, location does indeed affect value. But the implications go beyond this to more practical suggestions. Namely, the spatial covariance structure of home values can be duly exploited to yield better predictions.

Another particularly notable fact is that the purely SAR model outperforms the purely hedonic one: an R^2 of 59.6% v. 48.1%. On the surface, the implications of this are rather astounding. If the only goal is to estimate a model with a high degree of predictive accuracy, then a dataset containing only the sale date, price and location of each property could accomplish this as well as models with considerable attribute data. However, ideally a wider range of attributes and spatial data should be included.

Of course, what this result actually points to is that there is spatial dependence in the hedonic attributes of homes. If your home is identical in size, design and age to your

Table 1. Hedonic v. SAR Regression Results: Yucca Valley, San Bernardino County Estimated Coefficients and (t-scores)

Variable	Specification 1: Hedonic Only	Specification 2: Spatial Only	Specification 3: Hybrid Hedonic-Spatial
Hedonics Variables Included?	Yes	No	Yes
Located in Joshua Tree Township	NA	0.01545 (0.09)	0.01497 (0.10)
Located in Landers Township	NA	0.13228 (0.81)	0.14311 (0.95)
Located in Morongo Valley Township	NA	0.02911 (0.18)	-0.11427 (-0.74)
Located in Twentynine Palms Township	NA	-0.15689 (-0.84)	-0.02911 (-0.17)
Located in Yucca Valley Township	NA	0.04325 (0.26)	0.02129 (0.14)
Log of Distance to Central Business District	NA	0.02984 (2.03)	0.01081 (0.78)
Log of Distance to Central Business District Squared	NA	-0.00207 (-0.22)	0.00371 (0.42)
Longitude	NA	0.48643 (2.35)	-0.06164 (-0.31)
Latitude	NA	-0.21872 (-0.83)	-0.81057 (-3.10)
Log of Average Price of Homes within ¼ mile	NA	0.62908 (30.98)	0.44096 (20.80)
Log of Average Price of Homes within ½ mile	NA	0.08851 (3.87)	0.0525 (2.43)
Log of Average Price of Homes within 1 mile	NA	0.03507 (1.37)	0.04856 (2.04)
Log of Distance to Fault Line	NA	-0.02034 (-2.49)	-0.00883 (-1.15)
Log of Distance to Fault Line Squared	NA	-0.00174 (-0.86)	-0.00041 (-0.22)
Adj. R^2	0.4814	0.5957	0.7111
F-Statistic	67.79	92.93	64.74
Durbin-Watson Statistic	1.740	1.844	1.941
Rho coefficient	0.130	0.078	0.030

neighbor's home, then the two values will be very similar. More generally, a market with a relatively homogenous housing stock (that turns over with a sufficient degree of liquidity) will generally yield models with higher predictive accuracy. In this sense, the change in the R^2 from 48.1% to 71.1% also yields insight into the magnitude of the stock's (local) homogeneity and level of liquidity. The increase is significant (although there are cases where this increase can be even larger). This may be due to the Yucca Valley markets' housing stock, which is highly variable and/or relatively illiquid. Indeed, the relative predictive accuracy of AVMs for areas that are older, more heterogeneous and less urban such as this one tends to be poor.

The coefficients on the spatial terms give insights into the market's geographic structure. While many of the coefficients may not be statistically significant for any reasonably low level, this is likely due to considerable multicollinearity amongst these variables. This multicollinearity represents a noteworthy weakness in this regression analysis. Future research could analyze this multicollinearity in greater depth in order to possibly eliminate some correlated variables. First, properties in the municipalities of Morongo Valley and Twenty-Nine Palms have discounts accorded to their value, relative to the other townships[3]. Since the quality of the housing stock has already been controlled for, this discount suggests that mixture of taxes and public services may be undesirable relative to the packages offered by other townships.

The coefficients on the variables measuring the distance to the CBD are not only positive, but convex as well. That is, all else being equal, home values not only become higher as you move away from downtown Yucca Valley, but the values increase at an increasing rate. This result may initially seem puzzling since, in the market, centrally located land is more valuable than peripheral land. Perhaps other omitted factors are likely at work here. For example, the downtown area of Yucca Valley may suffer from excessive traffic congestion and/or urban blight, thus making fringe areas relatively more attractive to households. If this is the case, then such variables measuring these factors must be added to the specification.

Another puzzling result is that the coefficients of the variables measuring the distance to the nearest fault line are both negative and statistically insignificant, but they were positive and significant when the regression was estimated for the entire county. They may be insignificant due to multicollinearity. If distance to a fault line is actually not significant, then it may be that the earthquake faults are not as active as the ones in other parts of San Bernardino, that the housing stock in Yucca Valley is more resilient to a seismic event, or that homeowners are not aware of the true threat and thus fail to capitalize the expected loss severity into the price of a new home. A simpler possibility could be that there is insufficient variation in the independent variable to identify the true relationship between distance and home value. Examining the previous map, we can see that this may very well be the case. Since several fault lines run through Yucca Valley, the distance to the "nearest" fault line may be relatively constant across the sample of homes. In the alternative case that distance to a fault line is significant, but negative, then the variable is likely proxying for something else. Namely, it may very well be (for whatever reason) that more expensive homes tend to be located nearer to fault lines. While the results suggest possible relationships, further research should attempt to discern the true dynamics.

A good case in point where spatial variables may actually be measuring the effects of omitted variables is latitude and longitude. In the western hemisphere, longitude is increasing from west to east, and latitude is likewise increasing from south to north. So in San Bernardino, as the values of both variables increase, you are moving from the southwest corner of the county to the northeast corner. The southwest corner of the county is the part that is closest to Los Angeles, and the northeast corner is mostly unpopulated desert. Since both these variables have negative coefficients, they are almost certainly capturing how distance to the Los Angeles metro area is negatively capitalized when all else is held constant.

The final spatial terms are the autoregressive variables measuring average home prices at distances of ¼, ½ and 1 mile. Across both the pure SAR and hybrid hedonic-SAR specifications they are typically significant and similar in value. The value of 0.44096 for the coefficient on Avg_1_4 suggests that, in a controlled setting, a doubling of the housing stock's value within ¼ mile of a home will cause its value to increase by 44%. This declines sharply to 5.25% for homes beyond a ¼ mile but within ½ mile, and then to 4.9% for homes beyond ½ mile but within one mile.

Further research could examine for the robustness of the SAR specification. Adding variables for additional distances (e.g., one to two miles) may improve further boost the R^2. And varying the categorization of the distance rings might also improve the regression; for example, replacing the three SAR terms with variables measuring average values at < 1/8 mile, 1/8-1/4 miles, 1/4-1/2 miles, 1/2-3/4 miles, and so forth. The actual structure of spatial dependence and covariance is difficult to observe directly, and often requires several attempts before arriving at the optimal specification of measurement.

Measure and Compare the Predictive Accuracy of the Models

The final step in the analysis is to explicitly measure and compare the relative predictive accuracy of both estimations. *Table 2* compares some summary statistics on average prediction error, defined as the absolute percent difference between predicted and observed value, for both the pure hedonic and hybrid hedonic-SAR specifications.

For the pure hedonic, the median prediction error across all observations is 16.7%. The interpretation is that for half of all homes in the sample, the predicted price "misses" the target of the actual price by 16.7% or less. In this same vein, 75% of all predictions have an error of 29.8% or less, and almost all predictions are within 86.6% of the actual price.

For the hybrid specification, prediction errors drop uniformly with greatest gain in accuracy in the tails of the distribution. Compared to the pure hedonic, the maximum prediction error declines by nearly 40%. In words, the distribution of prediction errors remains relatively centered around its median from the previous specification, but the tails contract inward dramatically, with a commensurate decrease in the standard deviation of prediction errors.

Table 2. Prediction Errors of Regressions

	Pure Hedonic	Hybrid Hedonic-SAR	Pct. Change
Median (50% Quartile)	16.7%	14.8%	**-11.4%**
75% Quartile	29.8%	25.3%	**-14.9%**
90% Decile	46.0%	36.7%	**-20.2%**
99% Decile	86.6%	52.3%	**-39.6%**

Having examined gains to the overall magnitude of prediction errors, it is also worth examining if prediction errors vary systematically across home values. That is, does the addition of spatial terms to the model also decrease any bias in the direction of prediction errors? To answer this question, we plot predicted values against observed values for both specifications in *Figures 1* and *2*. We also plot average error in *Figures 3, 4* and *5*.

What is desirable is that predictions be symmetrically distributed around the 45-degree line. But for low-priced homes, the hybrid model is biased upwards (predicted>actual), and vice-versa for high-priced homes (predicted<actual) (see *Figure 1*). This implies that the basic hedonic model is biased upwards for low-priced homes and biased downwards for high-priced homes.

Since only hedonic attributes serve as the independent variables in the specification, this result is not all that surprising. Very low-priced homes are often priced as such not just because they may be smaller and older, but also because they have serious structural problems (e.g., collapsing roof, crumbling foundations, water/fire damage) that are not reported in the data. In addition, these homes are often located in distressed neighborhoods where the quality of life is low and the delivery of public services is inferior. Conversely, very high-priced homes are not only larger and newer, but have many superior physical attributes (e.g., hardwood floors, detailed tilework/woodwork, skylights, nice views, etc.) that are also not reported in the data. Since there are no variables measuring these qualitative characteristics in the specification, the model is economically "regressive:" biased upwards for low-price dwellings and biased downwards for high-priced dwellings.

But the hybrid specification appears to considerably reduce this regressive bias. For low-priced homes (<$40,000), about half of all predictions appear equally distributed on both sides of the lines. For high-priced dwellings the downward bias still persists, but the magnitude of the bias has been reduced when compared to the pure hedonic. More predictions appear above the 45-degree line (over-predictions) than did previously, and the under-predictions appear to be reduced in magnitude and closer to the line. While a bias problem may still persist, the hybrid specification would certainly seem to make substantial improvements in the right direction (see *Figure 2*).

As shown in *Figures 3, 4* and *5*, average percent errors from the hybrid hedonic-spatial model are significantly lower in all price ranges than the errors from the hedonic model;

Figure 1. Predicted vs. Actual Home Values from Hedonic Model (Yucca Valley California, 1999 Sales Only)

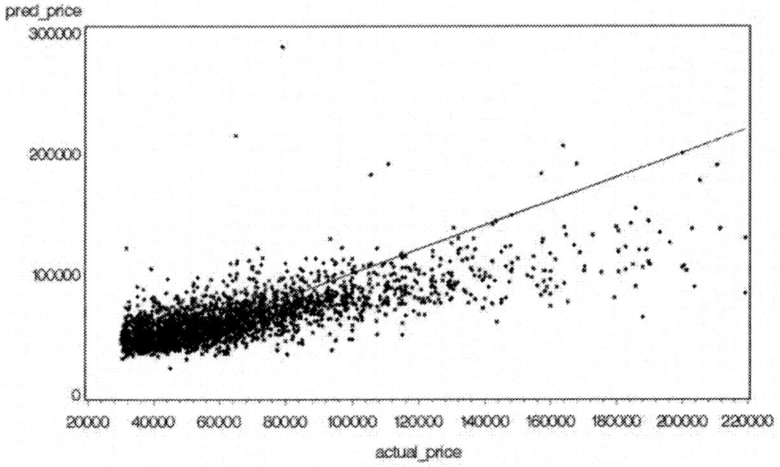

Figure 2. Predicted vs. Actual Home Values from Hybrid Hedonic — Spatial Model (Yucca Valley California, 1999 Sales Only)

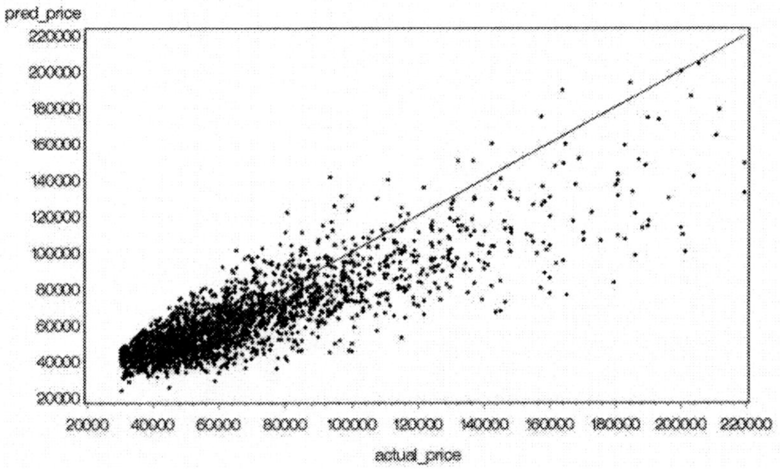

moreover, the spatial model alone performs better than the hedonic model alone. However, while smaller in absolute value, the errors still have the same signs across both models. Percent errors are positive for low-priced homes, while negative for high-priced homes. This indicates that a bias problem persists in the Hybrid Hedonic-Spatial Model, even if its effects have been reduced.

Figure 3. Average Percent Error by Price Range: Pure Hedonic

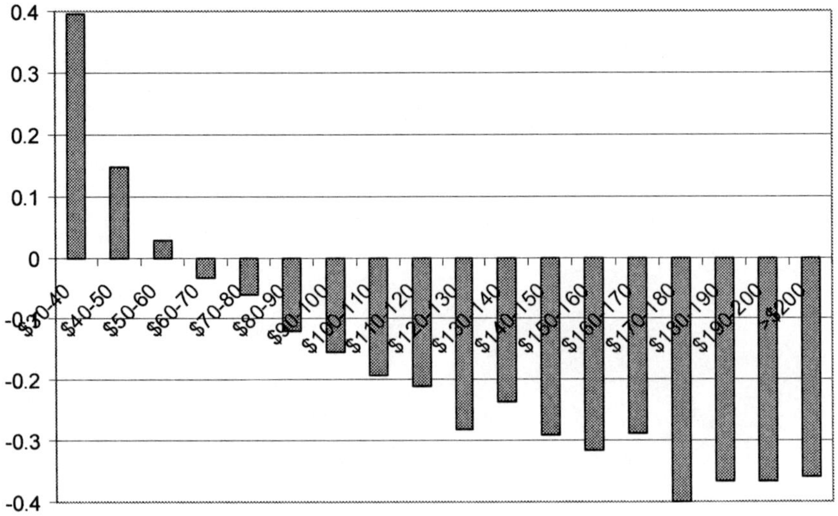

Figure 4. Average Percent Error by Price Range: Pure Spatial

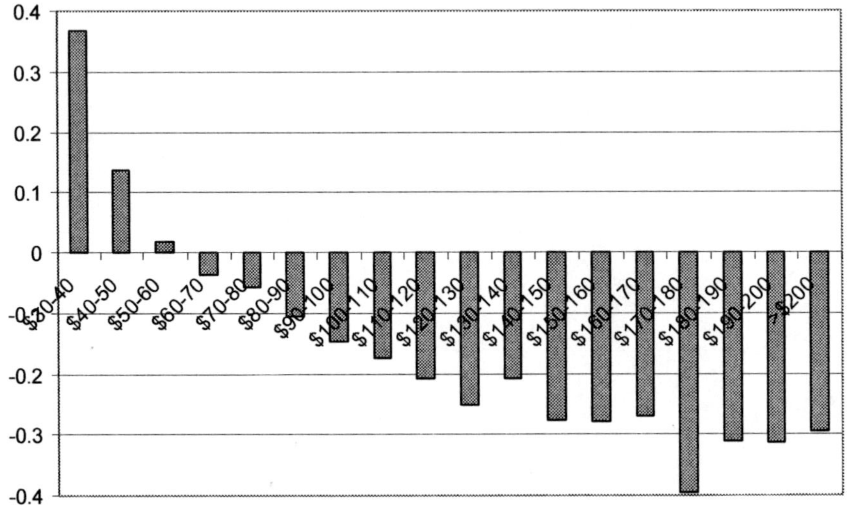

Figure 5. Average Percent Error by Price Range: Hybrid Hedonic-SAR

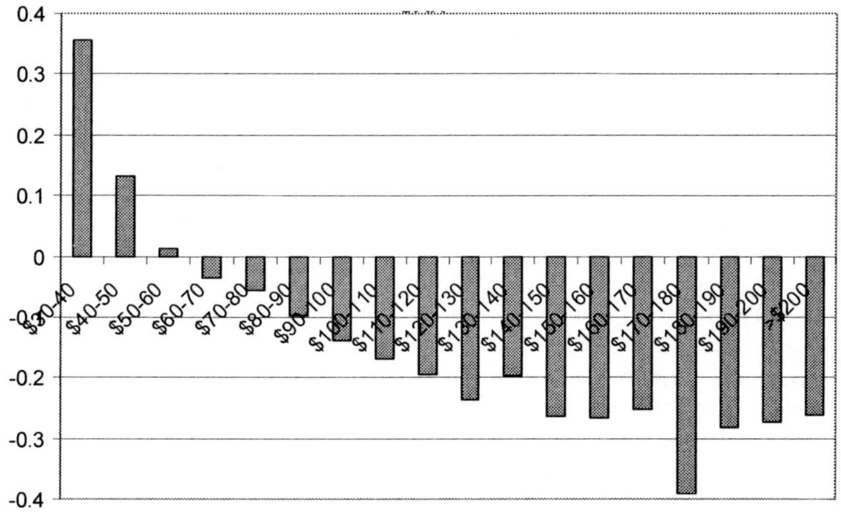

Conclusions

GIS has recently become one of the fastest growing mechanisms for data dissemination, retrieval and analysis. However, GIS — as with CAMA and AVM — has only partially entered business protocols, despite its power. In order for the usage of GIS to advance, education of mid to upper-level business managers is necessary. "The development of powerful personal computers, coupled with easy-to-use GIS products and widely available data, has created a new breed of GIS practitioner: the business professional" (Harder, 1997). However this professional may not yet be fully prepared for their role. A survey of courses and professors in business suggests that there still is limited discussion and development of case studies or hands-on courses using GIS technology. An integrated teaching model may well come out of the emerging geographical information science or GiScience (GISc) (Fotheringham et al., 2000) field but the concept and implementation are not yet developed.

While students in marketing, planning, geography, sociology and architecture may find access to GIS practice and protocol limited on campus, there is a slow but steady rise in their use for real estate investment decisions nationwide. Municipalities with a mandate to conduct accurate tax assessments, corporate offices with real estate departments that cannot solely rely on market analysis models for business expansion, and retailers interested in appropriate site planning are beginning to incorporate GIS.

Administrative challenges that face private companies and and municipalities are the same — time, money and capacity. In the recent past, data access was one of the main impediments to having a fully developed GIS. However, with the advent of data streams, including the Internet, the ability for managers to dedicate *time* for GIS integration, having the *money* to obtain state-of-the-art hard/software and a funding stream for continued use leads to an expanding issue of *capacity*. The ability to obtain, and sustain, a staff that will be able to perform routine to advanced valuation analyses is not simple. The skills of key business personnel who may have the managerial know-how may not also have the skill-set for technical or advanced spatial analyses. While public managers face this problem their private counterparts often find more favorable solutions due to financial incentives. The public manager, therefore, must find alternative means of breaking through the barriers that will keep the use of technology at the forefront although there remains multi-levels of resistance to the technology.

GIS opens the door to new ways of thinking and problem solving using spatial information. Practitioners consider GIS a "particularly appealing technology" because it provides business leaders with a multi-faceted product which is "visually oriented (which provides) compelling presentation of information and complex relationships," "facilitate(s) processing more information and implementing analysis more rapidly than can alternative, more conventional approaches," "can enrich the content of information presentations (to an audience which has a) proliferation of information readily available on the internet," and "(can provide their clients with) entertainment encroaching upon the traditional realms of communication" (Roulac, 1998).

Within the last two years, there has been a significant merging of GIS with tools for conducting spatial analysis. "GIS and spatial analysis have a longstanding association, and spatial analysis has often been seen as the ultimate objective of representing space in a GIS" (Goodchild, 2000). While many would consider the identification of properties using raster or vector analysis as "spatial analysis," this term now includes the means by which econometric model functions are fully integrated in the software. Such integration is still lacking for valuation technology; thus, valuation systems, which use hedonic-spatial-GIS techniques, are still in their infancy as is the ability to capture and analyze spatial information for improved valuation outcomes. Nonetheless, based on technology that is available today, it is clear that the combination of contemporary statistical hedonic and repeat sales models and emerging GIS technology has the potential to significantly increase the reliability and accuracy of valuation estimates.

References

Anselin, L. (1998). GIS research infrastructure for spatial analysis of real estate markets. *Journal of Housing Research, 9* (1), 113-133.

Appraisal Standards Board. (2003). *Uniform standards of professional appraisal practice*. The Appraisal Foundation.

Appraiser Central. (2003). *The BIG automated valuation model (AVM) lie.* Retrieved from: *http://www.appaisercentral.com/AVMs.htm.*

Belsky, E., Ayse, C., & Megbolugbe, I. (1998). A primer on geographic information systems in mortgage finance. *Journal of Housing Research, 9* (1), 5-31.

Bourassa, S.C., & Vincent, S.P. (1999). Hedonic prices and house numbers: The influence of feng shui. *International Real Estate Review, 2* (1), 79-93.

Bourassa, S.C., Hoesli, M., & Vincent, S.P. (2002). *Do housing submarkets really matter?* Manuscript.

Bureau of Labor Statistics. (1997). *Changes in calculating the consumer price indexes, the economic and budget outlook: An update.* Washington, D.C.: Bureau of Labor Statistics.

Can, A. (1998). GIS and spatial analysis of housing and mortgage markets. *Journal of Housing Research, 9* (1).

Case, B., Pollakowski, H., & Wachter, S. (1997). Frequency of transaction and house price modeling. *Journal of Real Estate Finance and Economics, 14*, 173-188.

Case, B., Pollakowski, H., & Wachter, S. (1991). On choosing among house price index methodologies. *American Real Estate and Urban Economics Association Journal, 19* (3), 287-307.

Castle, G. (ed.). (1998). *GIS in real estate: Integrating, analyzing, and presenting locational information.* Appraisal Institute in association with Adams Business Media/GIS World.

DiPasquale, D., & Wheaton, W.C. (1996). *Urban economics and real estate markets.* Prentice-Hall.

Dorsel, L. (2003). Environmental sampling and monitoring primer. Retrieved from the World Wide Web: *http://www.cee.vt.edu/program_areas/environmental/teach/smprimer/kriging/kriging.html.*

Dubin, R.A. (1992). Spatial autocorrelation and neighborhood quality. *Regional Science and Urban Economics, 22*, 433-452.

Ekeland, I., Heckman, J.J., & Nesheim, L. (2001). *Identifying hedonic models* (ceemmap working paper, CWP06/02). Centre for Microdata Methods and Practice, The Institute for Fiscal Studies Department of Economics, UCL.

Fotheringham, S.A., Brunsdon, C., & Charlton, M. (2000). *Quantitative geography: Perspectives on spatial data analysis.* Sage Publications.

Fotheringham, S.A., Brunsdon, C., & Charlton, M. (2002). *Geographically weighted regression: The analysis of spatially varying relationships.* New York: John Wiley & Sons.

Geho, M.L. (2003). *Prospects of applying computer aided mass valuation in Tanzania.* Computer Assisted Mass Valuation and Cost Management (TS21), FIG Working Week.

German, J. C., Robinson, D., & Youngman, J. (2000). Traditional methods and new approaches to land valuation. *Land Lines, 12* (4). Retrieved from the World Wide Web: *http://www.lincolninst.edu/pubs/pub-detail.*

Gillen, K., Thibodeau, T., & Wachter, S. (2001). Anisotropic autocorrelation in house prices. *Journal of Real Estate Finance and Economics, 23* (1).

Goodchild, M. (2000). Spatial analysis: Methods and problems in land use management. In M. J. Hill & R. J. Aspinall (Eds.), *Spatial Information for Land Use Management* (pp. 39-51). Gordon and Breach Science Publishers.

Graham, S. J. (1997). Products liability in GIS: Present complexions and future directions. *GIS Law, 4*(1), 12-16.

Harder, C. (1997). *ArcView GIS means business.* Environmental Systems Research Institute, Inc.

Houthhaker, H.S. (1952). Compensated changes in quantities and qualities consumed. *Review of Economic Studies, 19* (3), 155-164.

International Association of Assessing Officers. (n.d.). *Property appraisal and assessment administration. Mass Appraisal, 13*, 311.

International Association of Assessing Officers. (2003). *Member exposure draft: Standard on automated valuation models (AVMs)*. Retrieved from: *http://www.iaao.org/publications/avm_draft.htm.*

Jargowsky, P.A. (2003). *Stunning progress, hidden problems: The dramatic decline of concentrated poverty in the 1990s.* The Living Cities Census Series. The Brookings Institution, Center on Urban and Metropolitan Policy.

Kain, J.F., & Quigley, J.M. (1975). *Housing markets and racial discrimination.* Columbia University Press.

Kane, M.S., Linne, M., & Johnson, J. (2003). *Practical applications in appraisal valuation modeling: Statistical methods for real estate practitioners.* Rocky Mountain Valuation Specialists (draft; publication date via Appraisal Institute, November).

Klosterman, R.E. (1986). Educational strategies for land systems technology. In A.H.S. Lam & D.W. Robinson (Eds.), *Land Systems Technology: Background Papers Prepared for the Lincoln Institute of Land Policy Seminar on Land Systems Technology* (p. 7).

Le, L.H. (2003). *A security model for ArcIMS.* ArcUser: The Magazine for ESRI Software Users. Retrieved from: *http://www.esri.com/news/arcuser/0703/security1of2.html.*

Lee, C.M., Culhane, D.P., & Wachter, S. (1998). The differential impacts of federally assisted housing programs on nearby property values: A Philadelphia case study. *Housing Policy Debate, 10* (1), 75-93.

Lee, J., & Wong, D.W.S. (2001). *Statistical analysis with ArcView GIS.* New York: John Wiley & Sons.

Lopez, X. (1998). *The dissemination of spatial data: A North American – European comparative study on the impact of government information policy.* Ablex Publishing Corporation.

Malpezzi, S. (2002). Hedonic pricing models: A selective and applied review. In K. Gibb & A. O'Sullivan (Eds.), *Housing Economics: Essays in Honor of Duncan Maclennan.*

Mills, A.C. (1998). *The uniform residential appraisal report: Communicating the appraisal.* Appraisal Institute.

Overmyer, S.P. (n.d.). *The use of geographic information systems (GIS) in business education*. Retrieved from: *http://www.spatial.maine.edu/ucgis/testproc/overmyer/overmyer.html.*

Pace, K.R., Barry, R., & Sirmans, C.F. (1998). Spatial statistics and real estate. *Journal of Real Estate Finance and Economics, 17* (1), 5-13. Retrieved from: *http://finance.lsu.edu/academics/financ…scripts/web_sintro/htmel/web_sintro2.htm.*

Pace, K. R., Sirmans, C.F., & Slawson, Jr., V.C. (2002). Are appraisers atatisticians? In K. Wang & M. L. Wolverton (Eds.), *Real Estate Valuation Theory, Appraisal Institute and American Real Estate Society* (ARES). Kluwer Academic Publishers.

Rosen, S. (1974). Hedonic prices and implicit markets: Product differentiation in pure competition. *Journal of Political Economy, 82*, 34-55.

Roulac, S.E. (1998). Integrating, analyzing and presenting locational information. In G. Castle (Ed.), *GIS in Real Estate.*

Somers, R. (1998). Developing GIS management strategies for an organization. *Journal of Housing Research, 9* (1), 157-178.

Thrall, G.I. (1998). GIS applications in real estate and related industries. *Journal of Housing Research, 9* (1), 33-59.

Tse, R.Y.C. (2002). Estimating neighborhood effects in house prices: Towards a new hedonic model approach. *Urban Studies, 29* (7), 1165-1180.

Vogy, P.W. (1993). *Dictionary of statistics and methodology: A nontechnical guide for the social sciences.* Sage Publications.

Endnotes

1 The importance of considering spatial relationships in valuation models can be explained by location theory in the field of urban economics. The many individual parcels in a given metropolitan region produce a product- differentiated market where no two parcels, no matter how similar, are identical. Real estate may be similar in physical attributes, but it retains the unique quality of location. Therefore, house prices are dependent on households' valuation of related locational attributes, such as access to services, commuting time, distance to a CBD, and neighborhood quality. Households may seek out three-bedroom, two-story, brick homes, but apart from charm and rare structural attributes, the thing that differentiates the houses is location. Different locations provide different amounts of utility to each household based on their preferences, but some measures of neighborhood quality can appeal to consumers overall, such as a high-quality school district or low crime. GIS can identify and utilize these general locational attributes to measure a large portion of a parcel's value (DiPasquale & Wheaton, 1996).

2 The current hedonic model can estimate the value of unknown properties by using an iterative process (known as kriging) to estimate values on a smoothed surface.

Kriging "attempts to exactly interpolate.... and fits perfectly (0 error) all the points in the sample" (Pace et al., 1998). In this way, more information can be obtained about the relationship of known to predicted values by estimating the residuals.

In essence, kriging is simply the spatial version of generalized (or weighted) least squares. In the kriging process, the spatial covariance structure of the residuals is iteratively explicitly estimated and then applied to minimize spatial dependence (to zero if possible) to attain unbiased and efficient parameter estimates of a spatial stochastic process. If done correctly, the final model should have minimal, if not zero, spatial dependence in the residuals. This is difficult to implement. First, this is an iterative process that requires the researcher's active judgment and tuning at every step. Second, the spatial covariance matrix has dimensions of N-by-N, where N is the number of observations in your dataset. When this matrix is used in the application of GLS, it must be inverted. Computers cannot do this quickly if N is large, and often cannot do it at all.

For example, Dubin postulates that residuals are expected to be correlated across space: two houses which are near to each other should have similar neighborhood effects, and correlated residuals. As the distance between the houses increases, the likelihood that that house will be in the same neighborhood falls and the correlation declines (Dubin et al., 1999).

[3] The omitted category is homes in rural areas that do not lie in incorporated townships. Hence, most of their public services are delivered by the state or county.

Chapter XIV

Monitoring and Analysis of Power Line Failures: An Example of the Role of GIS

Oliver Fritz, ABB Switzerland Ltd., Switzerland

Petter Skerfving, ABB Switzerland Ltd., Switzerland

Abstract

This chapter presents a case study on the role of geographical information systems in power utility automation. It argues that geographical information systems are ideal software components for enterprise-wide application integration, data and asset management, and decision support applications. The authors believe that advances in industrial software applications are predominantly realized through stepwise integration and extension of existing systems. A case study on a monitoring application for power lines serves as example for this thesis. Furthermore, the authors hope that understanding technical and economic conditions governing the field of industrial automation helps readers in appreciating that software development and integration strategies are strongly linked to organizational and structural changes in the power utility industry.

Introduction

The supply of electricity to almost any inhabited part of our planet has been one of the cornerstones of social development in the late 19th and 20th centuries; its continuous extension and modernization still present a major challenge. Power utilities, i.e., the companies responsible for various parts of the process of generation, transmission and distribution of electricity, are showcases for political, economical, and technological successes and failures of our industrial history.

The first part of this chapter sheds light on the current state of automation in the utility industry. It presents a number of processes and applications and puts them into a structural and functional context. Current trends and ways to integrate key applications from different organizational units of typical power utilities and possible future strategies to come to more enterprise-wide approaches towards automation are discussed in the context of possible changes to the overall structure of the power distribution sector.

The second part of the chapter is devoted to a *case study* on the use of a geographical information system[1] as a support to network monitoring and surveillance. Although a rather focused application is described, the goal remains to exemplify and concretize a number of statements made in the first part.

The main thesis of this chapter is the view that GISs, through their *functionality* and *data management abilities*, are key components for *software integration strategies* in industrial automation. Monitoring and surveillance applications, such as the one that serves as an example to this chapter, are typically part of industrial processes carrying a potential for increasing efficiency through integration. However, such processes and tasks are often far from trivial and powerful software components are needed in order to achieve the required performance. GIS that provides functionality for such tasks will play an important role in this context, and ideas on the future of GIS within industrial automation in general are presented and related to the case study.

Although there is no "typical" utility, a look at common processes and activities at least allows establishment of a reasonably general model of the utility industry. References to "utilities" throughout this chapter should be understood within this context.

The chapter is of technical nature, although a number of business-related issues are dealt with. It is written from the perspective of a research and development center of a globally operating company offering products, systems, software and services for the automation and optimization of industrial and commercial operations.

Automation and IT in Power Utilities

The full treatment of the topic of automation and IT in the power utility industry goes obviously beyond the scope of this chapter. As it is a main goal of this chapter to show the importance of software integration as a part of a general IT strategy for utilities, an introduction to the topic is essential.

Automation can be roughly defined as the practice of reducing the necessity for human action or interaction within a given process. Apart from this broad definition, the concrete understanding of the role of automation in different industrial sectors varies a lot. Moreover, the actual levels of automation vary as much as their perception.

Automation is based on technologies, amongst which communication and information technologies are certainly most important ones. Traditionally, automation is still widely attributed to processes involving complicated workflows, sophisticated control algorithms, large amounts of data, or demanding constraints on production time. This leads to the view that progresses in highly elaborate technology and the resulting opportunities for automation are main drivers for more efficient processes.

From a business perspective, however, a rather lower-level form of automation has proven to become increasingly important. It can be called *decision support* automation. It often merely consists in making real-time information available to a person having to make a decision. Typical examples for this kind of automation can be found in business-to-business supply chain applications, where the relatively quick and accurate presentation of various offers or specifications is regarded as a helpful support tool with high business impact. As such, a decision support application automates the process of gathering and presenting the necessary knowledge to enable a person or group to take an informed decision.

Awareness of problems attributed to technological progress in automation is increasing, not only in a wider public, but also in industries and governments. Vulnerability of the infrastructure, security, issues around health, and safety have become important for customers of companies providing automation products and solutions. Some of these issues, e.g., the security of communication networks against intrusion, have not yet been resolved in a way that is acceptable for industries with high standards for reliability and availability of services to their customers. Power utilities certainly belong to this sector of the industry.

Historical development also has its influence on the degree of automation. Most utilities are organized in traditional structures. A large part of the current companies are or were until recently publicly owned or stem at least from strongly regulated market conditions. Although in recent years a political will to privatize and deregulate the utility market has been present, the infrastructure necessary for a safe operation of electricity generation, transmission, and distribution may be considered as a natural monopoly, and therefore public bodies will not withdraw in a foreseeable future from a very active role in this industry.

One of the consequences of these market conditions is the presence of a strong *vertical integration*. This means that many utilities provide a series of operations and services that, from a pure market philosophy, should not necessarily be offered by the same company, for example, companies operating, at the same time, large-scale generation, transmission, and trading of electricity, or providing distribution-network operation and end-customer oriented services. It is not the aim of this chapter to discuss the pros and cons of this situation. The actual processes found and IT systems used within utilities are, however, a consequence of the historical market conditions and the resulting organizational structures.

In a comparative study about the year-2000 bug, Jennex (2003) found that almost all modern communication and control systems in utilities rely on the use of microprocessors. They are found in hardware, where they replaced electro-mechanical devices, as well as in business related processes where they substituted manual or semi-automatic systems. In power generation plants, digital control systems control fuel and water flows, turbine and generator speeds, cooling systems, boilers, emission, and much more. Control centers use complex algorithms to run distribution and transmission networks and to satisfy the balance of supply and demand. Digital protection devices are continuously replacing older analog systems. At the enterprise level, software systems for customer management, billing, metering, and performance tracking, e.g., have made their way into the business infrastructure of utilities.

It is helpful to describe the functions of information and communication technologies in utilities in two ways, both from an operational and process perspective. The following section will order the most important software systems by their area of use, while the section subsequent will focus on software-supported processes.

Software Systems

Levels of Operation

Utilities typically operate a large number of widely dispersed assets. These installations collect, communicate and deliver a large amount of data. A variety of different, mostly proprietary software systems are used for this purpose. The bases for such systems are data-acquisition and control platforms (called SCADA, Supervisory Control And Data Acquisition) that come in a wide variety of sizes and scales. They represent the backbone of the communication structure delivering data and control signals from and to operation centers and local control and monitoring equipment.

A utility's assets can be grouped together in a number of *levels* that are distinguished according to their functionalities. An example of a functional entity is a substation where power lines of different voltage-levels are electrically connected through transformers and other equipment. At the substation level, automation implements protection applications, alarm handling and remote localization of errors. Substation automation comprises computers for local control, gateways for remote control from the network control centers, and logging printers. Intelligent electronic devices for protection and control are installed on the so-called *bay level* within a substation. The traditional terms *field level* or *local level* can sometimes be found in literature, they are used for the substation level or levels below it.

Large and costly Distribution Management and Energy Management Systems (DMS, EMS) govern operations at the network level. They may or may not draw on real-time information delivered from the substation level. Decisions on where to route the flow of electricity, when to adjust voltage or power levels at certain points according to the current demand are made with the support of such applications. Depending on the size of a utility and the market regulations, DMS and EMS can be extended by load-forecasting functionality and advances scheduling software for accessing power exchanges and transmission lines operated by third party companies or independent network operators.

Table 1. Components of Utility Automation (selection) Ordered by Level

<table>
<tr><td>Enterprise level</td><td colspan="2">Operation</td><td colspan="2">Maintenance</td></tr>
<tr><td>Network level</td><td>SCADA
EMS, DMS</td><td colspan="2">Wide Area Protection</td><td>Power System Monitoring</td></tr>
<tr><td>Station level</td><td colspan="2">Substation Automation</td><td colspan="2">Substation Monitoring</td></tr>
<tr><td>Bay level</td><td>Bay control
Load shedding</td><td colspan="2">Current-voltage measurements</td><td>Station Protection
Disturbance recording</td></tr>
</table>

On the left-hand side, operational control signals flow predominantly from the higher levels to the lower ones, whereas on the right-hand side, measurements flow back to the higher levels for monitoring and maintenance.

Demand and price forecasting do fall into the enterprise level, along with functionality like risk management, financial controlling, resource management, customer relationship management, or business-to-business and e-commerce applications. Software on this level has increasingly made the way into most commercial enterprises, independent of their specific sector. Some of these applications have enabled more innovative distribution utilities to offer a selection of new services to their customers. The chapter sections on asset management and system integration will discuss a number of possibilities to increase the use of the more resource-oriented enterprise-level applications through a stronger interconnection with the traditional industrial software systems used on network and field level.

Monitoring Applications

In order to prepare for the GIS case study presented later in the chapter we look at monitoring applications in more detail. Such applications can be roughly divided into two groups: Those used for continuous, planned, or scheduled actions and those specific for dealing with unexpected events. In this sense, monitoring applications have their main role in identifying faults, errors, and interruptions of the usual service, even if they are running constantly and provide the operations center with other useful data. They rely on data collected at field level, i.e., in protection and metering devices installed near the assets to be monitored, as well as on certain low-level analytical functions that identify an error based on the data collected.

The quality and level of detail of the information sent to a control center depend very much on the communication infrastructure available. High-performance communication structures (dedicated lines, wide-area networks, GSM or other wireless technologies) are hardly ever found in installations nowadays. The presence of a modem that calls the center once every two hours in order to transmit data is still regarded as an advanced solution. In spite of all predictions in recent years, Internet or mobile-telephony derived technologies have not quickly penetrated the market of utility communication systems.

The main workflow of many monitoring applications is based on events, often called alarms if signaling an undesired state. An alarm will trigger a certain sequence of actions, like notifying a repair crew or opening a circuit breaker. Some of these actions may happen automatically, others require human interaction. In order to follow the sequence of actions triggered by the alarm, state indications (e.g., new, acquitted, resolved) are attributed to alarms. The more information is available about an alarm, the narrower is the choice of possible decisions for an operator and the quicker a necessary decision can be made. In this sense, a monitoring application may be regarded as automation of decision support aiming to reduce costs occurring from non-availability of services.

Monitoring applications are also used for detecting the state of a device in a more continuous fashion. Data is collected in regular intervals and stored in suitable formats for analysis and post-processing. Such applications may enable a medium- or long-term asset management strategy, considering age and wear of certain assets. For continuous monitoring, requirements on the frequency of data transmission to a control center, particularly the amount of data used to describe the state of an asset, may be higher than for pure event or alarm monitoring. In addition, data must be stored for post-processing and analysis.

Software-Supported Processes

Power Availability Management

The main goal of any utility is to satisfy its customers with enough electricity of required quality in the most economic way. As electricity is for all practical purposes a non-storable good, this task cannot be fulfilled without an ongoing planning process that constantly ensures the availability of electricity. Moreover, this planning process has to take place on several time-scales: Strategic planning over years, medium-term planning over weeks to months, and finally the immediate planning process on halves or quarters of hours. It is clear that the information necessary and the decisions available are different for each of these time-scales. While on a long-term scale, new lines may have to be erected or new long-term delivery contracts closed, the monthly planning is more related to weather conditions or primary-fuel prices. The short-term planning process basically has to make sure that there are enough reserves available for satisfying the actual demand at the time of electricity.

These tasks primarily need forward-looking information. Major applications support the processes of generation and transmission scheduling, electricity trading, price forecasting and dispatching. These applications need customization for each individual company; they typically have to take into account individual capacities, marginal-cost structures, and redundancy policies dealing with reserve management and non-availability risks.

Maintenance Management

In order to perform maintenance on components of a power network, planned outages have to be scheduled. Those outages, however, should not affect the availability of electricity. The task of scheduling such maintenance intervals in an optimal way depends strongly on the structure of the network.

In principle, the electricity demand during a planned outage can be predicted from experience with sufficient accuracy. This requires the presence of applications that store and analyze historical data. This approach will usually lead to more or less regular service intervals without particular regard to possibilities of optimization. Increasing the flexibility of maintenance scheduling, however, is considered of high potential benefit to utilities (Shahidehpour & Marwali, 2000). A more advanced approach would therefore consider real-time data on the state of the assets to be serviced and the current load of the network. In this context, monitoring applications that are traditionally used for emergency or failure detection may become important tools for maintenance management as well.

Strategic Planning and Optimization

In order to perform medium- to long-term investment planning, a clear picture of the overall performance-function to be optimized by such a planning process must be defined. A conservative approach starts from fixed performance levels and well-known lifetimes of assets. Installations are used with reliable safety-margins; assets are renewed according to more or less fixed schedules. Planning takes place without much optimization. Only at the time of an actual investment decision, requirements and related costs are brought together and evaluations according to investment and risk policies are made.

More advanced possibilities are offered through a constant monitoring of the state and the performance of the installed assets. For example, a particular line or transformer can be run in a mode inappropriate for constant operation but acceptable for the satisfaction of a short-term peak-demand. Planning and monitoring of the ageing of assets represents an even more advanced concept. Assuming that a certain operation schedule for a given asset has a controllable influence on its deterioration and overall lifetime, an optimized pattern of the schedule can be calculated. In this context the term *lifetime optimization* has emerged (Draber et al., 2000).

Emergency Management

Alarms indicating unexpected events trigger actions and workflows. Depending on the type of alarm and its data contents these workflows run over certain timescales. Alarms indicating serious faults and outages are treated within minutes, analysis and diagnosis of the exact reasons within hours or a few days. Emergency management not only encompasses the immediate actions for restoring an error-free state, it also includes consequential processes for reducing error occurrence as far as possible.

Again, emergency management applications can make their way into the enterprise level if their functionality is extended in order to enable *learning* processes. This means, e.g., that experience from emergency handling is fed back into the maintenance and investment planning process in an automated fashion. The consequences of decisions made under consideration of lessons learned from emergency management can be measured and may bring substantial benefits to the whole enterprise.

Future Developments

It can be safely assumed that the general trend towards a further growth of IT-related spending also holds for utilities. The following sections present a selection of possible drivers and concepts for increased software use that are more specific to the utility industry.

Markets

Power networks and utilities exist in every country; in many places the level of automation on which they operate is far from the current state-of-the-art (Jennex, 2003). Market conditions, however, were relatively similar for most countries until the beginning of the last quarter of the 20th century: utilities were basically state-owned agencies with little exposure to competitive forces. Since then, some governments have started to liberalize market conditions, others have kept their rather regulated ones or are deregulating at a very slow pace.

The primary objective of *liberalized* markets is to give retail customers direct access to power producers. For this, generation has to be unbundled from transmission, to enable separate pricing and to allow deregulated generation to coexist with regulated wire commodities. Assuming that the use of such transmission assets is under the control of Independent System Operators (ISO), whose operation and pricing are accessible without discrimination to all market participants, local distribution companies may offer competitive power supply options to the retail customers, and perform meter reading, accounting, billing, and collection functions.

Under such market conditions, distribution utilities may adjust their economic priorities and resulting policies. Such adjustments are expected to affect

- Investment decisions on asset renewal, including upgrading and re-rating of existing ones,
- Optimization of operation and maintenance costs,
- Monitoring of the reliability of the network,
- Utilization of existing assets.

One can roughly argue that in the starting phase of liberalization, investments and maintenance costs are being reduced. Once the reliability of the operation suffers from these under-investments, spending in asset renewal and maintenance will surge again. An alternative is to use existing assets in an optimized way, i.e., reducing redundancies and safety margins or being able to allow a higher degree of flexibility of their use through increased monitoring of their actual state, which requires higher investments in automation (Roberts, 1996).

The effects on the four factors enumerated above can, however, also be shown in rather *regulated* markets. Here, utilities do not have the same economic incentives introduced by the specifics of deregulated or liberalized markets. Investments into equipment are varying periodically as devices are renewed, updated or upgraded, while maintenance budgets often remain more or less unchanged. Experience has shown at least the presence of willingness to review "old-fashioned" policies on investments and operations as well as an interest for higher degrees of automation allowing more flexibility. The incentives to invest in modernization may be less of economical origin, but rather of social or political origin, e.g., the demand for more detailed energy bills, narrower regulations on reliability, or the popularity of renewable energy amongst customers.

In the light of recent large power failures, unexpected rises of spot-market prices, and obscure manipulations by dominant market players, estimating the speed and rate of liberalization in the power utility sector for the near future is a risky task. Businesses should be prepared for a variety of scenarios under which advanced approaches towards automation in the utilities industry can be expected to thrive. The following sections give examples of such approaches on a conceptual level.

Asset Management

The term *asset management* originates in the world of finance where it is applied to the management of portfolios consisting of financial assets. In process industries it is often used as a synonym for plant management. For the power generation and transmission sector the view on the role of asset management is quite diverse: it encompasses strategic maintenance and investment policy, asset history recording and analysis, asset value depreciation, risk management, or financial performance (Kostic, 2003).

Looking at the cost structure of utilities brings some clarity. Apart from the bulk purchase of electricity or fuel, managing their existing assets is the main cost center for electric utilities, be it generation, transmission, or distribution companies. The management of these assets (e.g., lines, substations, turbines, generators, transformers, cables, switches) can be called asset management and has been defined by the Victorian government (1995) as "the process of guiding the acquisition, use and disposal of assets to make the most of their service delivery potential and manage the related risks and costs over their entire life."

In the case of electrical utilities, these assets are geographically widely dispersed, and have been installed over many decades. The management of these assets depends on reliable information regarding the age, condition, performance, criticality, and reliability of the assets in the electrical network and, at the same time, on a decision framework to ensure that the system is operated in the most effective and efficient manner. Asset management software solutions typically provide utility asset managers with an integrated view on relevant data, supporting their decision making process.

Figure 1. Enterprise View of Maintenance and Utilization Oriented Units and Respective Processes

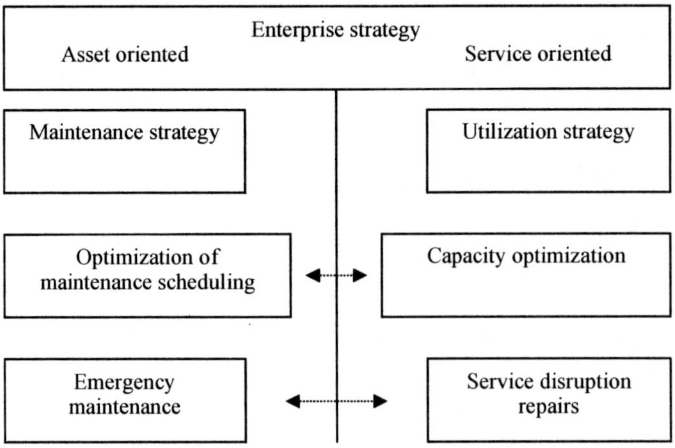

This means that a whole set of both technical and managerial activities form the process of asset management. Furthermore, it is unavoidable to combine and link activities from different traditional entities of a utility in order to make asset management more than just another layer of information collection, i.e., more than databases and tables showing well-known data in a slightly different format (Kay, 1998). Consequently, reorganization of businesses may be needed in order to exhaust the full potential of an enterprise-wide asset management strategy.

Figure 1 shows a simplified diagram of the dependencies of processes and units of a utility from the point of asset maintenance and service provision. While the top-down planning and management processes are focused on the actual assets and their maintenance (left-hand side), and the provision of electricity (right-hand side), horizontal links between tasks are often missing although both sides could profit from each other. For example, optimized maintenance scheduling without regard of electricity provision obviously lacks a crucial parameter.

On the software side, integration of applications is the strategy to support asset management. The following section treats this issue.

System Integration, Standards and Platforms

Unlike the home user, enterprises rarely ever exchange or upgrade completely their software applications. Rather, a strategy of steady improvement, enhancement, and incremental upgrade including the integration of legacy systems is predominant. For mission-critical systems, i.e., protection and control applications in distribution utilities, the attitude of waiting for proven solutions before adopting a new technology is enhanced even further by the fear of high migration costs and associated risks.

When integrating two or more applications two main challenges are seen. They can be exemplified using well-known analogies from the office-software world. The first challenge consists in the requirement that the user can "navigate" from one to the other. In the office-world this means, e.g., that a user writes a letter with a word-processing application and then enters the addresses from a database application without the need of writing macros. The user is able to see the addresses, sort or filter them without having to interrupt the actual process of writing the letter. This approach can be translated into industrial software applications conceding the substantial difference that the user writing a letter and the one entering addresses is often not the same person. However, assuming changes in organizational behavior and workflows, cross-navigation between traditionally separated applications becomes one of the most important requirements (Barruncho & Vidigal, 1995).

A major concept for software development addressing this issue of cross-application functionality from the developer's point of view is the notion of a *platform*. A platform is basically a combination of development tools and collections of basic, application-independent functional components that can be assembled and customized in a flexible way. Again the analogue in office software is well known: some years ago, word-processing, spreadsheet and database applications were rather unrelated to each other, in recent years a clear convergence has taken place and the customer is presented whole offices-suites combining traditional applications into complementary parts of a larger package. In addition, high-level programming languages allow automation of user-specific processes and implementation of non-standard functionality by directly accessing the object models of the various office applications. The advantage for the user is a common look and feel, the benefit of the developer a much more modular and cost-effective implementation strategy.

A second major challenge is posed by the necessity for common data models and data consistency. This means, in our example, that the notion of an address must be *abstracted* from its pure function to serve as a part of a letter. It must be defined as a set of data that describes a real-world object, the complete localization of a person's home or a company's site. Data consistency becomes important as soon as several applications with different purposes read and write data to a given data set. The requirement to meaningful data integration must be that no redundancy is present and that each application has access to the actual set of currently valid data through well-defined interfaces and safe access methods.

A key difficulty in older industries is obvious: a description or data sheet of a real-world object is usually made at the time of its delivery, which may have been be several years or even decades ago. At that point in time certain properties of the object may have either not been known or just not been measured and documented, as nobody thought they might become important at a later stage. If, at a later stage, a new application requires certain properties of an object to be known in order to fulfill its functionality, the need for extending the data model for the specific object and reconstructing the necessary data emerges. Given the numerous data formats and database systems present on the market, data handling and management has become one of the most demanding and costly parts of enterprise software systems.

A most important condition for handling data across various applications is the use of *standards* for data description[2]. Fortunately, a number of new standards have emerged in recent years. The International Electrotechnical Commission in its standards IEC 61850 (2003) defines rules and data models for monitoring and protection; CIM and IEC 61970 (2002) enable communication with control centers. IEC 61968 (2001) will allow more efficient engineering of interfaces and integrated applications.

On the side of IT technology, the success of component-based development and data description languages (e.g., XML) support a modular, step-wise realization of integration (Werner et al., 2000). Based on powerful platforms, high-level programming languages (Java, C++), and object-oriented interface- and component-libraries, development and customization of new and improved automation applications can take advantage of the full abilities of state-of-the-art hardware and communication performance (Charcosset & Cavailles, 1997).

Wrapping up these remarks, it can be safely stated that software integration represents a major, growing part of future investments in utility automation, but its growth will be rather steady and must include structural and organizational changes in business and industry.

GIS for Power Line Failures

GISs are not new to power utilities. They are commonly used in planning and construction workflows, service and metering processes, customer relationship management or load forecasting applications (Meyers, 2000). Each of these processes has different requirements on the display and analysis capabilities as well as the level of detail of the underlying geographical data.

The following sections describe as a *case study* an actual *pilot application* that the authors have developed within the framework of a customer-specific GIS project. The pilot application is based on an existing software package, Power System Monitoring (PSM), which takes care of data collection, fault identification and data transmission. A new software component integrating a GIS for the exact calculation of the geographical fault location and its visualization is the main outcome of the project. Particular care is taken to ensure that the design and development work does not limit the resulting component to its specific task within the project. Therefore, a number of more general requirements and design principles are considered in order to allow the finished component to be reused in similar projects.

In this case study, publicly available geographical data for political boundaries, roads and buildings is used alongside customer data on substations and power lines. The customer has been using PSM for a number of years; the business motivation for installing a GIS mainly resides in an expected reduction of the power-line restoration time.

PSM

Power System Monitoring (PSM) is an application that connects substation-based disturbance recorders to a central computer analyzing the collected records. Disturbance recorders typically store amplitude and phase information of currents measured at certain points in the power network. A number of well-known algorithms allow the detection of failures and faults within the network from the analysis of these measurements. Through the analysis of records collected at multiple points of the network, e.g., both ends of a power line, a higher accuracy of the result of the analysis can be achieved. When a fault is detected through these algorithms a corresponding alarm is created and sent to the control center.

It is important to stress that the information on the location of the recorded fault is initially relative. This means that a fault is indicated by its distance to one end of the line relative to the line's length. A typical alarm therefore contains the information shown in *Table 2.*

If we assume that the length of the line called "Main Link" is 20 km, this record indicates a fault at a distance of 15.26 km from the substation called "Main Station."

The further workflow is then driven by the actions of maintenance personnel (see *Table 3*). Analysis of other messages or indicators on a possible failure is performed and then a

Table 2. Alarm Data Delivered by PSM (schematically)

Time stamp	Line	Substation	Distance
2003-05-22	Main Link	Main Station	0.763

Table 3. Typical Workflow of Line Restoration

Role	Action(s)			
Protection system (PSM)	Detects fault and triggers alarm			
Protection engineer	Verifies fault	Dispatches repair work order		
Maintenance engineer		Receives Work order	Repairs line	Reports repair
Protection engineer				Releases line back to operation

decision on the restoration of the malfunction is taken. The elapsed time between the initial fault alarm and the restoration is typically of the order of hours.

PSM holds its data in a separate database. The most important entries in this database are the names and locations of the disturbance recorders, the lines and substations that are monitored, and the impedance and length of the power lines. These are necessary parameters for the correct calculation of the location of the line fault. The disturbance records and results from their analysis are stored in the same database.

GIS Extension

Functional Goals

The most immediate benefit from a GIS extension is the ability to represent the location of an event delivered by PSM on a suitable map. Furthermore, a GIS is an additional, attractive user interface, particularly during certain phases of the decision-making process triggered through an alarm. A better overview over the whole network and its accessibility through roads and landscape is achieved as well.

In addition to these main goals a number of further requirements related to the possibilities of cross-navigation between the human-machine interfaces[3] of PSM and the GIS are part of the pilot project described in this section. The reason behind this lies in the fact that PSM is an established application, and protection and maintenance engineers are familiar with its abilities. The GIS extension can then serve as an additional option without replacing any known features of the base application. It is an ideal feature to be extended to a number of pilot customers in order to evaluate acceptance and collect feedback. Such a pilot-application strategy is a typical one in the context of the extension of industrial software packages through integration.

The main goals of the extension are the following:

1. The transformation of the relative location information as described above into absolute geographical coordinates (longitude, latitude),
2. The display of a marker indicating the location of the failure on a map of the network,
3. A number of standard visualization options (zooming, selection of map layers) for identifying the surroundings of the fault location (nearest roads, service centers),
4. The ability to perform post-analysis tasks on possible relations between environmental influences (weather conditions, lightning) and failure occurrence.

The first goal is the most formal one: all that is required for the transformation of the relative fault-location information into an absolute one is a geographical representation of the faulty power line. This means that a data layer needs to be added to a set of geographical data sharing a fixed coordinate system. This layer contains sets of data consisting of measurable directed lines and an identifier relating the line to its counterpart in the PSM database.

The algorithm uses the information delivered by PSM (*Table 2*), and calculates the geographical coordinates (longitude and latitude) of that single point on the corresponding line in the GIS database that has the same relative distance to the indicated substation. After this calculation the entry of the alarm looks like this:

It should be noted that this calculation is the single most important feature of the entire pilot application. It seems somewhat obvious or even trivial, but it is the condition to all further steps. One such next step is the display of the calculated point on a map view. In the chosen implementation, a right-click on the line will display further information on the fault in table format. Standard zoom operations allow the easy identification of the surroundings of the location, roads or other access paths and enable a much faster decision on what further action to take. Options to transfer the coordinates to car-navigation systems based on GPS or automatic notification of the next available service engineer are possible depending on the availability of suitable equipment.

Post-Processing

Faults can have a number of reasons. Trees touching a line, e.g., can trigger a fault alarm. Although the line can be used immediately again after the tree has moved away, a later analysis may show that the tree should be removed altogether. A structured collection of disturbance records and their analysis with further tools like a GIS enables the use of a protection application within the context of a planning and optimization tool. It can become part of an asset management process for power lines that is not traditionally conducted by maintenance departments of utilities. Therefore, the enhanced information (*Table 4*) is stored in a history database for post-processing. This makes it possible to visualize fault occurrences on top of weather maps containing rain, wind, or lightning data. Such analysis may indicate causes for unusual failure frequencies and support a continuous maintenance planning process through access to statistical evaluations.

Geographical Data

A main technological challenge of the pilot application just presented is the modeling of the relation between real-world objects, i.e., power lines and substations, real-world events (line faults) and their representation in the two software components (PSM and GIS) each of which is based on their proper database formats and object models. Suitable custom entities are chosen that can be represented in both applications and the actual instances of which form one-to-one relationships.

Table 4. Alarm Data including Absolute Geographical Coordinates

Time stamp	Line	Substation	Distance	Longitude	Latitude
2003-05-22	Main Link	Main Station	0.763	118°33'12''	41°22'34''

Figure 2. View on a (fictitious) Power Network (on the line between San Francisco and San Jose, a fault is indicated)

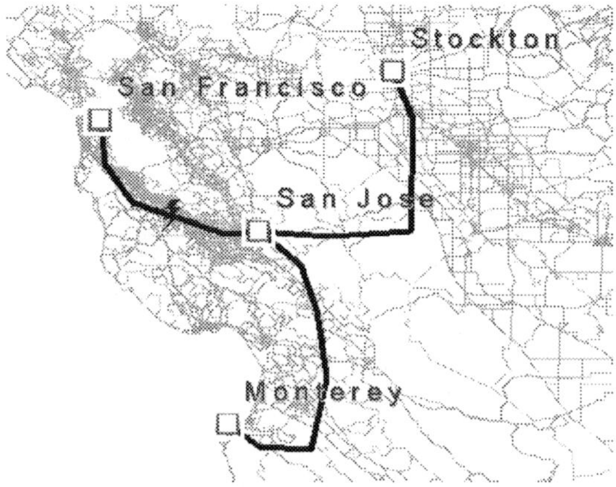

Those necessary custom objects are *network*, *substation*, and *line*. The network at the top level of data modeling contains the necessary information for identifying the PSM installation running on that network, and information on the geographical database containing layers of features that can be displayed on a map. A substation serves, in our context, always as a start- or an end-point of line. The line, finally, holds the records (*Table 4*) of all faults occurred on that line and information on the substations located at its start and end.

Substations and lines have direct counterparts in the geographical database. Substations are points; lines are measurable composite lines with a defined start and end. Such geometrical objects are standard in any GIS. In the present case they are copied along with the substations from a map provided by the customer of the pilot-installation.

The level of detail for the data should be at least as good as the overall accuracy of the PSM application. Practically speaking, lines should have its vertices placed as accurately as possible at the real vertices of the corresponding power line (e.g., major towers). Data of this accuracy is often available to utilities from the initial planning and construction process.

While substations have no particular functional role in the pilot application, line objects serve as containers for the whole fault location functionality. They receive the alarm generated by PSM, trigger a query for their representation in the GIS, perform the transformation of relative to absolute coordinates, and finally add the result to its list of fault events.

Below the data layer of custom objects, geographic data containing political boundaries, settlements, roads and rivers are used. Generally, the availability of geographic data varies strongly from country to country and costs for good data must not be underestimated (Grimshaw, 1999, p. 96). In the case of the pilot-customer this data is available in sufficient quality at low cost. Data import, aggregation, and map design are performed with the standard layout tool of the GIS, i.e., the full functionality of these tools is available for the configuration process of an individual pilot installation. Later on, the end user of the installed application is offered a limited choice of layout options, basically the possibility of hiding certain layers of geographical data if desired.

Value and Further Possibilities

The assessment of the possible impact of this case study and its relation to the remarks in the first part of this chapter is made in two stages. The perspectives of functionality, analysis, user behavior, and data management are addressed, and an integrated view on the present and future role of GIS in application integration is given.

Functionality

The clear benefit of the GIS from a functional perspective is the ability to locate a power-line fault in absolute geographical coordinates. This new feature relies purely on a geometrical algorithm from the GIS that enables the calculation of a point from a given relative distance along a line. Strictly speaking, this is a geospatial analysis function (Grimshaw, 1999, p. 209). Furthermore, this calculation takes place dynamically, i.e., when triggered through an alarm received from PSM.

Although there are many further examples of specific functions of a GIS being used in a similar sense, the combination of converting real-time information (the alarm) and static, user-defined asset data (the geographical description of the utilities' power lines), is somewhat novel for the utility industry. The consequential workflow is reminiscent of emergency-oriented applications (traffic, police, ambulance).

The authors are convinced that using such specific functionality is the main argument to justify to management the impact of a GIS for increasing the degree of automation of a given process. This reflects the remarks made in earlier sections: upgrades in control and monitoring automation are made step-wise, and each extension needs to be evaluated through its proper value. General cases on "new" platforms, concepts or technologies that do not address a specific improvement of a concrete workflow are usually not accepted.

Analysis

The post-analysis of information received from monitoring applications is of increasing interest (Draber et al., 2000). The more flexible usage of assets in critical conditions is particularly beneficial in sectors of the industry where investment costs are high, hardware extensions are expensive, and assets are used over decades. The ability to relate weather data to failure occurrence is the main case for integrating GIS into the process of post-analysis. Again, this circumstance is quite specific, as utilities do operate assets over widely dispersed areas, under varying climate conditions and in remote locations, it is exactly this ability to relate the geographical position of an asset to its actual state and condition that is beneficial. Maintenance planning can draw on this additional information given, and a GIS-enhanced monitoring application therefore becomes a valuable decision-support tool. On a strategic level, information available through GIS supports the process of identifying the best possible locations for a planned investment (Baban & Parry, 2001).

Data Management

GIS can support the task of managing the massive amount of data present in utility automation. Although most data is completely unrelated to the geographical location of assets, at least another layer of data presentation can be added through the display and layering abilities of typical geographical databases. This remark should not be misunderstood: it is not the mere browsing on a map that gives new insight; the data stored in the geographical database should rather be used as a master-reference during certain maintenance, service and planning processes (Dieterle et al., 1995). Geographical databases implement particularly detailed object hierarchies and allow operations such as advanced ordering and sorting.

User Roles

New components or applications generate shifts in user behavior. In the analogy made in an earlier section on office software, the change of behavior is obvious: writing letters and managing basic address data were not only performed with different applications, often the same person was not even able to operate both programs or to have access to them. Today, it is absolutely commonplace that the same person performs various standard and non-standard tasks with their address data, e.g., send letters or e-mails to selected groups, keep a worksheet on their birthdays, and update addresses and telephone numbers regularly.

We do not assume that workflows and processes within utilities will change as quickly as those in our analogy, but the increased accessibility to various applications across traditional department boundaries is indeed assumed to expand. Particularly the availability of technical real-time information to the management level is a requirement that is increasingly desired. Obviously, the presentation of this data and information must lose some of its technical look and feel and become more flexible and adaptable to the preferences of the new users. Here, GIS indeed offers a major benefit. We say again that it is not the mere browsing on colorful maps that is needed, but much more the ability to use the map with its various ordered layers of assets to *navigate* to the desired application and find the desired information. It is functionality like this that will, from a user's point of view, make the perspective of increased application integration more attractive.

An Integrated View

In the last sections we have highlighted different levels of potential benefit of GIS in utility automation. Summarizing these points, we formulate to our main conclusion from this case study: *GIS is an ideal tool for enabling application integration and integrated asset management.* In our case, it "promotes" an established monitoring application into a more enterprise-oriented level of usefulness. In a more general view, its various functional levels make it attractive to be a key component within a platform strategy for utility automation. Specific analysis functionality, data management, navigation, and visualization of the state of dispersed assets on various levels of detail can be exploited for the development of applications that are accessed from different departments of an enterprise. It thereby facilitates changes in user-behavior and the increases the willingness to introduce more enterprise-oriented processes with improved efficiency.

The issues surrounding our conclusion have been extensively discussed in a wider context in literature and various similar case studies (see Meyers, 2000, p. 793, on the specific capabilities of GIS in application integration; or Grimshaw, 1999, on the strategic role of GIS on enterprise level). It is widely assumed the awareness of the strategic value of GIS is on the rise. Our case study, although focused in scope, nevertheless exemplifies one possible migration path of a traditional, process-driven protection application into the realms of strategic planning and decision support.

Table 5 summarizes the relation between the capabilities of a GIS and various process levels found in utility automation. It also tries to summarize the possible impact on each of these levels.

Conclusions

The first part of this chapter has offered an overview on typical processes in the utility industry and their state of automation; the second part has presented a pilot application

Table 5. Relation of Functional Levels of GIS to Process Levels within the Utility Industry

Functional level	Process level	Impact
Functionality	Monitoring and control	New information available for emergency handling Faster reaction times Precise routing of service personnel
Analysis	Maintenance planning	More insight into possible reasons for unusual equipment failure Improved asset use in critical situations
	Investment planning	Availability of additional level of information
User behaviour	All	Enabling cross-department use of applications and access to real-time information Promoting integrated approach to asset management

using a GIS to enhance the functionality of an established monitoring application. The main purpose of this combination is to exemplify a subset of current trends in utility automation that are regarded as carrying a high potential. Our findings can by summarized as follows:

- Utility automation is strongly related to the *organizational structure* of the whole sector of power utility industry.
- Future developments are inevitably linked to changes in enterprise strategies towards *integrated* processes. These strategies will define the needs for integrated software applications.
- Changes in market structure define the concrete customer benefit of novel automation solutions.
- The workflows suited for increased IT-support are complex and need powerful functionality offered by advanced software components.
- The process of renewal and modernization in utility automation is *step-wise and incremental*. Modular, component-based approaches and the introduction of a *platform* concept for application development are key to successfully offering products to the utility industry.
- Data engineering and maintenance in integrated applications rely on *standards* and *open interfaces*. GIS, through its tight integration of object-oriented, relational data management is ideally suited to play an important role within this context.
- GISs are key components able to deliver crucial functionality for application integration. They bridge the traditional gap between asset-oriented monitoring and control applications and more enterprise-oriented analysis, planning, and optimization processes.

Finally, we would like to state that, independent of the specific further evolvement of automation in industry, continuing investment in research and development in the fields of power technologies and system and control engineering will remain the most important factor for success for any enterprise in the utility sector.

Acknowledgments

The authors are indebted to Ralph Nilsson (ABB, Sweden) and John Meza (ESRI, USA) for their crucial support during the realization of the pilot application.

References

Baban, S.M.J., & Parry, T. (2001). Developing and applying a GIS-assisted approach to locating wind farms in the UK. *Renewable Energy, 24* (1), 59.

Barruncho, L.M., & Vidigal, A.M. (1995). GIS and distribution management: System design and integration issues. *Proceedings of the Stockholm Power Tech International Symposium on Electric Power Engineering, 4*, 282.

Charcosset, G., & Cavailles, J. (1997). Experimentation of a SCADA/DMS-GIS interface development. *Proceedings of the 14th International Conference and Exhibition on Electricity Distribution, 4*, 35.

Dieterle, H., Roquet, A., Vierheilig, N., & Weil, W. (1995). Linking distribution management system, geographical information system and network planning system. *Proceedings of the 13th International Conference on Electricity Distribution, 4*, 8.

Draber, S., Gelle, E., Kostic, T., Preiss, O., & Schluchter, U. (2000). How operation data help manage life cycle costs. *Proceedings of Cigré 2000*, Paris. Paper No. 23/39-02.

Grimshaw, D.J. (1999). *Bringing geographical information systems into business* (2nd ed.). New York: John Wiley.

International Electrotechnical Commission. (2001). *System interfaces for distribution management*. Draft IEC standard 61968. Geneva: IEC.

International Electrotechnical Commission. (2002). *Energy management system application programming interface*. Draft IEC standard 61970. Geneva: IEC.

International Electrotechnical Commission. (2003). *Communications networks and systems in substations*. IEC/TR 61850. Geneva: IEC.

Jennex, M.E. (2003). IT use in the utilities of Ukraine, Armenia, and Georgia. *Communications of the Association for Information Systems, 11*, 413.

Kay, M.J. (1998). A business driven approach to asset management through the strategic use of information. *IRR conference on asset management*, London.

Kostic, T. (2003). Asset management in electrical utilities: How many facets it actually has. *Proceedings of the 2003 IEEE power engineering society general meeting*, Toronto. Available on CD as IEEE paper No. 0-7803-7990-X/03.

Meyers, G.R. (2000). GIS in the Utilities. In P.A. Longley, M. Goodchild, D. Maguire, & D. Rhind (Eds.), *Geographic information systems* (2nd ed.). New York: John Wiley.

Roberts, B.F. (1996). *Load and revenue analysis and forecasting for restructured electricity markets*. Presented to the electric utility forecasters forum, Santa Fe. Retrieved from: *http://www.econsci.com/euar9603.html.*

Shahidehpour, M., & Marwali, M. (2000). *Maintenance scheduling in restructured power systems*. Boston, MA: Kluwer Academic Publishers.

Victorian Government. (1995). *Asset management series: Principles, policies and practices*. Management improvement initiative by the Victorian government. Retrieved from: *http://home.vicnet.net.au/~assetman/welcome.htm.*

Werner, T., Vetter, C., Kostic, T., & Lohmann, V. (2000). Data Exchange in asset management applications for electric utilities using XML. *Proceedings of the 5th international conference on advances in power system control, operation and management* (APSCOM 2000 Hong Kong), *1*, 220-224.

Endnotes

[1] The abbreviation GIS is used throughout the chapter for singular form

[2] These should not be confused with standards for data access, e.g., ODBC

[3] This term (HMI) has been replacing the term GUI (Graphical User Interface) in recent years.

Appendix

Glossary

CIM: Common Information Model

CIGRÉ: Conférence Internationale de Grands Réseaux Électriques

CMMS: Computerized Maintenance Management System

CRM: Customer Relations Management

DMS: Distribution Management System

EMS: Energy Management System

ERP: Enterprise Resource Planning

GIS: Geographic Information System

GPS: Global Positioning System

GSM: Global System for Mobile Telephony

HMI: Human Machine Interface

IEC: International Electrotechnical Commission

IEE: Institution of Electrical Engineers

IEEE: Institute of Electrical & Electronics Engineers

ISO: Independent System Operator

IT: Information Technology

O&M: Operation & Maintenance

ODBC: Open Database Connectivity

PSM: Power System Monitoring

SCADA: Supervisory Control and Data Acquisition

T&D: Transmission & Distribution

XML: Extended Markup Language

Chapter XV

GIS in Agriculture

Anne Mims Adrian, Auburn University, USA

Chris Dillard, Auburn University, USA

Paul Mask, Auburn University, USA

Abstract

This chapter introduces the use of geographic information systems (GIS) and global positioning systems (GPS) in agricultural production. Precision agriculture is a catch-all term that describes using GIS and GPS technologies to manage specific areas of fields. Precision agriculture technologies use information from multiple sources to assist farmers in making crop production and management decisions based on the variability of production potential within fields. In this chapter, we describe the technologies used in production agriculture and we review some of the research associated with the use and future trends of these technologies. The purpose of this chapter is to define and explain GIS and GPS technologies used in agriculture and some of the economic benefits, impacts, and challenges of using these technologies.

Introduction

Farmers have long known variation existed within their fields, but did not have the tools to properly quantify, view and manage that variation. Geographic information systems (GIS), along with global positioning system (GPS) enabled technologies, have given farmers the ability to make management decisions with more precision and information than ever before. Fields no longer have to be managed on an across-the-board basis, but rather fields can be managed on the production potential of each area of the field. Agricultural crop production can be limited by soil chemical or physical characteristics, topography, crop variety, or a number of other variables. GIS allows farmers to determine where deficiencies exist, search for the cause of the deficiencies, and make management decisions necessary to improve productivity in the problem areas. With GIS and GPS systems, farmers can have access to a tremendous amount of information on yield variation, soil properties, topography, water absorption, plant health during the growing season, and records of chemical use.

The catch-all term, precision agriculture, means to use GIS and GPS technologies that allow farmers to manage specific areas of fields. GIS serves as one component of decision support systems (Grupe, 1990), which lends itself to the working definition of precision agriculture that is published by the National Research Council (National Research Center, 1997) as, "a management strategy that uses information technology to bring data from multiple sources to bear on decisions associated with crop production." Farmers see that their production is being squeezed with higher input costs and tougher international competition and some have considered adopting precision agriculture as a way to lower production costs, protect the environment, and to manage large farms (Olson, 1998).

Precision agriculture tools are used to monitor crop yields, apply inputs at a variable, rather than constant rate, and to guide equipment. Other tools are used to determine soil electrical conductivity, manage soil on a site-specific basis, and to monitor crop growth and health from satellite or aerial images. All of these tools utilize GIS to acquire, process, analyze, and transform the data collected into information that farmers can use to better manage production and improve profitability. The incorporation of precision agriculture tools began in the mid 1980s (NRC, 1997) and the initial adoption has been slow (Swinton & Lowenberg-DeBoer, 1998). While economic benefit is the deciding factor for sustained use of a precision agriculture technology, other reasons, such as attitudes toward technology, may possibly affect adoption (Cochrane, 1993). Although the research on the economic benefits is mixed, (Malcolm, 1996; Sawyer, 1994, Swinton & Lowenberg-DeBoer, 1998), farmers are primarily investing in them sequentially (Isik, Khanna, & Winter-Nelson, 2000; Dillon, 2002).

The purpose of this chapter is to define and explain the technologies used in precision agriculture, to explain some of the economic benefits, the impacts of producers' decisions, the challenges of using these technologies, and to review some of the precision agriculture technology research. Benefits of precision agriculture technology include: reduced variable costs, increased yields, increased profits, and reduced environmental effects (Intarapapong, Hite, & Hudson, 2002; Sawyer, 1994). Increased yields result from increasing inputs in more productive areas of managed fields, thereby increasing yields. Sometimes fewer inputs are applied in areas of the field that have lower yield potential, which reduces variable costs (Intarapapong et al., 2002). Fewer environmental effects are possible because of the more precise application of inputs (Kitchen, Hughes, Sudduth, & Birrell, 1995; Intarapapong et al., 2002; Sawyer, 1994). However, farmers cannot vary the input until they have the information that shows how yields or soil properties, such as pH levels, vary across the fields.

Generally, the first precision agricultural tools farmers implement are yield monitors or grid soil sampling. They learn the GIS and equipment technologies, gather data, and begin to make production decisions that may not require variable aspects of the field, such as changing cropping sequence. Additionally, information is obtained by targeted soil sampling to gather nutrient and physical characteristics of the soil.

Agriculture has always been a driving force in the growth of the United States and its economy. The agriculture sector contributes over $200 billion annually to the U.S. economy and employs almost one million workers (U.S. Census Bureau, 2002). Globalization of trade, decreased commodity prices and reduction in farm subsidies increase the importance for farmers to utilize their resources as efficiently as possible. GIS and GPS technologies provide farmers with techniques to help maximize production and efficiency and increase the information available to make sound business decisions. Precision agriculture technologies are provided by farm equipment manufacturers, agrochemical companies, pharmaceutical/biotech companies, data management firms, and high-tech Pentagon and intelligence community contractors (Marrero, 2003). We will describe many of these techniques and their economic benefits, the impact of the additional information on the producers' decisions, and some of the challenges that producers face with the new technologies.

GIS-Enabled Precision Agriculture Tools

Yield Monitors

Yield monitors serve as information gathering tools and decision support systems for the precision agriculture practitioner. The information gathered by yield monitors serves as the basis for many production and management decisions. Examples of these are determining management zones, selecting crop varieties, and applying inputs such as fertilizer or nitrogen. Creating zones within fields of different yield productivity levels is one of the primary techniques for managing fields on a site-specific basis. The delineation of yield zones based on yield variances would not be possible without GIS analytical tools.

Farmers have always known that variability existed within their fields, but had no way to quantify that variability until the advent of the yield monitor. Yield monitors are devices installed on crop harvesting equipment, such as a combine or cotton picker. Yield monitors use GPS, GIS, computer, and sensor technologies to accurately measure the amount of crop harvested at a specific location and time. In addition to measuring yield, monitors allow for the recording of crop moisture, elevation, variety, and a number of other harvest variables. It is estimated that 48 million of the 160 million acres of corn and soybean harvested in the United States in 2000 were harvested with a yield monitor-equipped combine (Lems et al., 2003).

Yield monitors are used on a variety of crops including corn, wheat, soybeans, sugar beets, potatoes, and cotton. Yield monitors utilize sensors to measure the crops' mass or volume and are found to be accurate to +/- 3% of actual harvested amount, but require routine calibration to maintain accuracy (Lems et al., 2003). The mass or volume measurements are recorded in the on-board computer along with harvester travel speed, crop moisture, and harvester width to produce once per second indirect yield measurements. GPS provides the field location for each measurement. The location and yield data are recorded onto a storage card and transferred to a desktop GIS for processing, viewing, and analysis.

Yield monitors provide a large amount of valuable data while harvesting, but they do not provide critical decision-making information until the data are processed in a GIS. The data gathered from yield monitors are transformed with GIS into the information required for subsequent management decisions, such as management zone creation, variable rate application, and targeted soil sampling. The management zones delineated in a GIS are used to manage the fields according to variation. Each zone may be sampled for deficiencies in order to manage them in a targeted manner, rather than applying inputs in a uniform application. The separately managed zones allow farmers to accurately diagnose problems, and compare management records on a year-to-year basis. Farmers are able to use their GIS to produce detailed harvest reports, determine trends from harvest to harvest, and compare the production capabilities of different varieties and crop inputs. When properly implemented, yield monitors and GIS serve as valuable accounting, record keeping, and decision support tools for farmers.

Targeted Soil Sampling

The ability of farmers to produce high yielding crops is heavily dependent upon the soil in which the crops are grown. The soil type and its physical and chemical characteristics must be in proper balance in order to maximize production potential, and ultimately, return on investment. The more farmers know about the environment in which they are operating the more knowledge they have to make prudent business management decisions that will increase profitability and reduce adverse environmental impacts.

Until the industrialization and mechanization of farming occurred, farmers worked fields of relatively small size (Morgan & Ess, 1997). In the early 1900s, farmers were intimately familiar with each field and were able to manage each in a uniform manner. The advent of the tractor changed the farm structure dramatically by allowing farmers to farm larger fields and more acres. The small fields were primarily uniform in their soil properties, whereas the larger fields could contain significant variability. While mechanization allowed farmers greater efficiencies in production, it prevented them from managing the variability found within the larger fields. Today, more detailed information about the soil properties gives farmers the knowledge to make management decisions with more accuracy and economy. GIS is an integral part of the targeted soil sampling management technique.

Targeted soil sampling consists of two primary methods, grid and zone sampling. In each method GIS software is used in conjunction with GPS to create a boundary of a field and break down the areas within the boundary into individual segments for study. The field boundary and zones, or grids, are viewable on laptops or handheld computers and are used to assist farmers in navigating to the sampling site. Grids are normally square in shape and range in size from one-half to two and one-half acres. The smaller grid sizes provide a more detailed view of the field, but result in increased sampling costs. Larger grids require fewer samples, and hence less costs, but they do not provide as detailed a view. Zones are generally not uniform in shape, or size, and can be based on Natural Resource Conservation Service (NRCS) soil maps, areas of similar yield production, or any variable farmers are interested in using for delineation.

GIS provides the capability to collect and view soil sampling data, but its impact on economic return comes from its ability to reveal field deficiencies and correct those deficiencies in the most environmentally friendly and economic fashion. The benefits of utilizing GIS in targeted soil sampling are reduced costs by decreasing inputs and increased profits through maximized productivity.

Targeted soil sampling utilizing GIS is a valuable information gathering tool for making better business management decisions. Regardless of the technique employed, the result is a better understanding of the soil being used for crop production, and more information available for deciding how to manage farm resources. GIS allows farmers to gather the information that will help them increase productivity, lessen environmental impact, and increase profits.

Variable Rate Application

The high costs of pesticides, herbicides, fertilizer, and labor make it important to utilize such inputs as accurately and efficiently as possible. Farmers have traditionally taken a blanket approach to inputs by applying chemicals or nutrients at a constant rate across a field. Labor costs are greatly reduced, and the impact of inputs is maximized, by using machinery that efficiently and evenly applies product. The problem with the uniform approach is that not all areas within a field require deficiency correction, herbicide treatment, or pesticides at the same rate. Some areas of a field may require heavy treatments, while others require none at all. The purpose of variable rate application (VRA) is to utilize the information collected about a field to only apply inputs where necessary. Applying inputs at a variable, rather than constant rate reduces input and labor costs, maximizes productivity, and reduces the impact over-application may have on the environment.

Before inputs can be applied at a variable rate, farmers must determine what the application will be based upon. For example, nitrogen application decisions are generally based on the average yield from previous years. Fertilizer and lime application decisions are based on the information gathered from targeted soil sampling. Additionally, farmers may use aerial imagery or NRCS soil maps to break the field into management zones and treat the areas individually. Another option for farmers is to apply inputs based on soil physical properties as determined by NRCS soil maps. Regardless of the input being applied, VRA is dependent upon a GIS for data analysis and application map creation. All of the data by which application maps will be created will reside initially in the GIS.

Farmers can optimize resources and avoid damaging the soil and environment by only applying inputs at the rates that will maximize the productivity of each area. The idea is not to maximize the inputs for each plant, but to create small areas of similar productivity that can be managed on a site-specific basis. GIS gives farmers the ability to create the areas of similar treatment based on prior field productivity. By creating management zones of low, medium, and high productivity, farmers can create application maps that integrate with the application equipment on-board GIS and controller unit. As mentioned earlier, it is important for farmers to adhere to sound agronomic principles when developing VRA maps. Applying inputs at a variable rate will only maximize productivity if the individual areas receive the proper amount of input.

Given the high cost of inputs, such as pesticides and fertilizer, it is extremely important for farmers to utilize them as effectively and efficiently as possible to maximize the return on investment. GIS can help farmers determine if VRA is economically and agronomically feasible for their situation and it gives them the tools to incorporate it into their business operations. VRA requires a certain level of expertise and investment of both time and money, but the reward can be a maximization of return on investment, increased productivity, reduced labor costs, and a reduced environmental impact from over-application. Introducing technological solutions without basing them in sound agronomic principles will only exacerbate the existing field deficiencies and not maximize return on investment.

Equipment Guidance

Whether farmers are planting, applying inputs, or harvesting, it is important to operate the machinery in the most efficient manner possible. Labor, fuel and input costs, and potential for breakdown increase if the machinery is operating more than necessary. Overlapping areas of the field or skipping areas will result in over-application or under-application of inputs such as herbicides and pesticides. Over-application can result in damage to the crop and environment, and under-application will result in the input not achieving the desired crop effect. Equipment guidance and VRA are similar in that they both serve to maximize the return on investment of equipment, inputs, and labor.

Equipment guidance systems can be placed on any type of agricultural machinery that would benefit farmers to drive in a more concise pattern. Equipment operators have traditionally relied on visual cues such as a point on the horizon, a marking system consisting of foam emitters that mark the applied areas, tire tracks, or by counting over a certain number of rows to begin the next application pass. These methods work, but most lack the accuracy needed to avoid skips and overlaps, and they do not work in low-light conditions. Equipment guidance technologies serve to eliminate these problems by integrating GIS, GPS, on-board computing, and direction indicator devices to keep the machinery traveling in the most efficient manner across a field.

Reducing overlaps and skips are two primary economic benefits equipment guidance gives farmers. The high cost of chemicals, fertilizers, and labor make it essential to apply these items in the most economic and time-efficient manner possible. The equipment operators' time is best utilized if they are applying the product in question without overlapping. Eliminating overlaps also reduces the amount of time the machinery is running, resulting in lower fuel consumption and hours of operation. Guidance systems also extend business operating hours for farmers by giving them the ability to operate equipment and perform applications at night. Traditional guidance methods are only viable options during daylight hours.

The equipment guidance system that is the easiest and least expensive for farmers to adopt is the lightbar. The lightbar consists of a GPS antenna, on-board computer with built-in GIS, and a directional indicator device (lightbar) that provides navigational information to the operator. Lightbar is a generic term for a device that provides the equipment operator with navigation cues. Navigation cues are based on guidance information received from the GPS and on the desired driving pattern the operator enters into the on-board GIS. In addition to directional cues, lightbars may also display information to the operator such as speed, heading, number of pass, and warning signals to notify the operator that the machinery is traveling through an area that has already been covered.

The auto-steer systems work along the same principle as lightbar systems, but they actually steer the machinery instead of the equipment operator. Auto-steer systems utilize a real-time kinematic form of GPS that incorporates a base station located on the farm in a known location that sends GPS data that is accurate to the centimeter-level to the mobile antenna located on the equipment. The lightbar will guide the operator within a meter of the desired pass location, while the auto-steer will guide the equipment within two inches of the desired location. Auto-steer systems are much more expensive to adopt than lightbars, but they allow farmers to hire a less skilled seasonal driver, plant fields at the optimum spacing, reduce soil compaction, and record elevation data to the GIS for use in irrigation system layout. An auto-steer system has the capability to help farmers maximize return on investment in any area of operations where precise equipment use is important.

Equipment guidance is a viable agricultural business tool for farmers to implement. Integration with a desktop GIS allows farmers to track what inputs were applied, and exactly where those inputs were placed. The GIS-enabled lightbar and auto-steer systems maximize land utilization through precise row placement, and expand the hours of operation for expensive equipment. Labor costs are reduced by eliminating the need for an experienced driver, and the driver is able to perform at a higher level because of navigation assistance and reduction in fatigue. The problems inherent with over-application or under-application of inputs are also reduced because they are only applied where they are needed, and in the amount required. The reduction in land compaction, combined with the other equipment guidance benefits, helps farmers reduce environmental impact, efficiently utilize inputs, and ultimately maximize return on investment.

Remote Sensing

Remote sensing, like yield monitoring and targeted soil sampling, is an information gathering tool. Bird's eye views of farmers' fields provided by airborne sensors give farmers unique field perspectives. Aerial photographs have been available to farmers for many years, but they did not have the ability to capitalize on this business tool until the development of GIS. Integrating remotely sensed data into a GIS can reveal information about soil characteristics, such as organic matter or moisture content, and general crop health.

Remote sensing is the GIS-enabled precision agriculture tool least adopted by farmers, but it holds great promise as a way to gather large amounts of information in a very short time period and with very little labor. Barriers to entry for farmers considering utilizing remote sensing include economical data acquisition and the difficulty in determining correlations between the collected data and the crop.

Remote sensing involves obtaining information about an object without actually coming into contact with that object. In an agricultural environment, that would mean collecting information about the soil or crop from a plane-based or satellite-based sensing device. Data collected can range from a simple color photograph to the crop's emission of electromagnetic energy. Remotely sensed data can give farmers near real-time information regarding their crop, which can allow them to make corrective management decisions before the crop has deteriorated to the point of no return. Data can be collected for entire fields with remote sensing as compared to traditional scouting methods, and it can be gathered over multiple dates during the growing season. Data collection over multiple dates during the season allows farmers to monitor both positive and negative trends in crop progression. The challenge to farmers is to learn how to analyze the data collected and process that data into information that can be used to make management decisions that will increase productivity, maximize inputs and resources, and minimize risks.

Sensors that measure the amount of infrared energy reflected or emitted by the plant's interaction with sunlight concentrate on an area of the electromagnetic spectrum outside the visible range (Morgan & Ess, 1997). All objects emit electromagnetic radiation and most of it is emitted at wavelengths outside the visible spectrum. Light that hits an object, or plant in our case, is either reflected, transmitted, or absorbed. The way in which each plant reflects or emits the wavelength of light depends on the characteristics of the plant. This reaction to the light energy is known as the plant's spectral response. It is the differences in spectral responses that form the basis for remote sensing as an information gathering tool. Healthy plants have a different spectral response than less healthy ones. Remote sensing attempts to determine the characteristic responses that can be used to measure crop status by analyzing a small portion of the electromagnetic spectrum.

The problems associated with remote sensing data include the cost of data acquisition and the inability to acquire data in adverse weather conditions. Satellite imagery is expensive because of the inherently high cost of the technology utilized. Satellites are expensive to build, launch and maintain, and the volume of imagery acquired is not large enough to produce an economy of scale. When a satellite has been tasked for data acquisition, there is always the possibility of inclement weather blocking the view of the sensing equipment. More economical acquisition platforms are available, such as airplanes and helicopters, but the costs are still relatively high, and they face the same potential weather difficulties as satellite-based platforms.

The role of GIS in using remote sensing as an information gathering tool is to provide the analysis necessary to determine relationships between the remotely sensed data and the actual measured crop or soil data. When relationships have been determined that provide correlations between remotely sensed data and ground acquired data, farmers are able to analyze and make decisions concerning large areas of land without taking actual ground readings. Remotely sensed data will become a layer in farmers' GIS packages that can be compared against other data layers such as yield, soil chemical and physical properties, and fertilizer, herbicide, and pesticide applications. The information provided by remote sensing gives farmers an analysis of crop conditions while there is still time to rectify deficiencies revealed in the imagery. In addition to providing crop condition information, remote sensing and GIS assist farmers in delineating soil zones, predicting possible yield, efficiently placing drainage, and calculating field acreage. If properly collected, processed, and analyzed in a GIS, the information gathered from remote sensing can be a valuable agricultural business management tool.

A model of precision agriculture technologies is shown in *Figure 1*. The model shows the flow of data from the various sources into the GIS for mapping and analysis to variable rate prescriptions that go back into the GIS for further analysis and decision-making.

Precision Agriculture Technology Research Opportunities

GPS technology is used in combination with GIS to acquire, process, analyze, and transform the data collected into information that farmers can use as decision support systems to better manage production and thereby improve profitability. While economic benefit is the definitive factor for sustained use of a precision agricultural technology, other factors influence adoption, such as attitude toward technology (Cochrane, 1993), perceived benefit, spatial variability of the farm, land types, environmental impact (Intarapapong et al., 2002), and magnitudes of variance of yield response (Mahajanashettii, English, & Roberts, 1999). To adopt and use these technologies, not only do agricultural producers have to make large financial investments, but also investments in learning new skills — using GPS tools that integrate into existing farming implements, using GIS software, and understanding data associated with mapping, soil quality, topography, yield variance, and crop health. Most GPS and GIS in agriculture are used in various combinations, and the tools develop into customized information systems and these systems are not turnkey or black boxes (Olson, 1998; Lowenberg-DeBoer, 1996). Many

Figure 1. Example Model of the Use of Precision Agriculture Tools

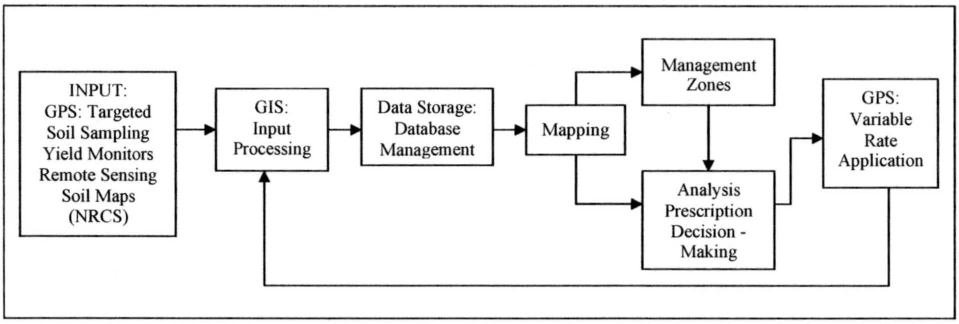

of these farmers are not clear on the exact value or benefit of these technology systems. Some researchers say that farmers are more likely to adopt an all-in-one technology (Byerlee & de Polanco, 1986; Olson, 1998) because of ease of use. While attempts have been made to create all-in-one technologies, these systems are expensive and do not allow for step-wise adoption.

Information Systems Research

GIS and GPS tools are seen as decision support tools for making better production and management decisions. Logically, better decision-making is possible with more information. Some researchers see precision agriculture as part of the larger context of information technology (IT) in agriculture (Sonka & Coaldrake, 1996). Just as IT is not a homogenous product (Markus & Robey, 1998), neither is precision agriculture technology (NRC, 1997). Rather, it is combinations of tools used for various applications for strategic, tactical, and operational improvement of agriculture production (NRC, 1997; Bouma, Stoorvogel, van Alphen & Booltink, 1999). Precision agriculture technology is used to manage specific areas of fields and also to achieve long-term goals of sustainability by providing historical information on the variations throughout farmers' fields. It can also provide accurate recordkeeping for government regulations.

Because precision agriculture technology is broadly applied to agriculture (Lowenberg-DeBoer, 1999), studying GIS and GPS technology uses in agriculture should follow some of the research practices used in the Management of Information Systems (MIS) field (Olson, 1998). Some of the same challenges that perplex Information Systems (IS) researchers are present in studying precision agricultural technology. Cost-benefit analysis is used to evaluate the efficiencies gained because the technology does not fully capture the impact of improved managerial decision-making. Also, precision agriculture could be expanded to a "holistic information intensive management strategy" that has a bearing on "agricultural production, marketing, finance, and personnel."

Some studies show that spatial decision support systems are more efficient in complex problems (Mennecke, Crossland, & Killingsworth, 2000). However, the farming industries lag behind in having reliable spatial decision support systems, which is the biggest failing of GIS systems (Fisher & Nijkamp, 1992). While the GIS farming systems offer a way to layer information and present the information in maps, farmers struggle with creating management zones and processing vast amounts of information. Few computerized spatial decision support tools are available. In the *Figure 1*, establishing management zones and data analysis are performed by farmers or by consultants and change agents. Therefore, farmers must be able to decipher and analyze complicated data and layers. There is a need for more development and research to determine methodologies to develop management zones, analyze data, and create queries and decision support systems that make using GIS more accurate and efficient.

IS research has established several streams of research that could be used in studying precision agriculture adoption, assimilation, and use. Human factors, such as experience (Mennecke et al., 2000), and visualization ability and the need for cognition — a willingness to engage in problem solving tasks (Speir & Morris, 2003) — are related to adoption and accuracy of decisions in using GIS technologies.

Diffusion of innovation theory indicates that perceived relative advantage, perceived complexity, and compatibility with values and operations will have an affect on adoption and use of technology, attitude toward change, exposure to technology, and compatibility of prior experience with technology. Interpersonal communication with other intended users (Rogers, 1983) will also affect adoption of GIS technologies (Nedovic-Budic & Godschalk, 1996).

Studies have indicated that users will use IT products when they see a perceived usefulness and perceived ease of use. The Technology Acceptance Model (TAM) developed by Davis (1986) could be instructive in studying farmers' intentions in using precision agriculture technology. Additionally, Sabherwal & Chan (2001) showed there is correlation between the business strategy profiles and the alignment of IT and business strategies. Using a similar method, the use of agriculture technologies could be correlated with the business strategies or goals of the agricultural producers. Adoption of agriculture practices is dependent on other characteristics, such as farm structure, farm size, education, and age of the farmer. The IS researchers can provide methods for measuring economic impact, precision agriculture success, and adoption.

Adoption

The initial adoption of precision agriculture has been slow and will not follow the traditional S-curve adoption because the technologies can be adopted in a step-wise fashion with different combinations of the technologies. Furthermore, the technologies are still new and each technology is being adjusted, transformed and refined; one exception is the yield monitor because the information can be used without adopting other technologies (Lowenberg-DeBoer, 1999). While research on adoption of precision agriculture exists, little has provided a holistic view of adoption barriers. Vanclay (1992), a rural sociologist, identifies ten barriers to farmers' adopting new technologies, some of these taken from the Diffusion of Innovation Theory (Rogers, 1983), the most cited theory in studying innovations in rural and agricultural sectors. These barriers include complexity, divisibility, congruence, economic risk and uncertainty, conflicting information, implementation economic costs, implementation intellectual costs, loss of flexibility, and physical and social infrastructure. Daberkow & McBride (2000) found that adoption rates varied by specific precision agriculture technologies according to region, size, type, and operator age, education and of time spent off the farm. Isik et al. (2000) studied the sequential adoption of the variable rate fertilizers after the adoption of the targeted soil sampling technique. Although the paper focused on the economic incentive to adopt sequentially with the variance in the soil properties, the authors warned that ignoring the sequential adoption of these technologies will overestimate the adoption rate of precision agriculture technologies.

Lowenberg-DeBoer (1996) compares the debates of economic benefits of precision agriculture technology to the debates of economic benefits of mechanization and the advent of the tractor. When tractors were beginning to be utilized, no one predicted the structural changes that would result from their adoption. The lack of involvement of social scientists in the early stages of precision agriculture technology indicates that we will not be much better at predicting the technological changes.

Research on adoption of precision agriculture technologies should incorporate its piecemeal technologies, whole farm effects, complexity of the technologies, and the investment in capital, time, and learning, as well as perceived support of these technologies. The IS field offers many differing views of technology diffusion which could be applied in studying precision agriculture adoption. The need-pull or the perception of need for improvement and the demand-push theories are rich grounds for research of the diffusion of precision agricultural technologies (Rothwell, 1982). The perception of benefit and ease of use should be studied. Information technology research indicates these two factors are indicators of technology use (Davis, 1986).

Economic Impact

Economic advantage proves to be the definitive factor for sustained use of a precision agricultural technology, but other factors, such as attitude toward technology, might affect adoption decisions (Cochrane, 1993). The characteristics and purpose of GPS and GIS tools use in agriculture vary greatly; therefore, it is difficult to define their effect on production, profit at the farm level, and the capability of making better decisions. Traditional farming requires farmers to apply inputs the same across fields, regardless of the inherent varying production characteristics of the soil and landscape. Precision agriculture applies GPS and GIS technology to provide massive amount of information about crop growth, health, yield, and topography and soil variability. This information is used to create management zones and apply inputs to each zone independently. Precision agriculture also allows holistic management of farm production that is not available in traditional farming (Olson, 1998; Lowenberg-DeBoer, 1996).

Research on the economic benefits of GIS and GPS technologies used in agriculture are mixed (Sawyer, 1994; Malcolm, 1996). Swinton & Lowenberg-DeBoer (1998) found that precision agriculture technology was profitable on a high value crop, possibly profitable on corn, and not profitable on wheat. Lowenberg-DeBoer (1996) suggests that mixed results are because precision agriculture economic research is based on a single aspect of farm production, rather than incorporating the effect of decision-making on the whole farm. Two reasons explaining the difficulty in measure economic impact of precision agriculture technology is that agriculture production success is not only dependent on good management, but also on uncontrollable conditions, including weather, pests, government programs, and economic climate. Measuring information technology (IT) economic benefits is difficult because factors, such as better decision-making, are not easily quantified.

Modeling the economic effects of agriculture production in the traditional farming system is complex and difficult and is confounded by the vagaries of the economic and environmental factors. It is difficult to model factors that are not under the direct control of the farmer, yet are very important in his/her decision-making. Weather, insects, plant diseases, environmental and governmental policies, and economic and international markets are a few examples. Additionally, growing conditions and climates require farm practices to differ from one region or environment to another. For instance, corn varieties are developed to offer better production for a given region, climate, and environmental conditions (to prevent particular disease susceptive for that region). Assessing and determining the most appropriate combinations of inputs and practices to maximize farming profit is a continuous problem that agricultural economists continue to research and debate and farmers have to address. Measuring the benefits of precision agricultural tools must take into account the complexities of measuring economic benefits that are not controlled by farmers, such as plant pests, plant diseases, climate, and economic environment.

The complexity of determining the economic effects of the use of IT in farm management is equally difficult. The mixed results of the economic benefits are unexpectedly similar to the results of economic benefits of IT in the IS research. Precision agriculture technology is still new, changing rapidly, and adjusting to meet the needs of agriculture producers and agribusiness. Similarly, IT use and development exploded in the 1980s. Prior to 1987, IT spending did not always demonstrate economic benefits; researchers called this the "productivity paradox." However, Brynjolfsson & Hitt (1996) developed one of the first convincing studies that IT investments were economically beneficial to the firms. They indicated that their findings were not because of their method of analysis — the use of econometrics and production functions — but rather, that because the measurement of IT was more current, organizations learned how to optimize the technology. Short-term economic effects may not be evident because organizations sometimes must change business processes and capitalize on experimentation in learning how to use technology effectively. Additionally, Brynjolfsson & Hitt (1996) used detailed financial information that was not available prior to their study. Access to detailed whole farm income may also prove problematic for researchers in studying the "whole" effect of precision agriculture.

Researching and modeling the effects of precision agriculture are equally difficult and complex. Researchers measuring the benefits of precision agriculture tools must be able to mediate these complexities of agricultural production and account for the holistic value precision agriculture technology offers. Additionally, Lowenberg-DeBoer (1996) calls for more research in studying both the microeconomic and the macroeconomic effects of precision agriculture. Little is known about the macroeconomic effects of IT, as well as the macroeconomic effects of precision agriculture technology.

Future Trends of GIS in Agriculture

GIS is becoming a more utilized business management tool for farmers because of current trends that will continue into the future. Farm computer usage has increased on a year-to-year basis according to the United States Department of Agriculture's National Agricultural Statistics Service (NASS) (2001). Although there are no data showing the percentage of GIS adoption, NASS data reveals that between 1999 and 2001, farmer computer access increased from 47% to 55%. The percentage of farmers using their computers for farm business increased from 24% to 29% over the same time period (NASS, 2001). Using this data as an indicator of future trends, farmers will likely continue to utilize computers as an integral farm management tool. Incorporating GIS is a natural extension of increased computer usage for farm business purposes. GIS will see improvements in hardware and software, changes in the way companies market and communicate to their clients, development of full-service precision agriculture providers, Internet-based data access, and the increased use of GIS to ensure environmental compliance.

Precision agriculture software packages have moved from general GIS applications to systems tailored to meet the needs of farmers with a minimum of training. Agricultural GIS developers must continue to upgrade and improve their software if they are to remain competitive. Precision agriculture has the potential of fundamentally changing the way farmers make production decisions (Sonka, 1998). The NRC (1997) report finds that the value of information will increase and this gained, detailed information will be more commonly employed and relied upon for future needs and uses.

More information about production should allow for better decisions. While the information obtained from precision agriculture in its current form increases operational efficiency, Sonka (1998) predicts that knowledge bases will develop and farmers will be able to link genetic capabilities to respond to consumer and societal needs.

If the general computing field is any indication, improved reliability and user-friendliness in both hardware and GIS software will continue into the future. For example, manufacturers have already made great strides in increasing the accuracy and reliability of yield monitors, and they have made them easier to implement and understand. Precision agriculture software packages have moved from general GIS applications to systems tailored to meet the needs of the farmer with a minimum amount of training. Hardware manufacturers will have to add features to their systems desired by farmers that will allow them to gain or maintain competitive advantage. Vendors will have to make their systems more reliable, user-friendly, and they will have to do a better job of clarifying the bottom line benefits precision agriculture tools and GIS can provide farmers.

Many farmers have postponed adopting GIS-enabled precision agriculture technologies because the potential benefits have not been demonstrated successfully. Integrating cutting-edge computer technologies into farming brings together two diverse fields. Farmers are traditionally slow to change and the world of information technology is continually changing at a rapid pace. The next form of IT for farmers is likely to be the development of decision support systems, which will help farmers to interpret the massive amounts of data and will help them to make better farm production decisions.

It will be the responsibility of participating companies to improve the way in which they demonstrate the benefits their products can offer. Precision agriculture practitioners have had success in adopting the new technologies when they hear the experiences of other adopters, and learn from those experiences, or when they are provided a strong commitment of both sales and support from precision agriculture companies. The trend for companies participating in the precision agriculture field will be to improve communication regarding benefits of their product, increase and improve the number of products available, and develop their abilities to make recommendations for farmers who have collected large amounts of data, but do not have the ability to properly analyze that data.

Utilizing GIS-enabled precision agriculture tools can be a daunting task even for computer literate farmers. The need for sales, service, analysis, and recommendations may foster the growth of new full-service precision agriculture companies. These new companies will serve as precision agriculture outsourcers who advise the farmer on the available technologies, assist in their implementation, and make sound agronomic recommendations based on the collected data.

Farmers who successfully collect data about their farm through yield monitoring, soil sampling, or remote sensing are often left with the problem of data analysis. The question arises, "I have the data — now how do I capitalize on it?" Recommendations will need to be tailored to the specific region of the country, crop, and soil conditions, among other variables, but a company could develop standard analyses and recommendations that incorporate these certain parameters. A company that provides quality service and assistance could convince farmers to adopt new methods that they previously thought were too costly and complicated.

Another potential trend for GIS in agriculture is the move to Internet-based data access. Whether the data is collected by the farmer, a third party service provider, or government agencies, the Internet can serve as a clearinghouse for a variety of data types of interest to the agriculture community. Farmers could obtain remote sensing data, soil type maps, or any other data type that could be incorporated as a data layer in their GIS. Fee-based and free maps and imagery are currently available to farmers online, and this trend is expected to continue. In addition to downloading pertinent information, farmers will also have the ability to submit data collected to a service provider for analysis, and receive the recommendation maps from the service provider over their website. Utilizing the Internet will decrease data collection costs, increase data availability, improve data analysis turnaround times, and serve as a valuable source of information on which to base farm management decisions.

An emerging trend in the precision agriculture industry is the use of GIS-enabled farming tools to assist in adhering to strict environmental standards. In addition to providing cost savings and improved crop production, precision agriculture tools give farmers the ability to closely track and monitor inputs applied that have an effect on the environment. GIS can serve as an accounting tool that is accurate, searchable, and capable of being analyzed for year-to-year changes in environmental compliance. The Food Quality Protection Act passed in August 1996 fundamentally changed the way EPA regulates pesticides (United States Environmental Protection Agency, 2003). The Act includes a new safety standard, which states that with reasonable certainty no harm must come from the application of pesticides on food. GIS gives farmers the ability to show when, where, and exactly how much pesticide was used during crop production. This ability will help ensure that they are in compliance with such environmental standards. The elevated concern about the environment and the impact agriculture has on it will only increase farmer necessity to track and monitor chemical inputs.

GIS-enabled precision agriculture tools are very much in their infancy, and they are being implemented in a primarily step-wise fashion. It will be important for each farmer to assess their situation, evaluate the technologies available, and implement those GIS-enabled technologies that can help reduce costs or improve productivity. Continued acceptance by farmers of these new technologies will depend on the hardware and software manufacturers' ability to communicate not only how to use these technologies, but also how they can provide benefits to farmers. Farmers who have implemented these tools are successfully collecting data about their farms, but the lack of instruction on how to convert the data into information for use in decision-making has limited further adoption.

A niche market may develop for analysis and recommendations to farmers who do not have the ability to conduct such analysis, or to determine how to use the results to improve management decision-making. Farmers of the future will have more Internet-based data available, and GIS will be an important tool not only for data analysis, but also for environmental compliance.

Conclusions

With falling crop prices and increased overseas competition, farmers need every option available to maintain profitability. Although requiring a considerable investment of time, money, and commitment, GIS-enabled precision agriculture technologies help farmers gather information that can improve management decision-making. From yield monitoring to applying inputs only where needed, precision agriculture incorporates many different tools for maximizing production. The benefits to the farm management system are not limited to production. Improved record keeping leads to more accurate, streamlined accounting, and the potential adverse environmental impact of over-application of pesticides, herbicides, and other inputs is reduced when applied in a site-specific manner.

While the potential for creating efficiencies are possible with GIS-enabled precision agricultural tools, the steep learning curve of these technologies and the initial investment of each of the tools complicate farmers' adoption decisions. Farmers must decide which tools and software will provide efficiencies for their situation. The complexities and holistic potential benefits of these technologies should be the incentive for researchers in both the hard sciences and social sciences to find ways to evaluate the adoption, use, economic benefits, and managerial benefits to both farmers and agribusinesses, such as machine manufacturers and third party software suppliers. It is Lowenberg-DeBoer's (1996) observation that the social scientists started evaluating precision agricultural technologies well after farmers were aware of them. Researchers have an abundance of opportunities to study the use, economic benefit, and managerial improvement at the microlevel (i.e., the farmer), as well as the overlooked macrolevel. The sociological changes, such as farm structure, privacy and policy effects need to be evaluated.

Agribusiness and IS practitioners have opportunities to build systems that streamline the information gathering process and make decision-making easier. Building systems that are simple to use and are helpful to a large number of different types of farmers with different growing conditions will be a challenge. In the meantime, third-party support businesses, such as consultants, may be the step some farmers take before implementing GIS-enabled precision agriculture technologies themselves. Farmers will have to assess their situations and the available technologies and determine which combinations will provide the information necessary for improved business management decision-making. The proper combinations of technologies will enable farmers to utilize their GIS not only for data processing and viewing, but also as a valuable decision support system.

References

Bouma, J., Stoorvogel, J., van Alphen, B. J., & Booltink, H. W.G. (1999). Pedology, precision agriculture, and the changing paradigm of agricultural research. *Soil Science Society of American Journal, 63* (6), 1763-1768.

Brynjolfsson, E., & Hitt L. (1996). Paradox lost? Firm level evidence on the returns to information systems spending. *Management Science, 42* (4), 541-558.

Byerlee, D., & de Polanco, E. H. (1986). Farmer's step-wise adoption of technology practices: Evidence from the Mexican Altiplano. *American Journal of Agricultural Economics, 69* (3), 519-527.

Cochrane, W. (1993). *The development of American agriculture: A historical analysis.* Minneapolis, MN: University of Minnesota Press.

Daberkow, S. G., & McBride, W. D. (2000). Adoption of precision agriculture technologies by U.S. farmers. Presented at *2000 International Conference on Precision Agriculture.*

Davis, F. D. (1986). Perceived usefulness, perceived ease of use, and user acceptance of information technology. *MIS Quarterly, 13* (3), 319-339.

Dillon, C. R. (2002). A mathematical programming model for optimal management zone delineation in precision agriculture. Presented at the *American Agricultural Economics Association Annual Meeting*, July 28-31, 2002. Long Beach, California.

Grupe, F. H. (1990). Geographic information systems: An emerging component of decision support. *Information Systems Management, 7* (3), 74-78.

Heermann, D. F. (2002). The promise of precision agriculture: Five-year, multidisciplinary project reveals potential and pitfalls. *Resource: Engineering and Technology for a Sustainable World, 9* (7), 6-8.

Intarapapong, W., Hite, D., & Hudson, D. (2002). *The economic and environmental impacts of variable rate application: The case of Mississippi* (Research Report 2002-2003) (pp. 1-41). Mississippi State University, Department of Agricultural Economics.

Isik, M., Khanna, M., & Winter-Nelson, A. (2000). *Sequential investment in site-specific crop management under output price uncertainty.* 2000 International Precision Agriculture Conference. Retrieved January 29, 2003 from: *http://precision.agri.umn.edu/Literature.*

Kitchen, N., Hughes, D., Sudduth, K., & Birrell, S. (1995). Comparison of variable rate to single rate nitrogen fertilizer application: Corn production and residual. In P. Robert, R. Rust, & W. Larson (Eds.), *Site-Specific Management for Agricultural Systems.* Madison, WI: American Society of Agronomy, Crop Science Society of America, and Soil Science Society of America.

Lems, J., Clay, D.E., Humburg, D., Doerge, T.A., Christopherson, S., & Reese, C.L. (2003). Yield monitors—basic steps to ensure system accuracy and performance. *Site-Specific Management Guidelines* (SSMG-31). Retrieved May 18, 2003 from: *http://www.ppi-far.org/.*

Lowenberg-DeBoer, J. (1999). Risk management potential of precision farming technologies. *Journal of Agricultural and Applied Economics, 31* (2), 275-285.

Mahajanashettii, S. B., English, B. C., & Roberts, R. K. (1999). Spatial break-even variability for custom hired variable rate technology adoption. Presented at *American Agricultural Economics Association Meeting,* August 8-11, 1999. Nashville, TN.

Malcolm, H. (1996). Technology helps farmers target and reduce chemicals. *Christian Science Monitor, 88* (248), 10.

Markus, M. L., & Robey, D. (1988). Information technology and organizational change, causal structure in theory and research. *Management Science, 34* (5), 583-598.

Marrero, C. R. (2003). Farming's brave new face. *E Magazine: The Environmental Magazine. 14* (4), 11-12.

Mennecke, B. E., Crossland, M. D., & Killingsworth B. L. (2000). Is a map more than a picture? The role of SDSS technology, subject characteristics, and problem complexity on map reading and problem solving. *MIS Quarterly, 24* (4), 601-629.

Morgan, M., & Ess, D. (1997). *The precision-farming guide for agriculturists.* Moline, IL: John Deere Publishing.

National Agricultural Statistics Service. (2001, July 30). *Farm computer usage and ownership.* Retrieved May 15, 2003 from: *http://usda.mannlib.cornell.edu/usda/reports/nassr/other/computer/empc0701.pdf.*

National Research Council. (1997). *Precision agriculture in the 21st century: Geospatial and information technologies in crop management.* Washington, D.C.: National Academy Press.

Nedovic-Budic, Z., & Godschalk, D. R. (1996). Human factors in adoption of geographic information systems: A local government case study. *Public Administration Review, 56* (6), 554-567.

Olson, K. (1998). *Sixth Joint Conference Food, Agriculture, and the Environment.* Center for International Food and Agricultural Policy, University of Minnesota, August 31-September 2.

Rogers E. M. (1983). *Diffusion of innovations* (3rd ed.). New York: The Free Press.

Rothwell, R. (1992). Successful industrial innovation: Critical factor for the 1990s. *R & D Management, 22,* 221-239.

Sabherwal, R., & Chan, Y. (2001). Alignment between business and IS strategies: A study of prospectors, analyzers, and defenders. *Information Systems Research, 12* (1), 11-33.

Sawyer, J. (1994). Concepts of variable rate technology with considerations for fertilizer applications. *Journal of Production Agriculture, 7* (2), 195-201.

Sonka, S. (1998). *Precision agriculture as a priority.* USDA NAREEE Advisory Board. 2nd National Stakeholder Symposium. Washington D.C. Retrieved May 23, 2003 from: *http://nsrl.uiuc.edu/Presentations/usdapre.htm.*

Sonka S. T., & Coaldrake K. F. (1996). Cyberfarm: What does it look like? What does it mean? *American Journal of Agricultural Economics, 78* (5), 1263-1266.

Speir C., & Morris, M. G. (2003). The influence of query interface design on decision-making performance. *MIS Quarterly, 27* (3), 397-423.

Swinton, S. M., & Lowenberg-DeBoer, J. (1998). Evaluating the profitability of site-specific farming. *Journal of Production Agriculture, 11* (4), 439-446.

United States Census Bureau. (2002). *Statistical abstract of the United States: 2002.* Retrieved September 5, 2003 from: *http://www.usembassy.de/usa/etexts/stab2002/agricult.pdf.*

United States Environmental Protection Agency. (2003, March 26). *Food Quality Protection Act of 1996.* Retrieved May 10, 2003 from: *http://www.epa.gov/opppsps1/fqpa/.*

Vanclay, F. (1992). Barriers to adoption: A general overview of the issues. *Rural Society, 2* (2), Retrieved July 8, 2002 from: *http://www.csu.edu.au/research/crsr/f_journal.htm.*

Chapter XVI

Isobord's Geographic Information System Solution

Derrick J. Neufeld, University of Western Ontario, Canada

Scott Griffith, University of Western Ontario, Canada

Abstract

This chapter presents a case study of Isobord[1], a Canadian manufacturer of high quality particleboard that uses straw instead of wood as the main raw material input. Isobord is facing critical operational problems that threaten its future. Gary Schmeichel, a biotechnology consultant hired by Isobord, must recommend how much straw collection equipment to purchase and what kind of information technology to acquire to help manage equipment dispatch operations. Schmeichel is exploring how geographic information systems (GIS) and relational database management systems (RDBMS) might help manage operations, but budget and time constraints and organizational inexperience seriously threaten these efforts. Decisions must be made immediately if there is to be any hope of implementing a system to manage the first year's straw harvest. Readers are challenged to put themselves in Schmeichel's shoes and prepare recommendations for Isobord.

Background

"... Like Rumplestiltskin, Isobord is spinning straw into a wealth of new opportunity!" The room erupted with applause as Gary Filmon, Premier of the Province of Manitoba, Canada, dramatically concluded his speech welcoming Isobord Enterprises Incorporated to the small town of Elie, Manitoba. The ceremonial ribbon cutting officially certified Elie as Manitoba's latest boomtown and home to the world's first large-scale strawboard production plant.

The new 215,000 square foot Isobord plant is designed to produce more than 130 million square feet of premium-quality strawboard per year. What makes the Isobord operation unique is its reliance on an annually renewable agricultural by-product, straw, as the primary raw material input. Most particleboard plants rely on wood as the primary input.

When it is completed, the Isobord facility is scheduled to process 200,000 tons of wheat straw per year to produce its high quality strawboard product. Initial runs of the product quickly earned great praise in consumer markets, due to the superior physical and mechanical properties of the straw-based board. Specifically, because Isobord uses straw fibers and non-toxic, environmentally-friendly isocyanurate resins in the manufacturing process, the final product performs better than standard wood-based particleboard in terms of water resistance, moisture swell, elasticity, internal bond, weight, density, strength, moldability, and screw retention. U.S. consumers of particleboard were so excited about Isobord's product that they agreed to purchase 75 percent of the output before the plant was even constructed!

According to Gary Gall, Isobord's president, "The beauty of the Isobord product is that it utilizes an annually renewable natural resource that was previously considered to be an agricultural by-product. By utilizing the straw we can simultaneously help to combat the negative effects of straw burning, and create a sustainable business in Manitoba." Until Isobord came along, Manitoba farmers were forced to burn straw after the harvest each fall. With the Isobord option, farmers can now sell the straw, reduce their workload, and cut down on air pollution in one fell swoop.

Setting the Stage

Isobord, a startup company headquartered in Toronto, Ontario, is in the process of developing a strawboard processing plant in the fertile Red River valley of Manitoba, a location some 2,000 kilometers away from the head office. Isobord's ability to create a sustainable operation in Manitoba is largely dependent upon the abilities of its management team. Unfortunately, while fund-raising, production and promotion have all received consistent attention by senior Isobord executives, the problem of harvesting straw has been essentially ignored. These executives seemed to believe that the straw would deliver itself to the new Isobord plant.

Feasibility studies were subsequently conducted by consultants to quantify the costs associated with collecting straw. Unfortunately, Isobord lacked an internal figure who understood both agriculture and the technical aspects required to coordinate the logistics of a massive straw harvest. Furthermore, Isobord had not yet retained any in-house information systems expertise. This lack of resources was paired with a technology-averse attitude of senior managers within Isobord, brought on by prior bad experiences with IT projects.

Case Description

Isobord's success hinges on the availability of straw. Without straw there can be no board. At full-scale production Isobord will consume 400,000 wheat straw bales per year. The quantity of straw demanded by Isobord far exceeds any previous efforts at straw collection. As such, Isobord must carefully devise strategies to undertake the collection of straw resources. If straw resources become depleted, Isobord will lose revenues of approximately $200,000 for every day that the plant is shut down.

The initial business plan was based on the assumption that if Isobord offered to buy the straw, farmers would be willing to collect and deliver it to the plant. After all, farmers typically had problems getting rid of straw after the harvest each year (with heavy crops, straw was particularly difficult to reincorporate into the rich soils of the Red River Valley). As a result, farmers often resorted to burning as the only viable straw removal alternative.

Isobord saw this as a golden opportunity: instead of burning, farmers could sell their straw. This would reduce the negative environmental/air quality effects of burning (smoke created by burning straw is particularly stressful to children and people with respiratory ailments such as emphysema or asthma). The plan was simple and logical. Isobord would help alleviate burning problems by purchasing the wheat straw from farmers.

Isobord offered farmers $30 per metric ton of straw delivered to the plant. By doing so, they placed an economic value on a material that previously enjoyed no real market. Wheat straw, which had been used in small quantities for animal bedding, could now be sold in large quantities for a guaranteed price. Isobord was certain they would receive a warm welcome from the local community: they were providing a vehicle for farmers to simultaneously dispose of their straw and to make money.

What Isobord officials failed to consider was the effort required to collect and deliver the straw. As they quickly learned from the farmers, it is far easier to put a match to straw than it is to bale, stack, and haul large loads of low value goods to a distant plant. Compared with a typical $160/acre wheat crop (assuming average yield of 40 bushels per acre and average price of $4 per bushel), Isobord's offer of $30/metric ton (effectively $30 per acre, assuming an average yield of one metric ton per acre) did not provide sufficient economic incentive to persuade farmers to collect and deliver straw bales. Farmers would need baling equipment, stacking equipment, loaders, and trucks — all different from the equipment required for harvesting and transporting grain. By the time all the work was completed, farmers claimed they would be losing money by selling straw to Isobord. The farming community quickly became disinterested in Isobord's proposals. They were not willing to act as the manpower, equipment suppliers and logistical controllers for Toronto's high rollers. Isobord's simple dream of "build it and they will come" had crumbled.

Plan A had failed: straw would not be baled and delivered directly by farmers. But the first endeavor was not a total failure. Isobord management had learned a great deal about straw collection through its interactions with the farming community. Clearly, obtaining straw would require more focused efforts.

Current Challenges/Problems Facing the Organization

If the project were to continue, Isobord would have to get very serious about straw collection. To make its business case to investors, Isobord had to demonstrate the availability, affordability and annual sustainability of the required 200,000 ton straw supply. The banks insisted that Isobord provide contracts stating that farmers would supply straw, at a stated price, for a five-year period. If Isobord could not produce proof of a guaranteed straw supply at a known price for a minimum time period, the banks would not continue to contribute money to the project. Once the straw supply was committed to Isobord by the farmer, it would have to be baled, stacked and transported to the plant. Management insisted that regardless of the chosen straw collection method, it must fit within the established budget of $30 per metric ton.

The firm encouraged a group of farmers to set up the Straw Producers Cooperative of Manitoba to help develop the straw supply base. The Co-op was made up of a board of ten directors, each of who was responsible for promoting the Isobord project to a district in southern Manitoba. The Co-op's chief objective was to convince farmers to sign straw supply contracts with Isobord.

Soon after the Co-op was formed, Isobord President Gary Gall decided that additional efforts should be focused on organizing the straw collection operations. Gary contacted Rudy Schmeichel, a motivated independent startup consultant from Toronto who had just completed a contract with Lifetech, a new biotechnology company. Rudy accepted the Isobord challenge and moved to Manitoba. He was met at the Winnipeg airport by Scott Griffith, Isobord's only Manitoba staff member. Scott drove Rudy to the Elie office, located on the main floor of an old three-story house that had previously been a convent. From the convent, Rudy and Scott began their crusade to organize the Straw Division.

Rudy quickly made contacts in the community. He spent considerable time in the coffee shops listening to the locals talk about the Isobord project. He began to get his bearings and soon established a place for himself in the community. From his "field research" Rudy began to devise a plan to bring straw from the field to the plant. Major operations included: (1) baling the swaths of wheat with a baling unit (i.e., a tractor pulling a baling machine that gathers loose straw into standard-sized 4'x4'x8' bales); (2) stacking the bales with a bale forwarder (i.e., a tractor that collects bales and stacks them at the edge of the field); and (3) dispatching a loader and multiple trucks to the field to pick up the bales and haul them to the plant.

Baling and Stacking Logistics

Each baling unit can process up to 36 bales per baling hour, travel (between fields) at a maximum speed of 30 kph, and will cost approximately $250,000. Each bale forwarder can process up to 90 bales per hour, travel (between fields) at a maximum speed of 90 kph, and will cost approximately $220,000. Since the capabilities of bale forwarders exceed those of baling units, it made sense to organize equipment into "baling crews" consisting of one bale forwarder and two or more baling units. Baling crews communicate with the central dispatch office using two-way VHF radios.

While the straw collection process may appear fairly simple initially, Isobord faces a number of serious constraints. First, the maximum time available for baling is estimated to be only 600 hours per season, based on expectations regarding typical planting and harvesting schedules (e.g., straw harvesting must fall within the August-October window), climatic conditions (e.g., rain or snow can make fields inaccessible, and straw cannot be collected if its moisture content exceeds 16 percent), available daylight hours, and anticipated equipment breakdowns. Rudy estimated that approximately 75 percent of the 600 available hours will be spent baling, although this could range anywhere from 90 percent to 60 percent, depending on how efficiently dispatch is able to route the baling crews.

A second constraint is that straw must be collected from a wide geographical area: approximately 6,000 targeted fields are randomly distributed over a 50-mile radius of the plant. Operators will be required to navigate an expansive network of unmarked municipal mile roads; finding an unknown field might be compared to finding a house in a city where the streets have no names — or perhaps to finding "a needle in a haystack." Furthermore, very few landmarks exist on the Manitoba prairies to help vehicle operators find their way. If operators get lost, they could waste hours of valuable baling time. Stephen Tkachyk, an Isobord baling crew chief, had experience navigating country roads at night. He stated that if you get lost at night, your best bet is to go to sleep in the truck and find your way in the morning!

Hauling Logistics

Several trucking companies were asked to provide quotes for hauling services. Because of the unique nature of this hauling job, all of them requested additional information. To provide quotations the trucking firms wanted to know the following:

- The number of loads that would be hauled from each location
- The distance to be hauled
- The location of the pick-ups
- The routes to be traveled

To accumulate the necessary information, Rudy suggested that "trip tiks," such as those offered by automobile associations, be created that defined the route from each field to the plant. Rudy needed the distance numbers as soon as possible so that the trucking companies could provide quotations, so Scott began developing handwritten trip tiks. However, with over 300 Co-op members, many of whom farmed multiple wheat fields, this process quickly became tedious. Furthermore, given the shifting nature of wheat field locations due to crop rotation, it was apparent that these "one-off" trip tiks would become outdated very quickly.

An Initial Solution

To organize routing of baling crews it was first necessary to determine where Co-op members were located. Scott purchased maps of Southern Manitoba and began marking the Co-op member locations using colored pins. Unfortunately, initial information provided by Co-op members only indicated locations of farmhouses, not wheat fields. Isobord would need to know specific wheat field locations in order to provide baling service (legal field descriptions are designated by section-township-range coordinates, e.g., field SW-12-07-02-W is on the South West quarter of section 12-07-02, West of the central meridian).

Over the next few weeks, Scott developed a system that would be easier to change as Isobord's needs changed. Using an Apple Macintosh computer and drawing software, he divided Isobord's straw collection area into a number of grid sections. Each grid section had a designated exit point, and the single best route between the grid exit point and the plant's straw storage site was identified. The best route between each particular field and the grid exit point was then determined on a field-by-field basis. The result of this exercise was a series of grid maps that included relevant roads, a section grid (i.e., one square mile sections each identified by a legal land description including section, township and range), towns, hydrography, and the locations of Co-op members' farmhouses.

A grid section reference identifier and distance-to-plant data were then added to each Co-op member's database record. The grid mapping process sped up the calculation of trucking distances, and provided a graphical reference base for all Co-op member locations. When the grid maps were complete Scott produced reports using Microsoft Access that summarized distance to be traveled, routes to be taken, and number of loads to come from each location (number of loads was determined as a function of acres under contract from each Co-op member).

When a field is ready to be baled, a Co-op member will now call Isobord's dispatch office to schedule baling. The farmer will provide his name, the number of acres to be baled, and the legal description of the field. Dispatch operators will then prioritize the calls and assign them to a specific baling crew. The baling crew will navigate to the new field using detailed grid maps. The goal of the dispatch operators will be to minimize each crew's travel time, in order to maximize productive baling time.

Isobord had minimal technology available to coordinate baling efforts. Isobord was currently only about ten employees strong, and Scott, a management student who was just starting to learn how to program, was the only employee with any knowledge of software development. Isobord's information technology architecture consisted of a PC running Microsoft Access database software, which Scott used to create prototype databases, as well as Scott's personal Macintosh computer that he used to create the digital maps. During the prototyping stage no formal systems methodology was used to guide development. Systems were developed based on theoretical concepts as to how a baling operation of Isobord's magnitude could be organized. During development the software was continually revised as new ideas surfaced. Based on the lack of in-house expertise and resources, Scott began to seek out a third party solution.

Information Technology Options

Scott saw the value of using computers to help solve Isobord's straw collection problems, and so he did further research to learn what information technologies were available to help administer dispatch operations. He learned that systems used for computerized dispatch typically utilize two technologies. The first, relational database management systems (RDBMS), are used to create the custom interface required for entering, manipulating and storing data. The other dispatch technology commonly used, geographic information systems (GIS), are used to store and manipulate digitized map data. GIS map data consist of real-world latitude and longitude coordinates, which may be input to the system using available paper map data, or via global positioning systems (GPS) receivers. When combined with RDBMS technologies, GIS can be used to attach meaningful attributes to map elements (e.g., a user could display a particular city on a map as circle, and then attach population, land area, and tax base data to the circle). The appendix contains a technical note describing GIS technology in more detail.

A RDBMS and GIS combination would allow Isobord to administer and control all information relevant to straw collection, such as:

- information pertaining to all Co-op members, including wheat field locations and summaries of accounts payable for harvested straw
- a record of all calls received requesting baling service
- activity records for all field operations
- inventory of all machinery, parts and supplies
- payroll for employees who provide baling services

The GIS component of Isobord's system would consist of two main parts: map data and analytical capabilities. To map out Co-op members' field locations and coordinate machinery, Isobord requires "quarter-section grid" data (which defines all sections of land in Manitoba by their legal description), and "road network" data (which details all provincial and municipal roads throughout Isobord's dispatch area). Other map datasets (e.g., provincial hydrography, rail lines, towns and cities) would be useful for navigation purposes. The ideal GIS would display incoming calls, color-coded to indicate priority, on a digital map of Isobord's straw collection area. Thus, dispatch operators would have a constant visual representation of current operations, and would be able to assign baling crews in such a way as to minimize travel time.

By using a true GIS, Isobord could track important information about baling operations on each farm field visually. The field would be displayed on a digital map and each field point could be tied to all the data pertaining to the field. Dispatch operators viewing a field location on a map could access data pertaining to that field, including when the farmer called to have the field baled, field size (number of acres), type of wheat straw planted on the field (spring, winter, durum), and type of combine (farm equipment) used to harvest the straw. This information would allow Isobord's dispatch operators to make decisions quickly based on current information. The power of easily accessible spatial and operational data through a GIS/RDBMS system would provide dispatch operators with the means to optimize the effectiveness of the baling fleet.

Isobord's GIS should also be capable of analyzing the shortest and/or fastest path between any two points on a map. During calculation, the GIS should take into account variables, such as the number of turns required to travel from one place to the next, the speed limit, and the road conditions for each segment of road. The user should also have options to classify roads (paved, gravel, or clay), define speed limits, or even remove a road from any route calculations (e.g., if it is unfit for travel due to flooding).

GIS Prototype

Scott believed GIS software would be perfect for managing Isobord's straw resources and maximizing baling crew efficiency. He began to develop an experimental prototype GIS application using Microsoft Access. The grid maps previously created on the Macintosh were imported into the Access database and attached to Co-op member records. For each Co-op member, the dispatcher could now click an icon to load up a map of the relevant grid section.

While the prototype system admirably demonstrated the benefits of using a GIS, its functionality was limited. The map "data" were actually static and discrete graphical images, rather than dynamic GIS datasets. As a result, any change in a Co-op member's field locations required a change to the underlying graphical image. In a true GIS, digital maps would be derived from data contained in an underlying GIS dataset; any changes to the data in a true GIS dataset would automatically be reflected in the map displayed. Furthermore, the prototype did not contain analytical functionality (thus, shortest route calculations could not be performed).

By this point Scott had developed a solid understanding of how straw collection logistics could be organized in order to maximize the utilization of equipment. However, he suspected that, as a new company, Isobord's ideas about how to manage operations would evolve significantly over the first few years of operations. As a result, Scott thought that retaining in-house control over software design would enable Isobord to implement a flexible solution that could be modified as the business of straw collection evolved. Scott therefore intended to internally develop database components in Microsoft Access, and then link that data to a commercial GIS tool. This solution would allow for flexibility in database design, while providing the analytical capabilities such as route optimization and thematic mapping offered by a full-scale GIS package. By performing some of the work in-house, Isobord would be able to control costs during the prototyping stage, and easily implement changes to software and processes as the art of a massive straw harvest became a science. At this point, Isobord approached three GIS consulting firms to discuss the possibility of creating a dispatch information system.

Firm #1: International Operating Systems (IOS)

With only three employees, International Operating Systems (IOS) was a small but innovative information systems consulting firm. IOS offered to sell Isobord a package called Bentley MicroStation (www.bentley.com), a geoengineering software application with GIS capabilities. MicroStation provided basic database and mapping capabilities, and IOS was willing to further develop the analytical capabilities required by Isobord.

While Scott was impressed with IOS, he had concerns about the complexity of the MicroStation proprietary programming language and the package's limited database functionality. To create a robust and user-friendly application, the MicroStation GIS would have to be linked to an Access database. The link was theoretically possible through ODBC (Open Database Connectivity — a standard database communications protocol), but IOS had never actually linked a MicroStation application to an Access database.

IOS developed a proposal for Isobord that included development of base map data in MicroStation, and a basic link to the Access database. IOS asserted that it would need to develop the road network dataset and portions of the quarter section grid dataset. IOS would also need to create and identify river lot locations. The total development cost would be $20,000, including the MicroStation application, dataset development, an initial MicroStation-Access link, and a two-day training session on MicroStation for one Isobord staff member.

Firm #2: DataLink Mapping Technologies

Further research revealed a second small GIS consultant, DataLink Mapping Technologies (www.granite.mb.ca/datalink/). DataLink specialized in developing GIS applications using MapInfo (www.mapinfo.com), a Microsoft-endorsed application. As with the MicroStation application, MapInfo could be linked to an external database via ODBC. MapInfo's documentation explicitly indicated that it could be integrated with Microsoft Access. It was also compatible with other database products through ODBC.

As with IOS, the DataLink proposal indicated that the road network data were not yet available from the Province of Manitoba. Thus, MapInfo could provide a solid base for displaying Isobord's geographic data, but could not perform shortest route calculations. Also, while integration of MapInfo with an external Access database was theoretically possible through ODBC, DataLink had not experimented with this MapInfo feature. DataLink produced an informal quotation to supply MapInfo, complete with the quarter section data, river lots, and hydrography, for under $10,000.

Firm #3: Linnet Systems

An official at the Manitoba Land-Related Information Systems department recommended that Scott contact Linnet Systems (www.linnet.ca), a data warehousing, surveying and systems development firm. After discussing the situation with Isobord, Linnet put together a detailed proposal that included acquisition of hardware, software, datasets, and all required development and training to provide a "complete business solution." Scott was surprised to learn that both DataLink and IOS would be purchasing all or part of their GIS datasets from Linnet.

Linnet wanted to aid in the development of an administration system to help Isobord monitor and control their operations. Linnet had already developed an extensive production management system for Louisiana Pacific, an oriented strand board (OSB) plan in northern Manitoba, to help monitor all aspects of its operations (e.g., resource planning, plant scheduling, hauling, paying invoices, tracking materials usage), and was eager to apply this knowledge to Isobord.

Linnet's proposal to Isobord was presented in standard systems development life cycle format. It included requirements and design, prototyping, application development, implementation, training, testing, and troubleshooting. The total price tag before taxes was $295,000. Throughout the quoting process, Linnet representatives repeatedly stressed that they only build systems that provide benefits to the client.

Scott further questioned Linnet about the possibility of using global positioning system (GPS) technology to monitor the movements of Isobord's baling fleet in real time. With GPS, Isobord could continuously monitor all of the equipment in the field; equipment operators would never get lost because the dispatch operators would always know their exact location. The GPS system would require a base receiving station equipped with a base-station monitoring device to relay all incoming GPS coordinates to a computer terminal. GPS units would cost approximately $2,200 per unit, plus $40,000 for a base station at the plant.

Linnet planned to develop the entire application using Power Builder to create the user interface and Oracle as the back end database. The GIS functionality would be developed within ArcView. All of Isobord's data would be stored in the Oracle database and accessed by both the Power Builder App and the ArcView App through ODBC. Linnet touted the benefits of Oracle and PowerBuilder as scaleable platforms capable of running systems throughout an enterprise. Their staff consisted of over 65 employees including several professional developers who consistently used Oracle/PowerBuilder and ArcView to provide solutions for clients.

Overall Evaluation of Technical Alternatives

Regardless of the solution chosen, Isobord would have to create a formal plan for the information system. The process of designing the structure of the database would be controlled largely by Isobord employees — and its success would hinge on their ability to communicate the system requirements to the chosen vendor. Through prototyping Isobord had created a working entity-relationship diagram for the project and had also documented the overall process flow for operations. At this point Scott and Rudy were the only individuals within Isobord who understood the intricacies of baling, stacking and hauling logistics. Unfortunately their knowledge was based on theory rather than experience. At this point they would have to invent the process and the supporting technology in parallel to complete an information system in time for the first straw harvest.

Time was required following initial development to test the system by simulating the conditions during a straw harvest. By populating the system with farm location data, and then running a series of test queries, Scott and Rudy hoped to verify that the functionality of the software met operational requirements. Typically farmers could provide a legal description of their land. The description would adequately identify field locations for about 75 percent of the Co-op members. For the other 25 percent of field locations, Isobord's dispatch office would require a paper map to ensure that the fields could be properly digitized and added to the GIS.

As a final provision, users would need to be trained in the operation of the dispatch information system. The software would have to be intuitive and user friendly if it was to be successful. This demanded that the software and hardware solutions had to be both stable, and not overly complex. Scott had heard mixed reviews about Microsoft Access. Professional developers he spoke with frequently said "it's a great place to start, but it has its limitations. It's not good for multi user systems beyond 10 clients or so, and it has a slow database engine so you need to watch your table size..." Beyond Access, the other technologies suggested by technical consulting firms were foreign to Scott. It was clear that new IT skills would have to be developed rapidly regardless of the solution chosen.

The Decision

Certain Isobord executives had become so enthused about the prototype system that they wanted to go ahead and use it for the first year. Other key players believed that it would be preferable to do things manually first, before investing in any information technology. Scott recommended implementing a real GIS system on a smaller scale during the pilot testing phase, when straw harvested would be used mainly as buffer inventory (i.e., a "backup" supply of straw that would help to prevent a stock out situation), so that the bugs could be identified and fixed before straw collection became mission-critical.

Budget constraints made the acquisition of expensive software problematic. The original budget had been prepared with the understanding that farmers would bale and deliver straw themselves. With Isobord undertaking straw collection operations, the realities had changed considerably. It was difficult to finance a software project that was not even anticipated during initial budgeting.

As Rudy paced back and forth through the upstairs chapel in the convent, he needed answers. He knew that equipment purchases were dependent on his ability to coordinate the logistics of straw collection, but how could he quantify this decision? How many baling and forwarding units should be purchased? How should dispatch operations be managed? What information technology solution should he recommend? What was the long-term, strategic value of a GIS to Isobord? He needed to come up with an action plan now to be ready for straw collection in five short months ...

References and Further Reading

Research Articles:

Dennis, A.R. (1998). Using geographical information systems for decision making: Extending cognitive fit theory to map-based presentations. *Information Systems Research, 9* (2), 194-203.

Fung, D., & Wilkes, S. (1998). A GIS approach to casino market modeling. *Journal of Applied Business Research, 14* (4), 77-88.

Smelcer, J.B., & Carmel, E. (1997). The effectiveness of different representations for managerial problem solving: Comparing tables and maps. *Decision Sciences, 28* (2), 391-420.

Watad, M.M. (1999). The context of introducing IT/IS-based innovation into local government in Columbia. *Journal of Global Information Management, 7* (1), 39-45.

West, L.A. (2000). Designing end-user geographic information systems. *Journal of End User Computing, 12* (3), 14-22.

Practitioner Articles:

Millman, H. (1999). Mapping your business strategy. *Computerworld, 33* (3), 77.

Reed, D. (1998). Mapping up customers. *Marketing Week, 21* (10), 47-50.

Pack, T. (1997). Mapping a path to success. *Database, 20* (4), 31-35.

Web Sites:

Interactive GIS demos, including JAVA and Shockwave applets, are available on the Web (e.g., *http://www01.giscafe.com/technical/Demos.php*; *http://www.mapinfo.com/free/index.cfm*).

Internet Guide to GIS. http://www.gis.com. Managed by ESRI, *http://www.esri.com*.

Company Web Sites:

Isobord (www.dow.com/bioprod), MapInfo (www.mapinfo.com), Linnet Systems (www.linnet.ca).

Trimble Navigation Limited. Trimble Global Positioning System (GPS) Tutorial, *http://www.trimble.com/index.html*.

Endnote

[1] Isobord was acquired by Dow Bioproducts Ltd. in June 2001 (www.dow.com/bioprod). The information in this case relates to Isobord's operations prior to the acquisition by Dow.

Appendix

A Technical Note on Geographic Information Systems

Geographic information systems (GIS) allow users to integrate databases with computerized maps. By integrating maps with traditional databases, GIS software can produce visual representations of complex data. The visual aspect of GIS is its key strength. Virtually any data can be mapped to create a picture that captures the nature and extent of the relationship between the data and its geography. For example, GIS is often used by marketing organizations to plot demographic data for regions within a given city or postal code. This form of analysis allows users to create maps that display regions that fall within required demographic parameters.

GIS have been embraced by many organizations as a tool for generating competitive advantage. It is most commonly used for strategic planning, natural resource management, infrastructure management, routing and scheduling, and marketing. Telecommunications companies in particular have embraced GIS solutions. For example, Ameritech Cellular and Germany's Deutsche Telekom have exploited the power of GIS software in comparing potential site locations for new communications towers. Factors such as physical characteristics of the land, interference from other communications devices, height restrictions, weather patterns, consumer demand, and even the probability of floods, tornadoes, and other natural disasters are all analyzed by the GIS model to help define optimal locations for new towers. John Smetanka, director of R&D for On Target Mapping, knows the importance of GIS in telecommunications planning. He commented that, "When you map out all the variables, sometimes there are only one or two places a tower can be built. Maps are critical to having plans approved."

GIS are also used for information management. Deutsche Telekom is converting all paper-based drawings of their telecommunications networks into digital data. The geography of their networks is currently shown on over 5 million paper drawings and schematics. The digital versions of these documents will be integrated into a GIS data warehouse that will eventually hold three terabytes (i.e., three trillion bytes) of information. By using a GIS rather than a conventional database, Deutsche Telekom is able to record precise spatial locations for every component within their communications infrastructure. This capability makes the maintenance of drawings and physical components much easier.

The potential applications of GIS may be limited only by the imagination of its users. Proponents believe GIS will become a universal business tool. GIS applications require powerful computers that until recently were not readily available. With cheaper, faster computing power infiltrating businesses, GIS is now accessible to mainstream users.

GIS Data

Map data are stored in two different formats. Raster data are used to store images such as aerial photographs, and are extremely demanding on computer systems in terms of storage and processing requirements. Vector data are used to represent spatial lines and points, such as the highways, roads and rivers found on common paper maps, and require less much storage space and processing power.

A major challenge currently facing GIS users is the acquisition of data. GIS software provides users with the power to perform complex data analysis, but map data (raster or vector) are unfortunately not always readily available, and developing new datasets is time-consuming and expensive. Data to build a GIS dataset usually come from three sources: paper-based maps, aerial or satellite photographs, and/or global positioning system (GPS) receivers.

- **Paper Maps.** Data from maps can be input using a digitizing tablet. Digitizing a map is a painstaking process of tracing each line with a digitizing tablet, clicking to create individual points, and then joining the points to create vectors that represent the lines on the map. The quality of a GIS dataset developed from paper maps is directly dependent on the quality of the original map, the number of points used to define each line, and the skill of the digitizer operator. It is also possible to scan the original map to create a raster image, and then use computer software to automatically generate a vector image. This approach is becoming more popular, but requires significant computer processing power.
- **Photographic and Satellite Images.** In addition to simple aerial photographs, satellites can take pictures of the earth using a wide array of imaging techniques. Data collected from satellites include pictures taken within the visible light spectrum, ultraviolet light (UV), and near infrared (NIR) light. Electromagnetic energy can also be measured using satellites. GIS software can use these different forms of spatial data to generate images.
- **Global Positioning Systems.** GPS receivers use satellite positioning to derive precise coordinates for a point on the earth. GPS receivers convert signals from orbiting satellites into latitudinal/longitudinal position, velocity, and time estimates. Data from a minimum of four satellites are required for these estimates. Coordinates for any type of landmark can thus be derived and then stored for translation into map data. Data can also be transmitted in real time from the GPS receiver to a base station. By joining these point coordinate data to create vectors or lines, a map can be created that displays the precise path traveled by the GPS receiver. This technique is often used to define roads. Collecting GPS data can be time-consuming because it requires collection of information at the physical location on the ground. However, the data collected using GPS is considered better than data collected by other means, since it relies on actual coordinate data rather than data interpolated from a paper map.

The Power of GIS

Once raw map data have been acquired, attributes are attached to map points and stored in an integrated database package for later reference. As many attributes as necessary can be attached to each line or point on the map. This allows users to create complex graphical structures such as road networks, and then keep track of details such as road names, speed limits, surface types, highway classifications, street signs, and number of accidents per segment.

Once detailed data have been attached to the GIS, the user can begin to unleash its analytical capabilities. For example, one add-on GIS application called Network Analyst allows users to determine the best possible route between any two points on the road network with two simple mouse clicks (taking into account variables such as speed limit, traffic patterns, number of turns required on route — and even assigning different times for left and right turns). The capability of GIS as a routing tool has been embraced by organizations such as emergency services, taxi companies, bus companies, package delivery companies, and trucking companies. FedEx now uses a GIS to plot carrier stops to optimize carrier efficiency, and save time and money. Routing is only one example of the analytical capabilities available through GIS.

Challenges

One of the greatest challenges facing users of GIS is the lack of data standardization. Different software vendors use different file formats, and problems arise when users try to integrate datasets with varying scales to create a digital map. For example, the dataset that defines all roads within a certain area may be of 1:50,000 scale, while the bridges on these roads may be at 1:20,000. When the two datasets are layered, the bridges may not appear on the roads. In this case, many users choose to adjust the bridge data so that the bridges appear over the roads. Unfortunately, this practice degrades the quality of the bridge data, resulting in a picture that is more visually appealing, but less precise than the original. Errors are introduced each time a dataset layer is added and altered, and the reliability of the data is increasingly degraded. Scale matching is one of the most challenging aspects for GIS data developers.

GIS critics have warned of possible abuses of this technology. In the marketing arena, for example, GIS have been criticized as counter-productive. When marketers use GIS to uncover specific target or niche groups, they can introduce greater costs by trying to reach small segregated populations. Direct marketing relies largely on economies of scale to achieve penetration. Chris Greenwood, Director of a GIS consulting firm, noted that when you drill down into a market, you may end up spending the same money as a mass mailer, yet you will reach a smaller population. Greenwood believes that "The trick is to get the geographical scale to match the potential business benefit ... In my experience, people try to buy better geography than the underlying business issues require."

Chapter XVII

GIS and the Future in Business IT

Joseph R. Francica, Directions Magazine, USA

Abstract

The development of mobile location technology, proliferation of web-based mapping applications, mandates by federal agencies, and lower cost of ownership have all spurred new interest in enterprise deployment of spatial information. Decisions taken today are already shaping the future of geographic information systems within corporations worldwide. The result is a greater reliance on spatial information, not just by executive management, but also by an increasingly mobile workforce that is using personal communications that are "location-aware." In short, cost, components, and connectivity will continue to be key business drivers for geographic information and spatial technology management.

Introduction

In the early 1990s, businesses slowly began to recognize the value of geographic information, even though both spatial technology and spatial data were used every day in support of critical investment decisions. Today, spatial information has become such an important strategic aspect of corporate planning that applications are deemed too proprietary for external discussions, and some of the more significant applications developed by companies go unreported in the literature.

The use of "business geographics" technology has gained acceptance by corporations because software developers like MapInfo, ESRI, Caliper, and now Oracle and Microsoft have released products that are critical to business processes. The proliferation of spatial information now goes beyond merely making maps for presentation, but has been embedded in enterprise systems such as customer relationship management (CRM), finance, logistics and route planning, human resources, and security of critical infrastructure such as in private utilities. These software development companies have moved beyond delivery of "shrink-wrapped" products to release components for the development of custom applications and, more recently, web services.

In the future, businesses whose core competitive advantage is rooted in geography will be using these tools to create a disruptive marketplace to gain market share and win customers. Overspending on technology in the 1990s has given way to the conservative, cost-conscious executives of the 2000s who are looking to both streamline workflows while providing exceptional customer service. For companies in retail, banking, and insurance, spatial information is a matter of exploiting location in tactical business processes, while those that have business models rooted in geography such as in real estate, utilities, and transportation, will continue to derive strategic benefits. The future business prospects of both groups depend upon systems, components, and services that consistently leverage spatial data.

In any chapter that attempts to predict or assume some likelihood of a future occurrence, the risk is to tempt obsolescence following its publication. Yet, reasonable prognostications are feasible in view of planned product releases that are likely to shape business decisions in the coming months and years. This chapter will explore why certain catalysts are already in place to move the early adoption phase of technology acceptance into maturity and productivity. From enterprise systems to in-vehicle navigation systems to mobile handsets that are "location-aware," spatial information will continue to find its way into more mainstream IT systems.

Background

During the decade of the 1990s, three significant factors contributed to the success in bringing spatial information technology deeper into the corporate framework for developing business strategy. The first was a decrease in the total cost of ownership of spatial information technology. Software prices, specifically for personal, desktop computers, and data costs, both demographical and geographical (streets, areas, points of interest), declined significantly in price. There was also a consistent drop in the price of personal computers and memory, both random access and hard disk, during the past ten years and a corresponding increase in processing speed, which was fundamental for the ability to rapidly analyze complex spatial phenomenon.

Desktop GIS systems decreased in price and increased standard functionality. Software providers had to add basic boundary or demographic data to entice buyers. Engineering, environmental, and government users gravitated toward systems that provided better data capture and maintenance functions, while business users sought software that was easy to combine demographic data with standard political geographies. While many GIS companies (ESRI, Intergraph, MapInfo, Autodesk, Tactician) offered solutions below

$5,000* and many near $1,500*, others enticed buyers with sometimes-equivalent functionality in the $250 to $500* range (Caliper, Scan/US, Microsoft) [*Prices as of 2003].

The second factor was the move from complete desktop systems to component technology. As a result of moving to a component object model (COM) framework for many software systems, spatial information software "applets," such as for conversion of file formats or projection systems, could be "embedded" within other information technology systems. Hence, the door to the spatial attributes of corporate databases was now open and geographic relationships could now be discerned.

The third, and perhaps most significant factor was the Internet. As mapping technology was moving onto the Internet, more corporations were taking advantage of exposing its staff to information stored in its databases. Internet mapping software made it easy to visualize these data and to distribute this information to remote office locations.

Now consider the first few years of this century. The rapid development and affordability of personal mobile devices and handsets has created yet another opportunity to both capture and broadcast location-based content. "U-computing" ("U" for ubiquitous) is engaging more people who can easily transport business data with mobile devices and stay connected with the more "non-mobile" enterprise data systems of corporate headquarters. Some of this has been driven by the United States Federal Communications Commission that has required cellular telecommunications carriers to eventually identify the location of emergency cellular telephone calls to within 50 meters for 67 percent of calls and 150 meters for 95 percent of calls. These location-aware devices also provide the carriers with the opportunity to market services for helping to locate your nearest "buddy" as well as the nearest restaurant or movie theater. These "smartphones" and Personal Digital Assistants (PDA) can display maps as well as directions to events, entertainment, and other destinations.

The convergence of affordable devices, software, and a demand for better mobility has made "location" an essential attribute of any information service. The desire for more mobility has driven the move from desktops to laptops; from laptops to PDAs; and maybe from PDAs to more robust TabletPCs. In so doing, the emphasis is now on communicating with these mobile devices, wirelessly, and hence the desire to know "where" these devices are located. Support for wireless technology has become a standard component of mobile devices. For example, the Intel "Centrino" microprocessor has built in 802.11b Wi-Fi capabilities. Mobile devices that function as phone, camera, and PDA are common and many also build in slots for "add-on" Compact Flash cards to support Global Positioning Satellite (GPS) sensors.

The rise in importance of spatial information can also be attributed to the parallel rise in stature of the chief information officer (CIO) within corporations. The CIO, with a broad mandate, had to assess the effectiveness of "all" information systems that contributed to the day-to-day business activities within the company. For companies that relied on location-based data as a critical part of their business, such as those mentioned previously, the CIO often had to become more aware of GIS as a component of enterprise resource planning (ERP) and allocation systems. As such, an important challenge remains: how will CIOs enable the integration of spatial information and technology with existing and critical enterprise systems (financial, human resources, and marketing) to more fully develop location-based business intelligence.

This fact has not been lost on database software suppliers. In the mid-1990s Oracle and Informix (later acquired by IBM) began to provide the ability to incorporate functionality to understand spatial primitives (points, lines, and areas) as attributes with their database products. Users could pose spatial queries of the database without the need for a separate geographic information system. Modules for geocoding and map display have been added as well, and most basic spatial functionality is now being offered as a standard benefit. The result is built-in functionality that database administrators can leverage to expose location information in any enterprise system, such as human resources, financial management, and marketing.

These factors contribute to the future of GIS in the private sector. Vertical industries that were either slow to adopt spatial information or had relegated it to a single department's domain will find the workflow easier to manage because functionality is incorporated within the database.

Future Trends of Spatial Technology in Business

Innovative product functionality has driven the adoption of geographic information system software. Some functionality has migrated from the desktop system to the web-based mapping system, allowing adoption among a greater number of users and the introduction of new applications. There are a few trends noted here that merit particular attention, especially where Internet connectivity meets a wireless mobile application, databases, and the development of new tools and data that support them all.

Web Services

Web services constitute the ability to construct spatial solutions using compatible application interfaces in conjunction with geographic data to create an Internet-based application. Web services are a clear evolutionary path for Internet mapping software solutions. The Open GIS Consortium (OGC) has provided guidelines for the development of web services solutions. They state, "OGC Web Services will allow future applications to be assembled from multiple, network-enabled geoprocessing and location services. This capability will be possible because rules will be established for these services to advertise the functionality they provide and how to send service requests via open, standard methods. In this manner, OGC Web Services will provide a vendor-neutral interoperable framework for web-based discovery, access, integration, analysis, exploitation and visualization of multiple online geodata sources, sensor-derived information, and geoprocessing capabilities."

The vision for web services holds that interoperable components can be rented or purchased from vendors to create an Internet mapping solution. Geospatial data will also be available through a subscription service, which resides in a format compatible with the service. The intended result is for application developers to have the ability to quickly assemble a solution from a library of services and data. The OGC goes on to state that, "OGC Web Services will allow distributed geoprocessing systems to communicate with each other using technologies such as Extensible Markup Language (XML) and Hypertext Transfer Protocol (HTTP). This means that systems capable of working with XML and HTTP will be able to both advertise and use OGC Web Services" (OGC, 2003).

Why is XML so critical to the development of web services? Steve Lombardi of Microsoft's MapPoint product division explains it this way: "XML allows computing machines to share data regardless of the operating system or programming languages used by their peers. At their core, XML Web services are small applications that use XML to share data and functionality amongst themselves across the Internet. As XML is an open standard supported by all major operating systems, development tools, and platforms. XML Web services enable communication between previously disparate systems- A Linux server application can interact with a Windows Server application, a Pocket PC device can programmatically access services hosted by a Solaris server, etc." (Lombardi, 2002). Therefore, it matters little if the geocoding service a developer wishes to use has been written in Microsoft Visual Basic and resides on a Windows 2000 server, and the map rendering engine written in C++ resides on a Windows XP computer, or the thematic mapping functionality written in Java resides on a Linux server. XML provides the protocol for communicating with any service that allows the most appropriate components to be joined into single web-based application. Therefore, barriers that once prevented rapid prototype development are quickly being removed through a standard programming interface.

Wireless Location-Based Services

Wireless location-based services (LBS) will build on present innovation and deliver future benefits. Jim VanderMeer (2003) stated that, "LBS scenarios such as 'send me a coupon from the nearest coffee shop' and 'provide me turn-by-turn routing to avoid traffic on the way from the coffee shop to work,' were the hype behind LBS which continued into 2002. Now, as we move into 2003, LBS will start to mature through realistic implementations of location revenue models. To fulfill these realistic implementations, however, the location-based services industry will require consolidation by the various mapping, infrastructure, and service providers. Despite minor setbacks and continued slow consumer uptake, 2003 will mark a turning point for LBS as a market segment... The location-based services market segment is no longer in its infancy. Location determining technologies combined with mature GIS solutions are available today, and are ready for horizontal mass markets" (VanderMeer, 2003). As the demand for location technology increases, the revenue model used by wireless carriers and LBS providers is not likely to take on any single revenue model. Some will charge monthly fee specific services. Others may charge by the amount of minutes or storage consumed: i.e., the more you use, the more you pay.

More GIS vendors are targeting enterprise users to deliver real-time and "sometimes connected" information services. As more wireless networks are deployed, more information can be transmitted to mobile professionals in real-time. However, for mobile workforces in the utilities and telecommunications sectors, spatial data is being deployed on PDAs and now TabletPCs. After information is gathered onsite, it can be reconnected with the primary data warehouses and updated as necessary. VanderMeer also said that, "As wireless networks and mobile platforms evolve, the infrastructure needed for second-generation LBS capabilities will slowly be put into place, setting the stage for larger offerings at the start of 2004. In addition, consumer uptake for telematics and personal location services will be slow as wireless carriers experiment with the right mix of location-enabled service offerings. Once this evolution has transpired, location-centric applications and Location Determining Technology (LDT) will play a key role in defining mobile devices, wireless services, and backend infrastructure implementations" (VanderMeer, 2003).

When the Web Meets Wireless

So, what might be expected from the combined resources of web services and wireless LBS? Much of this question lies in how much more productive a mobile workforce can become or what a consumer is likely to pay for a location-based service. The answer can be found in the ability of the service to provide relevant, real-time information. Look for two areas that are well known to begin to thrive, and two relative unknown areas to emerge: Weather and Traffic. Fleet managers and commuters know them very well. They are always changing and both greatly factor into route planning and resource allocation. Although not necessarily new to those familiar with websites that provide both, the delivery of this information to in-vehicle navigation systems and other wireless communication devices will be the norm. Vehicle manufacturers are expanding the number of vehicles on which "nav" systems are available. Solutions for modeling weather and traffic patterns will determine new routes to save fuel and time in order to serve the customer better. Companies such as Westwood One and Meteorlogix for gathering traffic and weather information, respectively, will support companies who provide asset tracking and fleet management solutions.

Two emerging technologies will factor significantly into the use of geospatial intelligence: Radio Frequency Identification (RFID) and location-aware WiFi.

RFID can be defined as: "Involving the contactless reading and writing of data into an RFID tag's nonvolatile memory through a radio frequency (RF) signal. An RFID system consists of an RFID reader and an RFID tag. The reader emits an RF signal and data is exchanged when the tag comes in proximity to the reader signal. The RFID tag derives its power from the RF reader signal and does not require a battery or external power source" (Atmel, 2003). So, once an item passes an RFID reader, its location is immediately known; instantaneous tracking at various delivery points provides feedback to dispatchers and consequently, the customer.

RFID provider Checkpoint, puts it this way: "RFID has unlimited potential to enhance the way that materials are moved, wherever item-level management accuracy and efficiency are important." RFID applications include:

- Airline baggage handling
- Automatic manifesting
- Automatic truck location
- Distributed self-checkout
- Document management
- Industrial laundry sorting
- Inventory tracking and logistics
- Medical supply/device tracking
- On-shelf inventory systems
- One-on-one loyalty marketing
- Parcel shipping
- Pharmaceutical manufacturing
- Point of Sale (POS) auditing
- Verification of shipment integrity
- Warehousing/distribution

Work-in-Progress Management

The exciting thing about this technology is that it is all location centric at some level. Marketers, manufacturers, and shippers need to know location as a critical means of tracking the efficiency of product throughput. The result is a better understanding of your business processes, otherwise known as "business intelligence" (Checkpoint Software, 2003).

Location-enabled Wi-Fi (wireless fidelity) is also emerging as a means of providing "small area" delivery of information. Dartmouth College (Hanover, NH) is installing LocaleServer™ a software platform from Newbury Networks that is a supplier of an 802.11b-based Location-Enabled Network™ (LEN). This new network will be used to "push" class materials to students with 802.11b-enabled PDAs or laptops. "Dartmouth's Thayer School of Engineering is currently using Newbury's Digital Docent™ application for content provisioning, e.g., to 'push' class notes, course information and other relevant materials to students in classrooms using 802.11b supported handheld or notebook computers — based on their location. For example, Digital Docent ensures that only students in that particular classroom receive this timely, location-aware information," according to a statement from the company (Newbury Networks, 2003).

The emergence of this technology indicates that "where" you receive information is just as important as knowing where you are in the first place. And so, professors who deem it necessary to release information to students only when they enter a certain "area of significance" can broadcast data for "just-in-time" delivery of information capital. For example, perhaps tests are distributed only in certain classrooms at the appropriate time. Maybe bluebooks turn into blue "tabletPCs" for exams. Many other applications can take advantage of this technology. For example, retailers can offer discounted merchandise through special promotions by sending a message to buyers who hold credit cards or frequent purchaser programs with the company as they enter with proximity to a store. These "push marketing" techniques have the benefit of reaching a very targeted audience that is receptive to company information and promotions.

Business Models of Spatial Information Companies

Open Source GIS

There is a growing movement toward open source geographic information systems that could expand as other open source operating environments begin to take hold in corporations because of the lower cost of ownership. Open source software is software code that is freely distributed and generally comes without much documentation or support. Users obtain the software to then develop customized user interfaces and other functionality to meet their specific needs. Some professionals also feel that commercial GIS software vendors have not addressed some of the challenges associated with database technology that supports spatial data (see databases). Gilberto Câmara, director for Earth Observation for the National Institute for Space Research (INPE) in Brazil believes that, "An important challenge for the GIS community is finding ways of taking advantage of the new generation of spatially-enabled database systems to build 'faster, cheaper, smaller' GIS technology. One of the possible responses to this challenge would be to establish a co-operative development network, based on open source technology. In a similar approach as the Linux-based solutions, the availability of GIS open source software would allow researchers and solution developers access to a wider range of tools than what is currently offered by the commercial companies" (Câmara, 2003). There are clearly pros (faster, more reliable, open source, programming flexibility) and cons (security, quality assurance testing, support) with respect to open source GIS. Yet, there is no economic incentive for GIS companies to develop open source products or for users to swap existing installations of current solutions. Some systems, such as the Geographic Resources Analysis Support System (GRASS), have been available as an open source tool for some time. GRASS is a freely distributed GIS system that supports both raster-based data and topologically structured sources. Those in the natural resource industry particularly favor the software.

More recently, MapServer from the University of Minnesota has gained popularity as a tool for creating web-mapping applications. However, more coordination of open source code would have to be established, most likely by a university or government entity. There is a movement toward pursuing such resources, and they will likely be developed in the next three to five years.

With respect to the work by the OGC, open source and interoperability should not be considered mutually exclusive topics. Interoperability specifications and open source should support integration not only between GIS solutions, but also other enterprise systems where geospatial information is considered vital to business survival.

Databases

The ability of an enterprise database to manipulate, query, and manage spatial data will be the single biggest factor in the adoption and proliferation of GIS at major corporations. Since enterprise databases manage the data for many disparate functional areas of a company, the ability to query the geographic component of that data will facilitate its use. Enterprise databases are further explained in Chapter V of this book by Julian Ray. In reports issued by International Data Corporation in 2002 (IDC, 2002), the market research company had this to say about the expected influence of spatial database technology:

"The Spatial Information Management (SIM) industry is being reshaped by broad spatially-enabled database management and data access capabilities that are emerging across the IT industry. We expect basic spatial information management (SIM) capabilities to be ubiquitous within data access and database management layers within three to five years. SIM market structure will be shaped by the following four factors during this period:

- *Open availability of basic spatial functionality in data access and database management.*
- *Substantially lower cost of entry for SIM vendors that want to enter new vertical markets.*
- *Substantially lower costs for IT vendors that want to include location-specific functions in their applications.*
- *The delivery of SIM functionality both as software and services."*

Specifically, Oracle Corporation has been the leader in deploying spatial analysis functionality within their standard database products and advanced functionality modules. The IDC report (IDC, 2002) summarize Oracle's influence in this market this way:

"Oracle's SIM functionality is now available to their 200,000 client organizations and their developer community as an integral, no-cost feature set. This fact drives significant new opportunities for SIM vendors and other technology vendors that want to spatially-enable business applications.

Oracle has tightly integrated spatial capabilities into their database and application server technology. This moves geospatial technology from being a specialty application to being a part of base-level information infrastructure. This simplifies the use of spatial data in business applications and removes much of the cost of using spatial data.

Oracle's spatial feature set is now available to the Oracle and Java developers' community. This means that developers can integrate spatial features directly into business and location-based applications at relatively low costs and with minimal training."

Another significant development with respect to the spatial functionality of enterprise databases, such as Oracle, is their ability to serve location-based information to a wireless application. Both Internet and wireless applications can be supported by the same database and Oracle9i was intentionally developed with those users in mind. Oracle10g will build upon that fundamental premise with support for a topological data structure, geocoding, and other capabilities that introduce more functionality that is both complementary and competitively positioned with respect to current GIS software solution providers.

Xavier Lopez of Oracle (Francica, 2003) commented that, "Oracle Spatial has been fundamental in launching a number of very successful wireless LBS deployments worldwide. For example, in early 2002, *JPhone*, Japan's number three wireless carrier, was the first to offer the ability to deliver color maps to General Packet Radio Service (GPRS) phones as part of their LBS service. The *J-Phone J-Navi* LBS applications were written in Java and run on Oracle Spatial. This particular deployment runs nearly all of its LBS functions (geocoding, map rendering, and location capability) directly from the spatial database and is able to achieve scalability requirements of 30,000 user sessions per hour. The result is the ability to deliver over 1 million color vector and raster maps per day to a new class of GPRS and Universal Mobile Telecommunications System (UMTS) enabled multimedia handsets" (quoted in Francica, 2003).

However, database functionality that includes the management of spatial data has not been lost on the other major information technology providers, IBM and Microsoft. Both companies have customers that use DB2 and SQL Server, respectively, to manage spatial information. IBM has three modules that support spatial data: DB2 Spatial Extender, Spatial Databade Module (based on IBM Informix Dynamic Server), and Geodetic Datablade Module. These products can function with Linux, Solaris, or HP-UK operating systems as well as AIX and Windows. Spatial Extender conforms to the OGC's simple feature specifications for SQL and the ISO SQL/MM spatial standard.

Microsoft's SQL Server has been popular with many customers because of its lower price point. It stores spatial features as Binary Large Objects (BLOB). However, BLOB support will be improved with the inclusion of User Defined Datatypes in an upcoming release of SQL Server. This new release will also provide support for the .NET framework. For many types of customers, this solution is adequate and sometimes preferred to solutions from IBM and Oracle, which come with the additional cost of hiring a database administrator and potentially others.

Each of these database software providers has close relationships with the GIS software companies. Both groups look at opportunities that support existing customers and which try to leverage strengths of the total software solution. However, the spatial functions in the database sometimes overlap that which is being provided by a client-side desktop GIS. Database managers, especially those who support large spatial database, will be looking to see which spatial functions are better suited for processing on a database server, and those that can be handled at the desktop.

New Tools

Look for two types of display technologies to invade the domain of GIS mapping software currently on the market in the near future: Scalable Vector Graphics (SVG) and Macromedia Flash.

What is SVG? George Seff of Limbic Systems, Inc. has this explanation: "Scalable Vector Graphics (SVG) is an XML language for describing two-dimensional graphics. SVG allows for three types of graphic objects: vector graphic shapes (e.g., circles, rectangles, and paths consisting of straight lines and curves), images and text" (Seff, 2002). And Mr. Seff elaborates on the problems with today's web mapping solutions: "Most mapping systems employ two approaches when delivering interactive maps on the Web. The first approach is the familiar Java applet. The second approach involves generating map images on the server and delivering them to the user in either GIF or JPEG image format. Java applets perform well concerning interactivity, but have suffered from browser compatibility and firewall issues. Image maps are compatible with web tools and environments, but interaction with the map such as zooming, panning, layer control and thematic shading necessitate a round trip to the server in order to re-render the image. Today, SVG solves all of these problems. SVG is an open, HTTP compatible standard that allows fully interactive mapping applications — without the need for applets or a round trip to the server every time the map presentation is tweaked" (Seff, 2002).

Macromedia Flash is widely used on websites today for animation and graphically appealing advertising. Mapping, as a visual tool, can take advantage of this technology as a means of rapidly distributing spatial information to a greater audience. Chris Goad of the MapBureau comments that, "The Macromedia SWF (SWF is the file format used by Macromedia Flash technology) authoring tools are particularly effective in three areas: support for creating animation and interaction without programming, integration of sound and video, and integration with server-side applications and development tools, both those of Macromedia and others... SWF has the compelling advantage of needing no download of a plugin for most viewers. Requiring a 2.3MB download of an SVG player to view a map is not a realistic way to reach the web masses" (Goad, 2003).

The reason these technologies are an effective alternative to current tools supplied by GIS software companies is programmability and data representation. Goad (2003) continues, "SWF and SVG are in many respects similar technologies, and both satisfy the basic technical requirements needed to support a rich GIS experience on the Web. These requirements are:

(1) Vector (rather than raster) representation of 2D geometrical objects

(2) Availability of event handling primitives that allow flexible design of interactivity

(3) Full access from a programming or scripting environment to the geometric and event models, so that arbitrary kinds of animation and interaction can be programmed, and

(4) The ability to query the server as needed for incremental updates to the map. SWF and SVG are similar not only in the fact that they both meet these requirements, but in how they represent geometrical data: their primitives for modeling and manipulating the geometrical world correspond closely, though not exactly."

Applications

Fleet & Field Service Management (FSM)

The promise of using spatial information by the mobile workforce of each will result in both bottom and top line revenue gains. In fleet management and FSM, the goal is cost containment. More enterprise solutions for routing and scheduling are incorporating mapping technology to more efficiently manage routes. A report from The Logistics Institute at Georgia Institute of Technology (2002) stated that in a transportation logistics survey, about 65% of the companies managing truck fleets indicated that with better planning they could save their organization 6% to 30% (Georgia Institute of Technology, 2002). However, looking at future growth for asset or vehicle tracking companies, Jim VanderMeer once again commented that, "2003 will be the year of saturation, followed by consolidation on the fleet and enterprise side of LBS. Asset tracking and Automatic Vehicle Location (AVL) solutions provide significant cost reduction and enhancements in efficiency through location-based offerings, but there are currently too many "me-too" hardware and service offerings bombarding the fleet and dispatch markets. As a result, the market is becoming saturated by an overwhelming number of GPS-enabled "black box" offerings. The slow economic climate, combined with this saturated market, will make it difficult for many asset tracking and AVL providers to attract and capture enough market share to sustain their business model." Still, the applications of location technology for fleet management offer a substantial return on investment (ROI). As the market consolidates, the companies that can substantiate the ROI stand to profit.

Customer Relationship Management (CRM) and Enterprise Resource Planning (ERP)

CRM systems from large vendors such as Siebel Systems are utilizing more spatial information, as are ERP solutions from SAP. MapInfo's products have been certified with Siebel 7, and SAP's products can be integrated with software from ESRI and data products from AND Products BV. Oracle's CRM suite can now leverage the spatial functionality included with the Oracle Enterprise database. Implementation of these solutions, however, has slowed as the technology marketplace has undergone significant retraction due to the economic conditions between 2001-2003. As companies look to upgrade systems that were installed during the later half of the 1990s, this market could once again see growth and a renewed interest in spatial information. Since the application of lifestyle segmentation systems and demographic analysis for consumer marketing by retailers was vital to the growth of franchise development by companies like Wal-mart, Starbucks, and Subway, more companies will take advantage of both marketing systems and network planning solutions as good sites become more difficult to locate.

Geospatial Data: Demographics & Satellite

Accurate and timely delivery of geospatial data is essential. The method of delivery, however, is evolving. Many demographic and satellite data providers are seeking solutions for better customer service. As such, the information to business clients will take the form of near real-time availability of products. For this, online access to an application service provider or a web service is an option for many corporations that do not want to host, maintain, or manage applications or data. Through subscription programs, demographic data products will be conveniently delivered by linking to a web service provider. The OGC and Intergraph (http://www.wmsviewer.com/) as well as ESRI, though its Geography Network (http://www.geographynetwork.com/webservices/index.html), have established websites that show how data from disparate sources can be linked to a single geographical visualization engine, without the need to import or convert the data. Web services that link data with visualization tools is one of the single most important initiatives that could propel the adoption of spatial information within businesses. These services eliminate the complexity associated with developing spatial data warehouses, visualization, and integration tools. Web services will provide the interoperability necessary to allow spatial technology to be integrated with other information systems. The application service provider (ASP) business model is apt for delivery of data, especially demographics that can be segmented and packaged as a suite of data for a specific industry. However, some companies are fearful of the ASP model due to privacy and security issues and want all data and software solutions to remain within the corporate firewall.

Prices for data will continue to fall, as some data products become commodities. Bundled with software, the entire United States street centerline database of roads and highways, and 2000 U.S. Census demographics can be purchased for less than $150. Demographic data companies are aggressively pricing products for niche data variables of consumer expenditures and lifestyle segmentation. Street centerline data vendors such as Tele Atlas, Navigation Technologies, and Geographic Data Technology are in a pitched battle to expand coverage, improve accuracy, and add international products in an effort to address both mobile location services and telematics. Because of downward price pressure from the less expensive desktop software solutions and free Internet services, these street data providers will find it increasingly necessary to enhance their products to show where they will provide a value-added benefit to justify higher price points.

However, all three companies offer different levels of higher quality data products, but at a substantially higher cost. The cost difference is related to the substantial improvements to the connectivity of the road network in order to provide a navigable data structure to companies in fleet and field service management. And a key differentiator for these street network providers is the ability to provide points of interest such as the location of hotels, restaurants, and gas stations as well as other attributes on the network, such as bridge heights, one-way streets, and other impedance factors that would affect routing solutions.

The satellite data providers will also feel the pressure to keep prices affordable as they look to expand into the business marketplace with ever improving spatial accuracy. In May 2003, the U.S. Government (Office of Science and Technology Policy, 2003) released a new policy on commercial remote sensing that allows satellite-imaging manufacturers to develop sensors that can recognize objects that are approximately 10-inches in size. These highly accurate data products may pose challenges in data storage because of the disk space they will occupy. In many cases, lower resolution imagery or less expensive aerial photography will suffice for business applications.

Conclusions

The catalysts that will shape the expansion of GIS within the mainstream corporate information technology infrastructure are in place. Web services, mobile location technology, and lower total cost of ownership will facilitate better adoption within businesses. In addition, the supporting tools developed for database management by Oracle, IBM, and Microsoft, as well as the server-based location processing tools for wireless applications will broaden the acceptance of spatial technology. If these companies continue to develop the spatial functionality of products and services, they will foster a greater demand for location-based services, more so than existing GIS software solution companies, because they have the financial ability to promote these tools more aggressively into business clientele. However, all firms will benefit by a raised awareness for geospatial information and applications. Software tools, components, and services for developing stand-alone applications are being used to move spatial information into enterprise systems that support corporate finance, marketing and human resource functions. Interchange protocols, such as XML, will allow a freer flow of spatial information in an increasingly inter-connected world. All of these developments will support the next phase of technology solutions that incorporate spatial information in the mainstream business IT environment.

References

Alien Technology Corporation. (2003). *Company literature.* Retrieved from: *http://www.atmel.com/products/RFID/.*

Atmel Corporation. (2003). *Company literature.* Retrieved from: *http://www.atmel.com/products/RFID/.*

Câmara, G. (2003, May 15). Why open source GIS software*? Directions Magazine.*

Checkpoint Systems. (n.d.). *Company product information.* Retrieved from: *http://www.checkpointsystems.com/content/rfid/default.aspx.*

Francica, J. (2003, January 23). Databases will become larger and global - Larry Ellison's keynote address - OracleApps World 2003. *Directions Magazine.*

Francica, J. (2003, February 11). The direction of oracle's spatial strategy. *Directions Magazine.*

Goad, Chris. (2002, June 10). Flash/SWF for GIS. *Directions Magazine*.

GRASS Development Team. (2003). *Home page and historical notes*. Retrieved from: *http://grass.itc.it/*.

International Data Corporation. (2002, October). *The structure of the spatial information management industry*. International Data Corporation #28246.

International Data Corporation. (2002, December). *Spatial information management: competitive analysis*. International Data Corporation #28348.

The Logistics Institute-GIT. (2002, January). *Transportation and logistics survey results*. The Logistics Institute at Georgia Institute of Technology.

Lombardi, S. (2002, November 19). XML web services, GIS, and location technologies - Part 1. *Directions Magazine*.

Newbury Networks. (2003, April 15). Press release. *Directions Magazine*.

Office of Science and Technology Policy. (2003, May). U.S. Commercial Remote Sensing Space Policy. Washington, D.C.: White House, United States of America.

Open GIS Consortium. (2003). *OGC web services phase 2 document (OWS-2)*. Retrieved from: *http://www.opengis.org/initiatives/?iid=7*.

Seff, G. (2002, May 7). SVG and GIS. *Directions Magazine*.

VanderMeer, J. (2003, January 29). LBS: Turning corners in 2003. *Directions Magazine*.

About the Authors

James B. Pick is professor in the School of Business at the University of Redlands and former department chair of Management and Business and former chair of the Business School Faculty Assembly. He holds a BA from Northwestern University, a MS from Northern Illinois University, and a PhD from the University of California, Irvine. He is the author of eight books and 110 scientific papers and book chapters in the research areas of management information systems, geographic information systems, environmental systems, population, and urban studies. He has published in such journals as *Information and Management, Simulation, Journal of Global Information Technology Management, Journal of Information Technology Cases and Applications, Socio-Economic Planning Sciences, Demography, Social Science Journal,* and *Social Biology*. His awards include several for faculty distinguished teaching and research from the University of Redlands, Senior Fulbright scholar award for Mexico, Thunderbird Award from the Business Association for Latin American Studies, and outstanding alumnus award from Northern Illinois University. He serves on three journal editorial boards.

* * *

Anne Mims Adrian is a PhD student in Management Information Systems, Department of Management, at Auburn University. She also co-leads the information technology efforts of the Alabama Cooperative Extension System, Alabama Agricultural Experiment Station, and the College of Agriculture at Auburn University. She has published in regional agricultural economics journals and has presented papers at the American Agricultural Economics Association meetings and the International Conference of Computers in Agriculture.

Arthur W. Allaway is a professor of Marketing and the M. Thomas Collins faculty fellow in Marketing at the University of Alabama in Tuscaloosa. His research interests include new product development, business-to-business marketing, spatial diffusion, site and market area analysis, marketing models, and e-business/commerce strategy. Dr. Allaway is extensively published, including *Journal of Marketing, Journal of Retailing, Journal* *of Business Research*, *Growth and Change*, *Economic Geography*, and the *Research in Marketing* series. While earning his doctorate at the University of Texas-Austin, Dr. Allaway trained with the pioneering David Huff on the application of spatial analysis to marketing. His business experience includes marketing research director for a large business-to-business manufacturer.

Luís Alfredo Amaral was born in 1960 and holds a PhD in Information Systems obtained at University of Minho in Portugal (1995). He is associate professor in the Department of Information Systems, the School of Engineering of University of Minho, where he teaches courses on information systems management and information systems planning for undergraduate and postgraduate degrees. He is also involved in research projects in the area of methodologies for organizational intervention activities such as: information systems management, information systems planning and information systems development. Other topics of interest are the adoption process of IT applications by organizations and the curricula for information systems professionals.

David K. Berkowitz is an associate professor of Marketing and director of the Center for the Management of Technology at the University of Alabama in Huntsville. He received his PhD in Marketing and Applied Statistics from the University of Alabama in Tuscaloosa. Dr. Berkowitz's research is focused on the new product diffusion and adoption process, retailing site selection and performance, brand management and international marketing. He has published his research in many leading journals including the *Journal of Retailing*, *Journal of Advertising*, and *Journal of Advertising Research*. At UAH, Dr. Berkowitz teaches marketing high technology products, managing technology development, marketing management, and buyer behavior. Before receiving his PhD, Dr. Berkowitz worked for 12 years with Hallmark Cards.

Subhasish Dasgupta is an assistant professor of Information Systems in the School of Business and Public Management, The George Washington University in Washington, DC. He received his PhD from Baruch College, The City University of New York (CUNY), and an MBA and BS from the University of Calcutta, India. His current research interests are electronic commerce, information technology adoption and diffusion, and Internet-based simulations and games. He has published in journals such as *European Journal of Information Systems, Decision Support Systems, Logistics Information Management, Journal of Global Information Management, Journal of Global Information Technology Management, Information Systems Management, Simulation and Gaming Journal* and *Electronic Markets: The International Journal of Electronic Commerce* and *Business Media*. He has presented his work at numerous regional, national and international conferences.

Chris Dillard is a Masters of Science student in Management Information Systems in the Department of Management, Auburn University. He is also an agricultural program assistant with the Department of Agronomy and Soils in the College of Agriculture. Chris has been involved with the Auburn University Precision Agriculture Program since 1998.

His involvement with the Precision Agriculture Program includes researching various GPS and GIS technologies and disseminating that information to Alabama and other southeastern farmers.

Joseph R. Francica is editor-in-chief and general manager of *Directions Magazine*, the leading online publication for GIS and spatial information technology resources. He is also the principal of Geodezix Consulting, which provides GIS support and analysis to market research, retail, public organizations, and consumer goods firms. He previously served as editor of *Business Geographics Magazine* and in executive management positions with both Fortune 1000 and start-up firms. He received a bachelor's in Geology (Rutgers University), a master's degree in Earth Science (Dartmouth College), and an MBA (Southern Methodist University). He has written 100 articles and editorials on business, marketing, and GIS topics. He is the conference chair for the Location Technology and Business Intelligence Symposium.

Oliver Fritz received his degree and PhD in Theoretical Physics from the University of Basel (1991 and 1995, respectively). As a post-doctoral fellow at Rutherford Appleton Laboratory (Oxford, UK) and Paul Scherrer Institut (Villigen, Switzerland), he mainly worked on static and dynamic properties of unconventional and low-dimensional magnetic systems. Since joining ABB Corporate Research (Baden, Switzerland) in 1999 he is concerned with a variety of problems from the fields of physics and information technology. His special interests lie in modelling and simulation of dynamical systems, decentralised power generation and distribution, and new concepts for integrated software applications.

Kevin C. Gillen is a sixth year PhD student in Applied Economics at the Wharton School. He has more than 9 years of experience in developing econometric models. At Wharton, Mr. Gillen has developed multiple applications of statistical forecasting models to real estate, using GIS technology. Several of these applications have resulted in scholarly publications in the field of real estate economics. He is currently funded by a HUD Dissertation Grant and a Lincoln Land Institute Dissertation Grant to study the market for urban land. Prior to Wharton, Mr. Gillen worked for the U.S. Department of Housing and Urban Development's Office of Federal Housing Enterprise Oversight (OFHEO) developing models of mortgage termination, and for the Federal Reserve analyzing the U.S. financial system. Mr. Gillen holds a BA in Economics and Mathematics (Canisius College) and has attended Georgetown University and Cambridge University (UK) on scholarship.

Richard P. Greene is an associate professor in the Department of Geography at Northern Illinois University (USA). Utilizing geographic information systems (GIS), global positioning systems (GPS), and remote sensing technologies, he is actively engaged in researching urban spatial structure. He has published in a variety of journals including *Landscape and Urban Planning*, *Forum of the Association of Arid Lands Studies*, *The Social Science Journal*, the *Annals of the American Academy of Political and Social*

Science, *Urban Geography*, and *Economic Geography*. He has taught a number of GIS workshops for the *National Science Foundation's* Chauatauqua short course series on the topic of GIS and the urban environment. He is a member of the Association of American Geographers, Arid Lands Studies Association, the Western Social Science Association (WSSA), and is currently serving a 3-year term on the WSSA Executive Council.

Scott Griffith graduated with a Bachelor of Commerce (Honours) from the University of Manitoba (1998). During his studies in Winnipeg, he developed an information system to manage the raw-material collection logistics for Isobord Enterprises. Since his graduation Scott has assumed the role of vice president for Tell Us About Us, Inc., an international firm specializing in Web-enhanced customer feedback services. Scott currently works on business development and marketing for Tell Us About Us, Inc.

Gary Hackbarth is an assistant professor of Management Information Systems at Iowa State University. His research interests include strategic applications of geographic and location-based technologies, electronic and mobile commerce, organizational memory, IT security, and IT and international relations. He has published in *Information Systems Management*, *Information & Management* and the *Financial Times*.

Brian N. Hilton received a PhD in Management Information Systems from Claremont Graduate University, School of Information Science. His current research interests lie in geographic information systems, spatial decision support systems, and open source software. His dissertation examines dynamic information system development using open source software and Web services for the development of an Internet-based spatial decision support system. He holds an MS in MIS (Claremont Graduate University) as well as a BA in Economics (Richard Stockton College of New Jersey). Dr. Hilton is currently a research associate at the Claremont Information and Technology Institute (CITI).

Thomas A. Horan serves as executive director of the Claremont Information and Technology Institute (CITI) and is an associate professor in the School of Information Science, Claremont Graduate University (CGU). As director of CITI, he is responsible for directing a wide range of applied technology research projects. His area of specialization is the development and deployment of advanced information technologies including geographic information systems. Prior to joining CGU, Dr. Horan was a senior fellow at George Mason University and a senior analyst at the U.S. General Accounting Office. Dr. Horan received both his master's and doctoral degrees from the Claremont Graduate School.

Esperanza Huerta is a professor at the Instituto Tecnológico Autónomo de México. She received a PhD in Management Information Systems from Claremont Graduate University. Her research interests include the use of GIS technologies, the cognitive implications of Web site design, and the theoretical bases of Web-based learning. Her writing has appeared in *Web-Based Education: Learning from Experience* and *Proceedings of the Americas Conference on Information Systems.*

Suprasith Jarupathirun is a PhD candidate in MIS at the University of Wisconsin-Milwaukee. His research focus is on the effectiveness of IS including DSS, SDSS, and expert systems. Suprasith has published several papers in proceedings of information systems conferences. Before joining the PhD program, he was a surveyor and made digital maps for several years in Thailand. He also served as a trainer in the use of survey instruments (e.g., GPS, Electronic Total Stations, Electronic Distance Meters, and Electronic and Optical Theodolites) as well as in the use of the mapping and designing software.

Peter Keenan has been a member of the Department of Management Information Systems at University College Dublin since 1990. Before joining UCD, he developed logistics and location software for a number of large Irish organisations. Currently his research interests include geographic information systems (GIS) in business, spatial decision support systems and the use of the Internet for decision support. He holds degrees from the National University of Ireland. These degrees include a Bachelor of Commerce (1984), a Master of Management Science (1985), and a PhD (2001).

Paul Mask, PhD, is a professor of Agronomy and Extension Specialist with the Department of Agronomy and Soils, Auburn University. He is recognized by NASA as the Alabama Space Grant Geospatial Extension Specialist. Paul is also the coordinator of the Precision Agriculture Program for Alabama, has conducted several grant programs pertaining to agricultural practices, and has published numerous extension publications in agriculture production journals.

W. Lee Meeks is a PhD candidate at The George Washington University in the Management Science Department of the School of Business and Public Management. His degrees include an MBA from The George Washington University and a BS from the U.S. Naval Academy. He works as a senior scientist within the Geospatial Systems Group of General Dynamics — Advanced Information Systems, located in Chantilly, VA. He has spent more than 7 years in the employ of or consulting to the National Imagery and Mapping Agency or its predecessor, the Defense Mapping Agency. His research interest is improving organizational decision-making through the innovative application of information technologies. His current focus is a data-centric view of the advances in geographical information systems as GIS and spatiotemporal data are applied to organizational decision-making. With Dr. Subhasish Dasgupta, he has in press, "Geospatial information utility: An estimation of the relevance of geospatial information to users" in *Decision Support Systems*. His previous publications have been for the National Imagery and Mapping Agency. He has presented at the Institute for Operations Research and Management Science (INFORMS) annual meeting.

Brian E. Mennecke is an associate professor of Management Information Systems in the College of Business at Iowa State University (USA). Dr. Mennecke earned his PhD at Indiana University in Management Information Systems and also holds master's degrees in Geology and Business from Miami University. His research interests include mobile and electronic commerce, spatial technologies and location-based services, data visualization and support systems, technology-supported training, small group decision making, and collaborative technologies. He has previously published a book on mobile commerce and articles in academic and practitioner journals such as *Management Information Systems Quarterly*, the *International Journal of Human-Computer Studies*, the *Journal of Management Information Systems*, *Organizational Behavior and Human Decision Processing*, the *Journal of Computer Information Systems*, and *Small Group Research*.

Lisa D. Murphy is assistant professor of Management Information Systems (MIS) at the University of Alabama in Tuscaloosa. Her research interests include geographic technologies, communities of practice, digital documents, knowledge management, and systems analysis and design. While earning her doctorate in MIS at Indiana University, Dr. Murphy completed in a fellowship at Environmental Systems Research Institute, the largest maker of GIS software, and was also a founding officer of the GeoBusiness Association (now a part of the American Marketing Association). She has conducted several workshops on introducing geographic technologies to business school faculty. Prior to earning her doctorate, Dr. Murphy was a systems developer at Boeing and a marketing manager at Cessna Aircraft Company.

Celene Navarrete is a PhD student of Management Information Systems in the School of Information Science at Claremont Graduate University. She holds a bachelor's degree in Information Systems from the University of Aguascalientes (Mexico). Her research interests include the evaluation of information technology programs and initiatives, e-government and geographic information systems.

Derrick J. Neufeld is a professor of Information Systems at the Richard Ivey School of Business, The University of Western Ontario (Canada). He holds a Bachelor of Commerce (Honours) from the University of Manitoba, a Certified Management Accounting designation from the Society of Management Accountants of Canada, and a PhD from the University of Western Ontario. Dr. Neufeld's research centers on virtual work, and he is presently investigating phenomena such as remote work, leadership at a distance, and computer-mediated communication.

Julian J. Ray is an assistant professor in the School of Business at the University of Redlands. Dr. Ray holds a BA from the University of Reading, UK, and an MSc and PhD from the University of Tennessee. He specializes in the application of geographic information systems to business problems focusing on spatial logistics, location based services, and spatial information systems. Julian has been an active practitioner and entrepreneur, having developed software for commercial GIS companies as well as founding two independent software companies. As a consultant and company founder, Julian has developed spatial information systems for a number of private, federal and state organizations including IBM, Intergraph Corp., The World Bank, Department of Defense, Department of Education, FEMA, FHWA and a number of state and local transportation agencies.

Terry Ryan serves as associate professor in the School of Information Science at Claremont Graduate University. He earned a PhD in Management Information Systems from Indiana University. He conducts research on the design, development, and evaluation of IS applications to support training and education, customer service, and government. He has published articles in *Communications of the AIS, Data Base, Information & Management, Interface, International Journal of Human Computer Studies, Journal of Computer Information Systems, Journal of Database Management Systems, Electronic Journal of Information Systems Evaluation*, and *Journal of System Management*.

Maribel Yasmina Santos is an assistant professor in the Information Systems Department at the University of Minho in Portugal. She has a degree in Informatics and Systems Engineering from the University of Minho (1991), an MSc in Informatics and a PhD in Information Systems and Technologies, both from University of Minho (1996 and 2001, respectively). Her research interests include business intelligence, data mining, geographic information systems, spatial reasoning and space models.

Petter Skerfving received his degree in Computer Science from the Lund Institute of Technology (Lund, Sweden) in 1996. Having worked with software design and implementation at SAAB Combitech Innovative Vision (Linköping, Sweden) and software design processes and software process improvement at Q-Labs (Lund, Sweden and Kaiserslautern, Germany), he joined ABB Power Automation (Baden, Switzerland) in 2000. As head of the software development group "System Tools and Applications," Mr Skerfving has concerned himself mainly with software architecture design, and system integration issues.

John C. Stager is an IT consultant and a PhD student in the School of Information Science at Claremont Graduate University. He is conducting research in the use of technology to support group collaboration. Additionally, he has been an information science practitioner for the last three decades. He is a member of the Association of American Geographers, the Association for Information Systems, the Western Social Science Association (WSSA), and is currently serving as the section coordinator for geography in the WSSA.

Michelle M. Thompson is a real estate education consultant focusing on community development and geographic information systems. Currently, Michelle works with the Lincoln Institute of Land Policy providing course development and valuation research services. Michelle recently taught ArcGIS8.2: Introduction to GIS in the Cornell University City & Regional Planning Department. As Principal of Thompson (RE) Appraisal and a licensed residential appraiser, Michelle performed fee appraisal services in Massachusetts (1992-1996). Michelle received a BA (1982) in Policy Studies from Syracuse University, a Masters of Regional Planning (1984) and a PhD (2001) from the City & Regional Planning Department at Cornell University.

Benjisu Tulu is currently a doctoral student in Management Information Systems at Claremont Graduate University, School of Information Science. Her current research interests lie in computer security, medical informatics, and geographic information systems. She is currently working on secure video conferencing and digital signatures in the health care domain. Ms. Tulu holds an MS in MIS (Claremont Graduate University) as well as a BS in Mathematics (Middle East Technical University, Turkey). Ms. Tulu is currently a research associate at the Network Convergent Lab at Claremont Graduate University.

Nanda K. Viswanathan is associate professor of Marketing and chair of the Department of Management, in the School of Management, Delaware State University. His research interests are focused in the areas of GIS, competitive theory, consumption, and branding. Prior to teaching at Delaware State University, Dr. Viswanathan taught at the University of California, Riverside and the University of Redlands. Dr. Viswanathan's research has been published and is forthcoming in journals and conference proceedings, including *Socio-Economic Planning Sciences*, *Social Science Journal*, *Academy of Marketing Science*, *Midwest Academy of International Business*, and *Global Information Technology Management Proceedings*.

Susan Wachter is Richard B. Worley professor of Financial Management and professor of Real Estate and Finance at The Wharton School at the University of Pennsylvania. Dr. Wachter served as assistant secretary for Policy Development and Research at HUD, a President appointed and Senate confirmed position, from 1998 to 2001, and was principal advisor to the Secretary responsible for national housing and urban policy. The Chairperson of the Wharton Real Estate Department from 1996 to 1998, Wachter is the author of more than 100 publications. Wachter served as president of the American Real Estate and Urban Economics Association and coeditor of *Real Estate Economics*, and currently serves on multiple editorial boards, and is visiting fellow at Brookings Institution and Fellow of the Urban Land Institute.

Fatemeh "Mariam" Zahedi is Wisconsin distinguished professor, MIS Area in the School of Business, at the University of Wisconsin-Milwaukee. She received her doctoral degree from Indiana University. Her present areas of research include IS quality and satisfaction, DSS, e-commerce and Web interface design, intelligent interface, IS design (for components, health networks, and maintenance), and policy and decision analysis. She has published more than 40 papers in major refereed journals, including: *MIS Quarterly, Information Systems Research, Decision Sciences, IEEE Transactions* *on Software Engineering, IEEE Transactions on Professional Communications, Decision Support Systems, IIE Transactions, European Journal of Operations Research, Operations Research, Computers and Operations Research, Journal of Review of Economics and Statistics, Empirical Economics, Socio-Economic Planning Sciences, Interfaces*, and others. She has numerous publications in conference proceedings, and is the author of two books, *Quality Information Systems* and *Intelligent Systems for Business: Expert Systems with Neural Network.* Dr. Zahedi serves on the editorial board of a number of journals and has many years of consulting and managerial experience in developing information systems and performing policy analysis.

Index

R

S

T

U

V

W

Y